REMOTE SENSING
AND GEOGRAPHICAL
INFORMATION SYSTEM

REMOTE SENSING AND GEOGRAPHICAL INFORMATION SYSTEM

A M Chandra
S K Ghosh

Alpha Science International Ltd.
Oxford, U.K.

A M Chandra
S K Ghosh
Geomatics Engineering Section
Department of Civil Engineering
Indian Institute of Technology
Roorkee, India

Copyright © 2006
Reprint 2007

Alpha Science International Ltd.
7200 The Quorum, Oxford Business Park North
Garsington Road, Oxford OX4 2JZ, U.K.

All rights reserved. No part of this publication may be reproduced, stored in a retrieval system or transmitted in any form or by any means, electronic, mechanical, photocopying, recording or otherwise, without prior written permission of the publisher.

ISBN 1-84265-278-8

Printed in India

PREFACE

Almost no field is left where the emerging technology of Remote Sensing combined with the technology of Geographic Information System (GIS) has not found its applications. This demands a comprehensive text at introductory level that explains the fundamentals of remote sensing and GIS, and their applications. This text primarily fulfills these demands, and, it will be of great help to those who have not studied remote sensing and GIS.

Since the primary aim of the text is to introduce about the remote sensing technology and GIS, it is not tailored to any single discipline. The text tries to stimulate students to investigate the subjects that they have not undergone through earlier. A special feature of this text is that the students will find the subject matter on remote sensing and GIS under one fold. The text is designed to cover the topics that are essentials to a beginner before specializing in the field of remote sensing and GIS.

The text has two sections; the first section, from Chapter 1 to 7, is devoted to remote sensing, and the second section, from Chapter 8-11, deals with GIS. Chapter 1 is Introduction about the book giving the details of subject matter covered in the text. Chapter 2 explains the fundamentals of Electromagnetic Energy, which is the prime source of energy used in remote sensing. Chapter 3 discusses the Sensors and Platforms, which that used to collect the raw remote sensing data. Chapter 4 and 5 present the Satellite Data Products and Image Interpretation technique. The Digital Image Processing, is an important part of the remote sensing technology, is discussed in Chapter 6. Chapter 7 gives a brief account of applications of Remote Sensing.

GIS starts from Chapter 8 and discusses 'What is a GIS', and its components. Since GIS has to handle a large amount of remote sensing data, various Data Models used in GIS are discussed in Chapter 9, and GIS Database and its Management is presented in Chapter 10. Spatial Analysis Techniques used in GIS, are given Chapter 11. Finally, Chapter 12 presents the Applications of GIS in various fields.

To enrich the text with valuable material, at some places in the text, illustrations and matters have been taken from already published works, and the same have been duly acknowledged at appropriate places, as far as possible.

The authors wish to acknowledge the moral support provided by the author's families during the course of writing this book. Thanks are also due to Sri Naresh Kumar Kaushik, who typed, edited the manuscript, and drew illustrations for this text.

A M Chandra
S K Ghosh

CONTENTS

Preface	v
1. Remote Sensing	**1**
1.1 Introduction	1
1.2 Remote Sensing System	2
1.3 Historical Development of Remote Sensing	3
1.4 Multi-concept of Remote Sensing	4
1.5 Advantages and Disadvatages of Remote Sensing	6
1.6 Some Applications of Remote Sensing	7
2. Electromagenetic Radiation	**9**
2.1 Introduction	9
2.2 Electromagnetic Energy	9
2.3 Energy Interaction in the Atmosphere	12
2.4 Energy Interactions with the Earth's Surface	15
2.5 Resolution in Remote Sensing	22
2.6 Pixel and Mixed Pixel	24
3. Sensors and Platforms	**25**
3.1 Introduction	25
3.2 Broad Classifications of Sensors and Platform	25
3.3 Land Observation Satellites and Sensors	26
3.4 High Resolution Sensors	40
3.5 Earth Observing-1 (EO-1)	48
3.6 Weather Satellites/Sensors	51
3.7 Other Weather Satellites	56
3.8 Marine Observation Satellites/Sensors	57
4. Satellite Data Products	**59**
4.1 Introduction	59
4.2 Data Reception, Transmission, and Processing	59
4.3 Remote Sensing Data	60
4.4 Data Products	63

4.5	Referencing Scheme	65
4.6	Standard Products	66
4.7	Digital Data Products	69

5. Image Interpretation — 73

5.1	Introduction	73
5.2	Interpretation Procedure	73
5.3	Elements of Photointerpretation	74
5.4	Image Interpretation Strategies	75
5.5	Photomorphic Analysis	76
5.6	Image Interpretation Keys	77
5.7	Equipment for Image Interpretation	77
5.8	Automated Approach to Image Classification	78

6. Digital Image Processing — 79

6.1	Introduction	79
6.2	Digital Image Processing	79
6.3	Overview of Digital Analysis Steps	81
6.4	Initial Statistics Extraction	82
6.5	Image Rectification and Restoration	83
6.6	Image Enhancement	88
6.7	Spatial Filtering	91
6.8	Image Transformations	92
6.9	Image Classification and Analysis	97

7. Application of Remote Sensing — 115

7.1	Introduction	115
7.2	Land use and Land Cover Mapping	115
7.3	Crop Inventory Studies	116
7.4	Ground Water Mapping	121
7.5	Urban Growth Studies	124
7.6	Flood Plain Mapping	125
7.7	Hydro Morphological Studies	128
7.8	Wasteland Mapping	130
7.9	District Level Planning	132
7.10	Disaster Management	135
7.11	Concluding Remarks	138

8. Geographic Information System — 145
  8.1 Introduction — 145
  8.2 Definition of GIS — 145
  8.3 Components of GIS — 146
  8.4 Geographical Concepts — 149
  8.5 Input Data for GIS — 150
  8.6 Types of output Products — 150
  8.7 Applications of GIS — 151

9. GIS Data — 153
  9.1 Introduction — 153
  9.2 GIS Data Types — 153
  9.3 Data Representation — 153
  9.4 Data Sources — 154
  9.5 Typical GIS Data Sets — 156
  9.6 Data Acquisition — 156
  9.7 Data Verification and Editing — 159
  9.8 Georeferencing of GIS Data — 160
  9.9 Spatial Data Errors — 164
  9.10 Spatial Data Models — 169
  9.11 Spatial Data Structures — 172
  9.12 Modeling Surfaces — 176
  9.13 Modeling Networks — 177
  9.14 GIS Database and Database Management System — 178

10. Spatial Data Analysis — 187
  10.1 Introduction — 187
  10.2 Data Analysis Terminology — 187
  10.3 Measurement of Length, Perimeter and Area — 187
  10.4 Queries — 190
  10.5 Reclassification — 190
  10.6 Buffering and Neighbourhood Functions — 190
  10.7 Data Integration - Map Overlay — 192
  10.8 Spatial Interpolation — 196
  10.9 Surface Analysis — 202
  10.10 Network Analysis — 207
  10.11 Digital Terrain Visualization — 208

11. GIS Application — 213
  11.1 Introduction — 213
  11.2 Problem Identification — 213
  11.3 Designing A Data Model — 214
  11.4 Project Management — 215
  11.5 Implementation Problems — 216
  11.6 Project Evaluation — 216
  11.7 Case Studies — 216

12. Future Trends — 241
  12.1 Introduction — 241
  12.2 Advances in Remote Sensing — 241
  12.3 Classification Accuracy Assessment — 255
  12.4 Advances in GIS — 260
  12.5 Internet GIS — 260
  12.6 Mobile GIS — 266
  12.7 Open GIS Consortium (OGC) — 271
  12.8 Decision Support System — 274

References — 281

Index — 287

REMOTE SENSING 1

1.1 INTRODUCTION

Remote sensing is defined as the process or technique of obtaining information about an object, area, or phenomenon through the analysis of data acquired by a device without being in contact with the object, area, or phenomena being studied (Chandra, 2002b). It consists of the interpretation of measurements of electromagnetic energy reflected from or emitted by a target from a vantage-point that is distant from the target (Mather, 1999). It is a methodology employed to study from a distance the physical and chemical characteristics of objects. Human sight, smell, and hearing are examples of rudimentary forms of remote sensing. Elements of photographic interpretation is considered a part of remote sensing, however, it is generally limited to study of images recorded on photographic emulsions sensitive to energy in or near the visible portion of the electromagnetic spectrum.

Remote sensing discussed in this book is mainly based on the electromagnetic energy sensors which are being operated from space borne platforms, and record energy in more quantifiable formats over a much broader range of the electromagnetic spectrum. Most of the remote sensing methods make use of reflected infrared band, thermal infrared band, and microwave portions of the electromagnetic spectrum.

Fig. 1.1 illustrates the schematic representation of remote sensing processes and its subsequent use in Geographic Information System (GIS) environment.

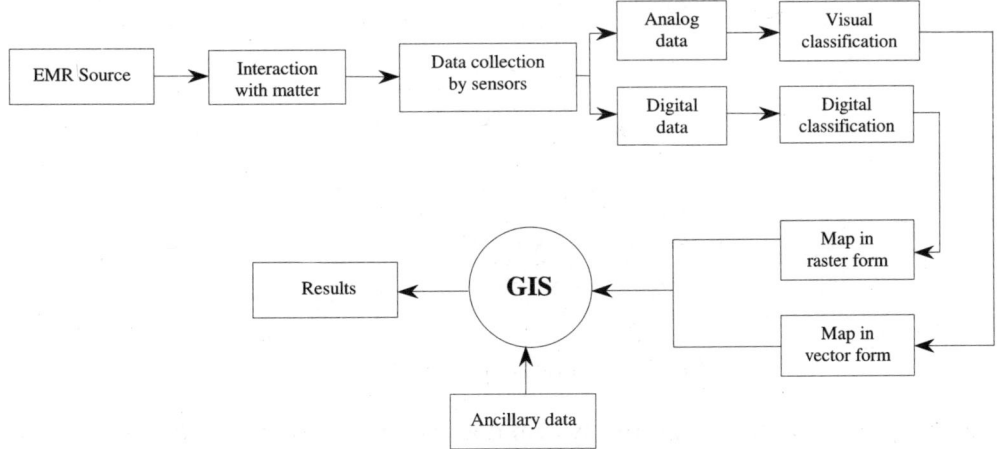

Fig. 1.1 Integration of remote sensing data in GIS

1.2 REMOTE SENSING SYSTEM

An ideal remote sensing system shown in Fig. 1.2, consists of source of electromagnetic energy, energy propagation, energy interaction, return signal, recording, and output for users (Chandra, 2002b). The source of electromagnetic energy provides a high level energy over all wavelengths at a known constant intensity. This energy passes through a non-interfering atmosphere where there is no loss of energy, and falls on a target. Depending upon the characteristics of the target, the incident energy interacts with the target, and generates unique and uniform reflected and/or emitted energy in all wavelengths.

The super sensor having the capability to record the reflected and/or emitted energy from the target in all wavelengths, records the spatial information in spectral form. The super sensor should be simple, compact, and accurate in operation having virtually no power requirement.

The information recorded by the super sensor is transmitted to a real-time data handling system on ground where it is processed instantaneously in an interpretable form to make possible the identification of all the features uniquely characterized by their physical, chemical, and biological characteristics. The interpretable data become available to users who are supposed to have in depth knowledge of making use of these data in their respective fields.

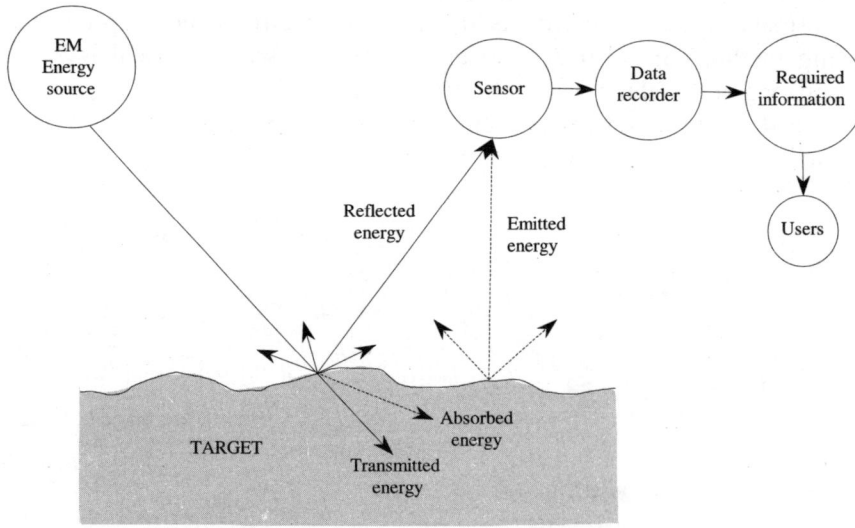

Fig. 1.2 An ideal remote sensing system

However, an ideal remote sensing system does not exist in real world as the components of an ideal remote sensing system have the following shortcomings.

(*i*) There is no energy source that emits uniform energy both spatially and temporally.

(*ii*) The constituent gases of the atmosphere and water vapour molecules and dust particles present in the atmosphere, interact with the energy leading to modification of strength and spectral distribution.
(*iii*) Same matter under different conditions may have different spectral response. Also different matters may have similar spectral response.
(*iv*) In reality there is no ideal super sensor which can accommodate all wavelengths of the electromagnetic spectrum.
(*v*) Due to some practical limitations, sometimes the data transmission and interpretation are not in real time. The transmitted data may also be not in the form which a user may desire, and thus again, the user may not receiving the data in desired form in real time.
(*vi*) All the users may not have sufficient knowledge of data acquisition, analysis, and interpretation of remote sensing data.

1.3 HISTORICAL DEVELOPMENT OF REMOTE SENSING

The concept of remote sensing was developed with the invention of photography using a camera in the nineteenth century, and astronomy was one of the first fields to exploit this technique. Subsequently, this technique was employed in military to obtain information about enemy, and to make strategy for war. Free-flying aircrafts were used during American Civil War to acquire information through remote sensing using aerial photographs for the movement of troops and supplies, reinforcement of fortifications, and in assessing effects of bombardments. Government-sponsored researches produced sensor technology for use in remote sensing for military use and subsequently for civilian applications. After the Second World War, the remote sensing technique was applied for environmental assessment, civilian mapping, and terrain assessment applications. Remote sensing was extended to space after 1960 with the development of space rockets and satellites.

The new era of remote sensing can be considered under military reconnaissance satellites, manned space flights, meteorological satellites, and earth resources satellites.

Space-monitoring capabilities of remote sensing using satellites for war were enhanced after 1960 by launching the reconnaissance satellites within the CORONA, ARGON and LANYARD programmes to provide intelligence data in the form of photographic data acquired at low altitudes. Later stereoscopic images were obtained with a resolution of 2 m. The early satellites were having the short duration missions of seven to eight days, but the later generations have longer duration of missions providing data for hundreds of days.

The manned space flight programmes were started in 1961 by USA, resulting into the first lunar landing in 1969. These programmes included Mercury programme, obtaining photographs of the Earth, systematic approach to remote sensing through Gemini project (1965-66), Apollo missions (1968-75) for lunar landing and remote sensing of the Earth, Skylab missions (1973-74) for earth resources experiments, space shuttle missions started in 1981 and multispectral images with a resolution of 100 m obtained in the visible and near infrared in nine separate bands.

Developments within the Russian space programs were also going on parallel to those of the United States. To name a few are lunar landing of Cosmonaut Yuri Gagarin, the first person to go in space on April 12, 1961, Vostak programme (1961-1963), Voskhod programme (1964-1965), Soyuz spacecraft, Salyut space station (first on April 19, 1971).

The first meteorological satellite was launched by the United States on April 1, 1960, for weather forecast, movement of hurricanes, and other associated uses. The first in a series of satellites which image large areas of the Earth with a high repeat cycle was TIROS-1 (Television and Infrared Observation Satellite).

The first satellite specially dedicated to resource management was launched in 1972. It was named as ERTS-1 (Earth Resources Technology Satellite), and it was well suited for agricultural purposes. These satellites are now referred to as Landsat. These satellites were aimed at to acquire data from the Earth's surface on a systematic, repetitive, medium-resolution, multispectral basis. Later in 1978, the first radar remote sensing satellite (SEASAT) was launched providing data to the public only for three months. First of the SPOT series having stereoscopic capabilities, was launched in 1985 by France. The first remote sensing satellite in India was launched in 1988, and was named as IRS (Indian Remote Sensing). Japan also launched its own Japanese Earth Resources Satellite (JERS) and the Marine Observation Satellite (MOS). Since 1975, China has been periodically launching its own satellites, the data from these satellites not freely available. The European consortium has launched radar satellites ERS (European Radar Satellite) in 1991 and 1995, and RADARSAT by Canada in 1995.

The operational dates for different remote sensing platforms are given in Fig. 1.3.

1.4 MULTI-CONCEPT OF REMOTE SENSING

Remote sensing may be viewed as an integral part of a larger Information system. In many applications remote sensing inputs are key component of a continuing cycle of decision making. Fig 1.4 shows a simple closed-loop cycle, unencumbered by the various feedback loops. The starting point, also the end point, is the set of panels labeled as Information requirements. This focuses on the ultimate driver of any information management system which is the user (his recurring needs). Various disciplines concurred with Earth observations and resources are represented. The terrestrial globe in the background symbolizes the worldwide scope of the system. Information requirements logically lead to user/customer demands.

The best current and future uses of most Earth observing data from satellites stem from correlating and interleaving this type of data with various other types that together are essential inputs to decision making and applications models. The multi-concept of remote sensing data consists of the following.

Multi-station images involve successive overlapping pictures along a flight path, using an aircraft or a spacecraft for better perception of three-dimensional features and improved signal-to-noise ratio.

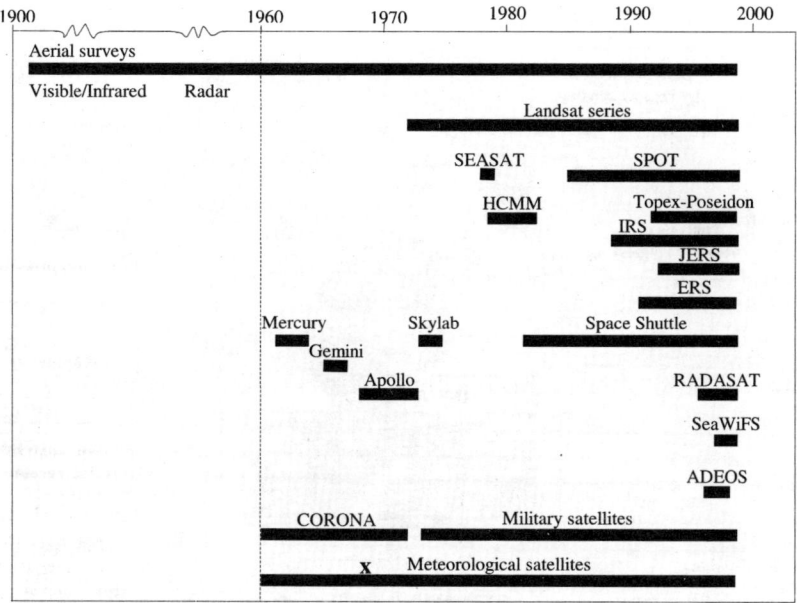

Fig. 1.3 Operational dates for different remote sensing platforms (*Source*: *Gibson, 2000*)

Multi-band images which exploit the fact that each type of feature tends to exhibit a unique type of tonal signature, thus, when brightness values seen in a series of imagery taken in different wavelength bands are suitably combined, it is possible to unambiguously identify specific terrestrial features.

Multi-date images that involve a comparative analysis of the images taken on a series of pre-planned dates, can provide an additional handle for identifying the signature since many features exhibit dynamic characteristics.

Multi-stage images involve a multistage sampling scheme. It gives progressively more detailed information from successively smaller sub-samples of the area being studied. A three-step process involving observations from space, aircraft, and ground is normally used.

Multi-polarization images enabling the delineation of features are based on the polarization of the reflected radiation. This approach exploits the fact that the some features such as a water body may reflect strongly polarized radiation whereas features such as vegetation or fractured rock may reflect weakly polarized radiation.

Multi-enhancement images involve the combination of multi-date, multi-band, and multi-polarization images to suitably generate composite images.

Multi-disciplinary analysis involves analyzing the data by two or more analyst from different disciplines to obtain a more accurate and complete information about the total Earth resource of an area. The results of such multidisciplinary analysis are usually presented in a set of multi-thematic.

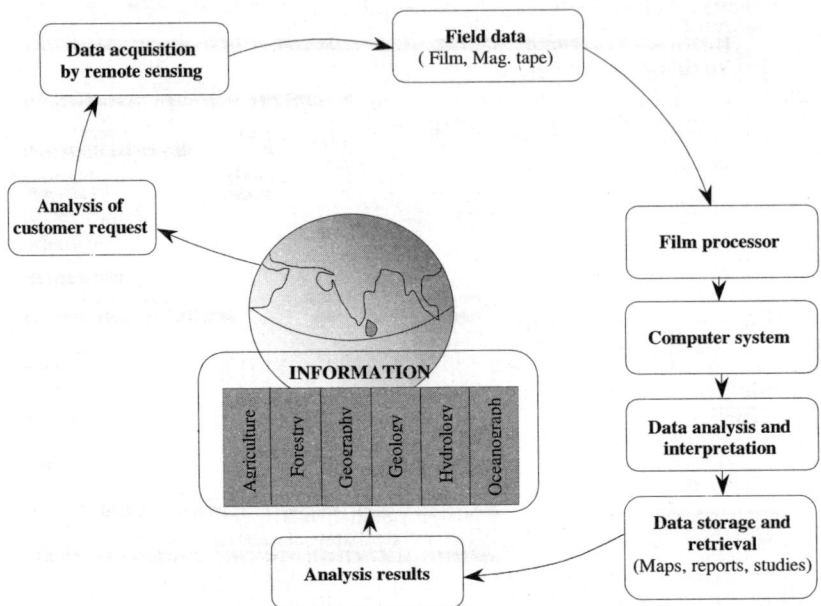

Fig. 1.4 Multi-concept of remote sensing

The bulk of the data in such systems have in common a geographical significance, and they are tied to definite locations on the Earth.

As vast amounts of spatial or geographically referenced data may have to be gathered, stored, analyzed in terms of their interrelations, and rapidly retrieved when required for day to day decisions, a Geographic Information System (GIS) that accepts these data must itself be automated (computerized) to make it efficiently utilized. GIS has a great role as a unifying means of handling geospatial data, including often mandatory inputs from remote sensing.

1.5 ADVANTAGES AND DISADVANTAGES OF REMOTE SENSING

The data acquired through remote sensing have several advantages. Satellite images provide useful information in various wavelengths, and they can be stored as permanent records for use in future. Since the imagery covers a large area, it becomes possible to make regional survey on a variety of themes and identification of large features such as landforms. Repetitive coverage has an advantage of monitoring dynamic themes like water, agriculture, land degradation, urban development, etc., which are encompassed under natural and human-induced effects. It also makes possible easy acquisition of data over inaccessible areas. Remote sensing provides data acquisition at different scales and resolutions. A single remote sensing image can be used for different purposes and applications. The data analysis can be performed in the laboratory which reduces the field work making the remote sensing data cost effective. Map revision at medium to small scales using remote sensing data is economical and faster. Colour composites produced from three individual band images, provide better analysis than using a single land images or aerial photographs. A three-dimensional analysis can be carried out using stereo-satellite data. The remote sensing data being in digital form, processing and analysis can be done faster using computer.

Remote sensing data, besides having numerous advantages, have some disadvantages. Remote sensing requires trained and experienced personnel for data processing and analysis. It becomes expensive affair if applied for a small area, particularly for one-time analysis. Satellite data cannot be used for preparing large scale engineering maps. Softwares used for processing the data are costly. Any interpretation based solely on remotely sensed data should be used with caution unless supported by ground verification.

1.6 SOME APPLICATIONS OF REMOTE SENSING

This section briefly describes some of the ways of using images acquired by remote sensing, by classifying these broadly into five categories.

(*i*) If accurate and/or up-to-date maps of an area are not available, the remote sensing image can be treated as rudimentary map for use as base map on which information obtained from other sources can be portrayed.

(*ii*) Remote sensing images can be used to delineate aerial extent or pattern of features for determination of their relative extent, or measurement of their respective areas. These require that images should be corrected for geometric errors.

(*iii*) Remotely sensed images provide means of preparing an inventory of different classes of features in the region.

(*iv*) Remotely sensed images can be used to assess the condition, or status of specific areas.

(*v*) Sometimes, the remotely sensed images can also be used for quantitative measurements of some properties of landscape surface.

Remote sensing technology offers the potential to produce a broadly consistent database at spatial, spectral, and temporal resolution, useful for resource management. Earth is a finite planet with limited resources. With growing population and rising standard of living, pressure on natural resources is increasing day by day. It, therefore, becomes necessary to manage the available resources wisely and prudently. Use of remote sensing data can be made effectively and economically for periodic preparation of accurate inventories of natural resources both renewable and non-renewable, and also for managing and monitoring the natural resources.

Since remote sensing is a technique which can provide information about the Earth's surface including the areas covered by water, it finds unlimited applications. It is being used to collect the information about agriculture, forestry, geography, geology, archeology, weather and climate, marine environment, water resources management and assessment, engineering, etc.

Analysis of land use and land cover provides a base for exploration of natural resources. Remote sensing also has application in study of natural hazards such as earthquakes, floods, land slides, and land subsidence. Table 1.1 broadly summarizes the use of remote sensing in various disciplines.

Table 1.1 Some application areas of remote sensing

Agriculture and forestry	Land use mapping	Geology	Water resources	Oceanography and marine	Environment
Discrimination of vegetation types	Classification of land use	Recognition of rock types	Determination of water boundaries	Detection of living marine organism	Monitoring of surface mining and reclamation
Measurement of crop species by acreage	Cartographic mapping and map updation	Mapping of major geological units	Mapping of floods and flood plains	Determination of turbidity patterns and circulation	Mapping and monitoring of water pollution
Measurement of timber acreage and volume by species	Categorization of land suitability	Revising geological maps	Determination of aerial extent of snow and snow boundaries	Mapping of shoreline changes	Detection of air pollution and its effects
Determination of range readiness and biomass	Distinguishing Urban and rural areas	Delineation of unconsolidated rocks and soils	Measurement of glacial features	Mapping of shoals and shallow areas	Determination of effects of natural disaster
Determination of vegetation vigour	Regional planning	Mapping igneous intrusions	Measurement of sediment and turbidity patterns	Mapping of ice	Monitoring environmental effects of human activities
Determination of vegetation stress	Management of transport networks	Mapping recent volcanic surface deposits	Inventory of lakes	Study of waves and eddies	
Determination of soil condition	Mapping of land water boundaries	Mapping of landforms	Delineation of irrigated fields		
Determination of soil association		Determination of regional structures			
Assessment of forest fire		Mapping of lineaments etc.			

ELECTROMAGNETIC RADIATION 2

2.1 INTRODUCTION

Remote sensing relies on the measurement of Electromagnetic (EM) energy. EM energy can take several different forms. One of the most important sources of EM energy is the Sun which provides energy at all wavelength. Some of these wavelengths are important to us, such as wavelength in the visible region which provides light and energy in ultraviolet wavelength which can be harmful to human skin.

Many of the sensors used in remote sensing measure reflected sunlight. Some sensors, however, detect energy emitted by the Earth itself or provide their own energy. To understand the principle of the remote sensors, a basic understanding of EM energy, its characteristics and interactions with matter are required. This knowledge is also necessary in order to interpret remote sensing data correctly.

In this chapter, EM energy, its source, and the different parts of the electromagnetic spectrum are explained. The interactions between EM energy with the atmosphere and Earth's surface have also been discussed.

2.2 ELECTROMAGNETIC ENERGY

Electromagnetic energy can be modeled by waves or by energy bearing particles called *photons*. In the wave model, electromagnetic energy is considered to propagate through space in the form of sinusoidal waves. These waves are characterized by electrical field (E) and magnetic field (M) both perpendicular to each other, and for this reason the term electromagnetic energy is used. The vibration of both fields is perpendicular to the direction of travel of the wave (Fig. 2.1). Both fields propagate through space at the speed of light c which is 299,790,000 ms^{-1}, and can be rounded off to 3×10^8 ms^{-1}.

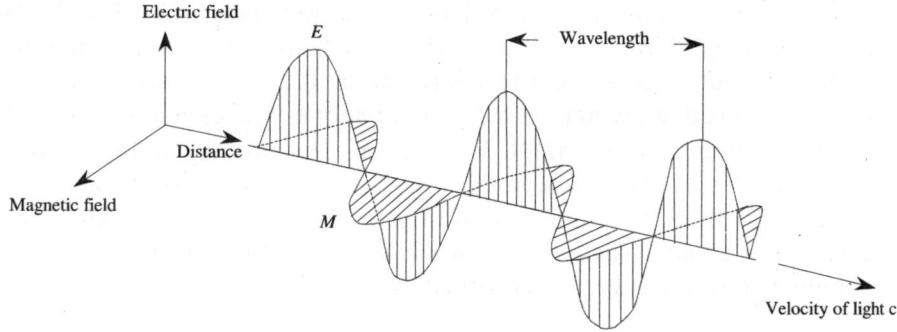

Fig. 2.1 The wave model of electromagnetic energy

The wavelength λ of electromagnetic waves, particularly important for understanding remote sensing, is defined as the distance between successive wave crests. Wavelength is measured in metres (m) or some fraction of metres, such as nanometers (nm, 10^{-9} metres) or micrometers (μm, 10^{-6} metres).

The frequency v of the electromagnetic energy is the number of cycles of a wave passing a fixed point over a specific period of time. Frequency is normally measured in hertz (Hz), which is equivalent to one cycle per second. Since the speed of light is constant, wavelength and frequency are inversely related to each other.

$$v = \frac{c}{\lambda} \qquad \ldots(2.1)$$

Most characteristics of EM energy can be described using the *wave* model. For some purposes, however, EM energy modeled by particle (photons) theory is more convenient to use. This approach is considered when quantifying the amount of energy measured by multi-spectral sensor. The amount of energy held by a photon of a specific wavelength is given by

$$Q = hv$$

or

$$= h\frac{c}{\lambda} \qquad \ldots(2.2)$$

where Q is the energy of a photon J and h is Planck's constant (6.6262×10^{-34} Js). From Eq. (2.2) it follows that the longer the wavelength, the lower is its energy content. Gamma rays (around 10^{-9} m) are the most energetic, and radio waves (around >1 m) are the least energetic. It may be noted that it is easier to measure shorter wavelengths than the larger wavelengths.

Electromagnetic Spectrum and its Characteristics

All matter with a temperature above Absolute zero (0° K) radiate EM energy due to molecular agitation in which movement of the molecules is taking place. This means that the Sun, and also the Earth, radiates energy in the form of waves. The matter capable of absorbing and re-emitting all EM energy is known as a *blackbody*. For blackbodies, both the emissivity (ε) and the absorptance (α) are equal to 1.

The amount of energy radiated by an object depends on its absolute temperature and emissivity, and it is a function of the wavelength. The radiation emitted by a blackbody at different temperatures is shown in Fig 2.2. The area below the curve represents the total amount of energy emitted at a specific temperature. It can be concluded that a higher temperature corresponds to a greater contribution of shorter wavelengths. The peak radiation at 400° C is around 4 μm while at 1000° C it is 2.5 μm. The emitting ability of a *real material* compared to that of the blackbody, is referred to as the *emissivity* of a material. In reality, blackbodies are hardly found in nature, and most natural objects have emissivity less than one. This means that only a part, usually between 80-98% of the received energy, is re-emitted, and the remaining part of the energy is absorbed.

Electromagnetic Spectrum

All matters at a certain temperature radiate electromagnetic waves of various wavelengths.

The total range of wavelengths extending from gamma rays to radio waves is commonly referred to as the *electromagnetic spectrum* (Fig. 2.3).

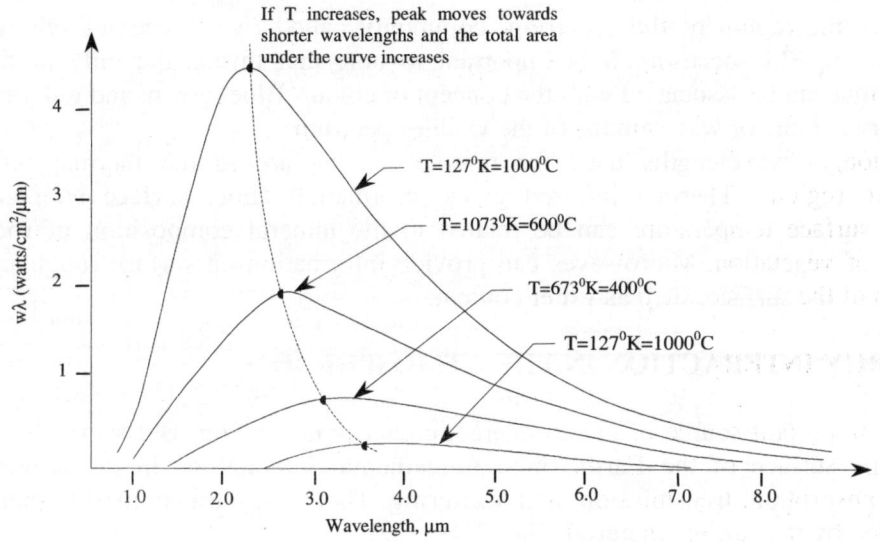

Fig. 2.2 Radiations from a blackbody

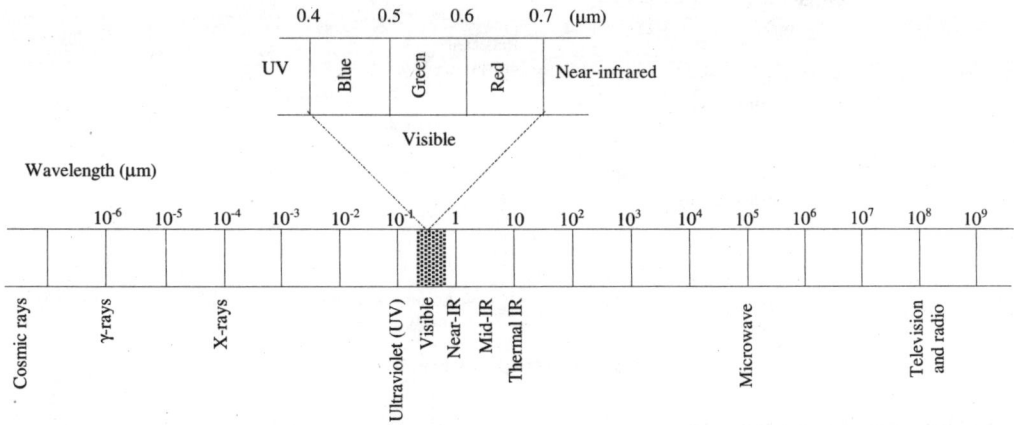

Fig. 2.3 Electromagnetic spectrum

Remote sensing operates in several regions of the electromagnetic spectrum. The optical part of the electromagnetic spectrum refers to that part of the electromagnetic spectrum in which optical laws can be applied. These relate to phenomena, such as reflectance and refraction that can be used to focus the radiation. The optical range extends from X-rays (0.002 μm) through the visible part of the electromagnetic spectrum including far infrared (1000 μm). The ultraviolet portion of the spectrum has the shortest wavelengths that are of practical use for remote sensing. This radiation is beyond the violet portion of the visible wavelengths. Some of the Earth's surface materials, such as primary rocks and minerals, emit visible light when illuminated with ultraviolet radiation. The microwave range covers wavelengths from 1mm to 1m.

The visible region of the spectrum, commonly called *light*, occupies relatively small portion in the EM spectrum. It is important to note that this is the only portion of the spectrum that can be associated with the concept of colour. Blue, green, and red are known as the primary colours or wavelengths of the visible spectrum.

The longer wavelengths used for remote sensing are in the thermal infrared and microwave regions. Thermal infrared gives information about surface temperature. For example, surface temperature can be related to the mineral composition of rocks or the condition of vegetation. Microwaves can provide information on surface roughness and the properties of the surface, such as water content.

2.3 ENERGY INTERACTION IN THE ATMOSPHERE

The most important source of electromagnetic energy is the Sun. Before the Sun's energy reaches the surface of the Earth, three fundamental interactions in the atmosphere are possible; absorption, transmission, and scattering. The energy transmitted is then reflected or absorbed by the surface material (Fig. 2.4).

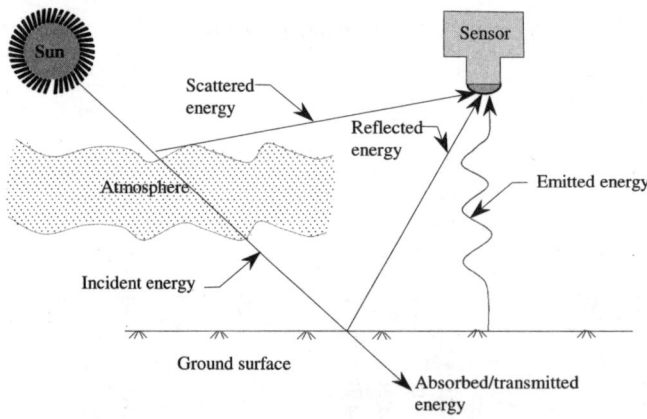

Fig. 2.4 Energy interaction

Absorption and Transmission

Electromagnetic energy travelling through the atmosphere is partly absorbed by various molecules. The most efficient absorbers of solar radiation in the atmosphere are ozone (O_3), water vapour (H_2O), and carbon dioxide (CO_2). Fig. 2.5 gives a schematic representation of the atmospheric transmission in the 0 to 22 µm wavelength region. It may be seen that about half of the spectrum in this region is useless for remote sensing of the Earth's surface, simply because none of corresponding energy can penetrate the atmosphere. Only the wavelength regions outside the main absorption bands of the atmospheric gases can be used for remote sensing. These regions are referred to as the *atmospheric windows*, and include a window in the visible and reflected infrared region, between 0.4 to 2.0 µm where the (optical) remote sensors as well as the human eye operate, and three windows in the thermal infrared region, namely two narrow windows around 3 µm 5 µm, and a third relatively broad window extending from approximately 8 to 14 µm.

Due to the presence of atmospheric moisture, the strong absorption bands are found at longer wavelengths. There is hardly any transmission of energy in the region from 33 µm to 1 mm. The more or less transparent region beyond 1 mm is the microwave region.

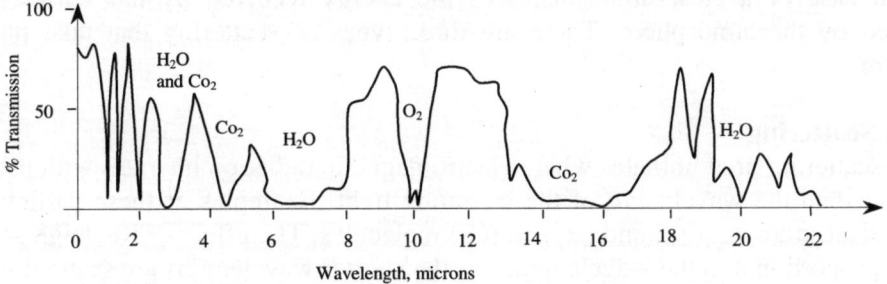

Fig. 2.5 Absorption and transmission (*Source: Lilliesand and Kiefer, 1998*)

The solar spectrum as observed both with and without the influence of the Earth's atmosphere is shown in Fig. 2.6. It may be found that the radiation curve of the Sun (measured outside the influence of the Earth's atmosphere) resembles a blackbody curve at 6000° K. Further, on comparing this curve with the radiation curve measured at the Earth's surface, the relative dips in the curve indicate the absorption by different gases in the atmosphere.

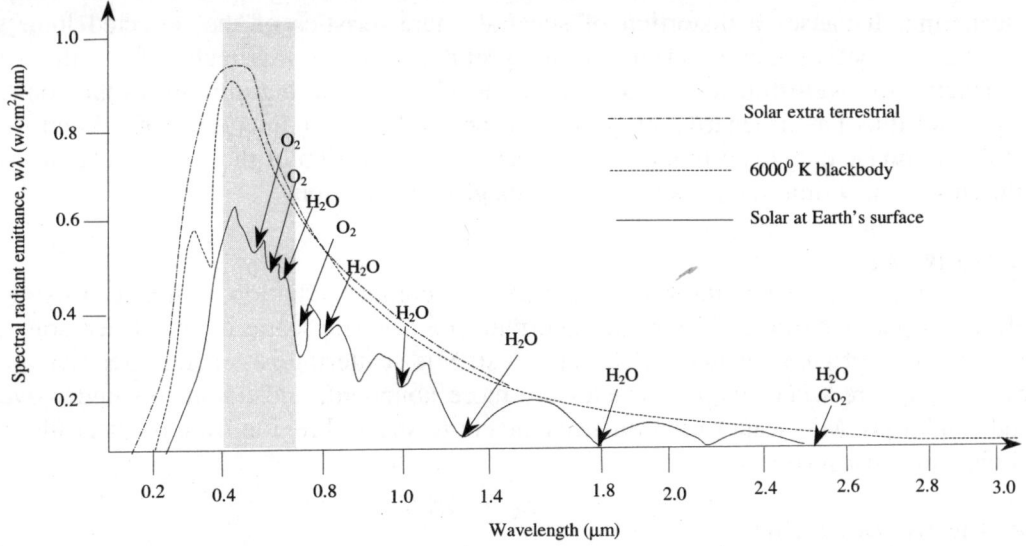

Fig. 2.6 Solar spectrum (*Source: Lilliesand and Kiefer, 1998*)

Atmospheric Scattering

Atmospheric scattering occurs when the particles or gaseous molecules present in the atmosphere cause the electromagnetic waves to be redirected from their original path. The amount of scattering depends on several factors including the wavelength of the radiation, the amount of particles and gases present in the atmosphere, and the distance traveled by the radiation through the atmosphere. In the visible wavelengths, 100% (in case of cloud cover) to 5% (in case of a clear atmosphere) of the energy received by the sensor is directly contributed by the atmosphere. There are three types of scattering that take place in the atmosphere.

Rayleigh Scattering

Rayleigh scattering predominates where electromagnetic radiation interacts with particles that are smaller than the wavelength of the incoming light. Examples of these particles are tiny specks of dust, nitrogen (N_2) and oxygen (O_2) molecules. The effect of Rayleigh scattering is inversely proportional to the wavelength, *i.e.*, the shorter wavelengths are scattered more than the longer wavelengths.

In the absence of particles and scattering, the sky would appear black. In daytime, the sun rays travel the shortest distance through the atmosphere. In that situation, Rayleigh scattering causes a clear sky to be observed as blue because this is the shortest wavelength the human eye can observe. At sunrise and sunset, however, the sunrays travel a longer distance through the Earth's atmosphere before they reach the surface. All the shorter wavelengths are scattered after some distance, and only the longer wavelengths reach the Earth's surface and as a result, the sky appears orange or red.

In the context of satellite remote sensing, Rayleigh scattering is the most important type of scattering. It causes a distortion of spectral characteristics of the reflected light when compared to measurements taken on the ground. Due to Rayleigh effect, the shorter wavelengths are overestimated. In general, the Rayleigh scattering diminishes the contrast in photos, and thus has a negative effect on the possibilities for interpretation. When dealing with digital image data (as provided by scanners), the distortion of the spectral characteristics of the surface may limit the possibilities for image classification.

Mie Scattering

Mie scattering occurs when the wavelength of the incoming radiation is similar in size to that of the atmospheric particles. The most important cause of Mie scattering is the aerosols which are the mixture of gases, water vapour, and dust. Mie scattering is generally restricted to the lower atmosphere where larger particles are more abundant, and dominates under overcast cloud conditions. Mie scattering influences the entire spectral region from the near ultraviolet including the near infrared.

Non-selective Scattering

Non-selective scattering occurs when the particle size is much larger than the radiation wavelength. Typical particles responsible for this effect are water droplets and larger dust particles. Non-selective scattering is independent of wavelength, with all wavelengths scattered about equally. The most prominent example of non-selective scattering includes the effect of clouds (consisting of water droplets). Since all wavelengths are scattered equally, a cloud appears white. Optical remote sensing cannot penetrate clouds.

2.4 ENERGY INTERACTIONS WITH THE EARTH'S SURFACE

When electromagnetic energy is incident on any given feature of earth surface, three fundamental energy interactions with the feature are possible. Various fractions of the energy incident on the feature as shown in Fig. 2.4 are reflected, absorbed, and/or transmitted energy. Considering that all energy components are function of wavelength, from the principle of conservation of energy, the interrelationship between the three energy integrations can be written in the form of an energy balance equation as below.

$$E_I(\lambda) = E_R(\lambda) + E_A(\lambda) + E_T(\lambda) \qquad \ldots (2.3)$$

where E_I = the incident energy,
E_R = the reflected energy,
E_A = the absorbed energy, and
E_T = the transmitted energy.

The proportions of energy reflected, absorbed, and transmitted vary for different earth features, and depend on the material type and condition of the feature, and these variations permit to distinguish different features appearing on an image. Further, even within a given feature type, the proportion of reflected, absorbed, and transmitted energy vary at different wavelengths. Thus, two features may be indistinguishable in one spectral range and be different in another wavelength band. These spectral variations within the visible portion of the spectrum result in the visual effect called *colour*. For example, objects appear blue when they reflect more of the blue portion of the spectrum and green when they reflect more highly in the green spectral region, and so on. Spectral variations in the magnitude of reflected energy is utilized to discriminate between various objects by interpretation.

Due to the fact that many remote sensing systems operate in the wavelength regions in which reflected energy is dominant, the reflectance properties of earth features play very important role in distinguishing different features. In view of this, the modified form of Eq. (2.3) written as Eq. (2.4) in which the reflected energy is taken equal to the energy incident on a given feature reduced by the energy that is either absorbed or transmitted by that feature, is often found very useful.

$$E_R(\lambda) = E_I(\lambda) - [E_A(\lambda) + E_T(\lambda)] \qquad \ldots (2.4)$$

The geometry of reflected energy from an object is also an important consideration which is primarily a function of the surface roughness of the object. Specular reflection is obtained from flat surfaces that manifest mirror-like reflection where the angle of reflection equals the angle of incidence. Diffuse (or Lambertian) reflection is from rough surfaces that reflect uniformly in all directions. From most earth surfaces, a mixed reflection is obtained having characteristics somewhat between the two extremes.

Fig. 2.7 illustrates the geometric character of surface giving specular, near-specular, near-diffuse, and diffuse reflections. The category that characterizes any given surface is dictated by the surface roughness in comparison to the wavelength of the energy incident upon it. For example, in the visible portion of the electromagnetic spectrum a sandy beach appears rough whereas in the relatively long wavelength of the energy it can appear smooth to incident energy. In short, when the wavelength of incident energy is much smaller than the surface height variation or the particle sizes that make up a surface, the reflection from the surface is diffuse.

Diffuse reflections contain spectral information on the colour of the reflecting surface, whereas specular reflections do not. Hence in remote sensing, measuring the diffuse reflectance properties of terrain features is an important factor.

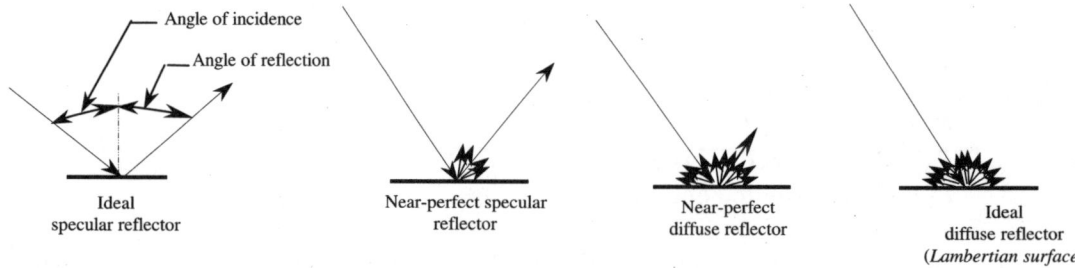

Fig. 2.7 Specular and diffuse reflectance (*Source: Lilliesand and Kiefer, 1998*)

The reflectance characteristics of earth surface features may be quantified by measuring the reflected portion of incident energy as a function of wavelength, called *spectral reflectance* ρ_λ expressed as percentage as below:

$$\rho_\lambda = \frac{\text{Energy of wavelength }(\lambda)\text{ reflected from the object}}{\text{Energy of wavelength }(\lambda)\text{ incident upon the object}} \times 100$$

$$= \frac{E_R(\lambda)}{E_I(\lambda)} \times 100 \qquad \ldots(2.5)$$

A graph showing the spectral reflectance of an object as a function of wavelength is termed as *spectral reflectance curve*. Its configuration gives an insight into the spectral characteristics of an object, and has a strong influence on the choice of wavelength region in which remote sensing data are to be acquired for a particular application. Fig. 2.8 illustrates a highly generalized spectral reflectance curves for deciduous *versus* coniferous trees. The curve for each of these object types is plotted as a ribbon (or envelope) of values and as a single line due to the fact that neither the spectral reflectance of one deciduous tree species and another will be identical nor will the spectral reflectance of trees of the same species be exactly equal.

Spectral Reflectance Curves

The energy reaching surface of a material is called *irradiance* and the energy reflected by the surface is called *radiance*. Each material has its own specific reflectance which is the fraction of the incident radiation that is reflected as a function of wavelength. From such a curve, the degree of reflection for each wavelength can be determined. Most remote sensing sensors are sensitive to larger wavelength bands (*e.g.*, from 400 – 480 nm), and the curve can be used to estimate the overall reflectance in such bands. Reflectance curves are made for the optical part of the electromagnetic spectrum (up to 2.5 µm).

Reflectance measurements can be carried out in a laboratory or in the field using a field spectrometer. The reflectance characteristics of some of the common land cover types are discussed below.

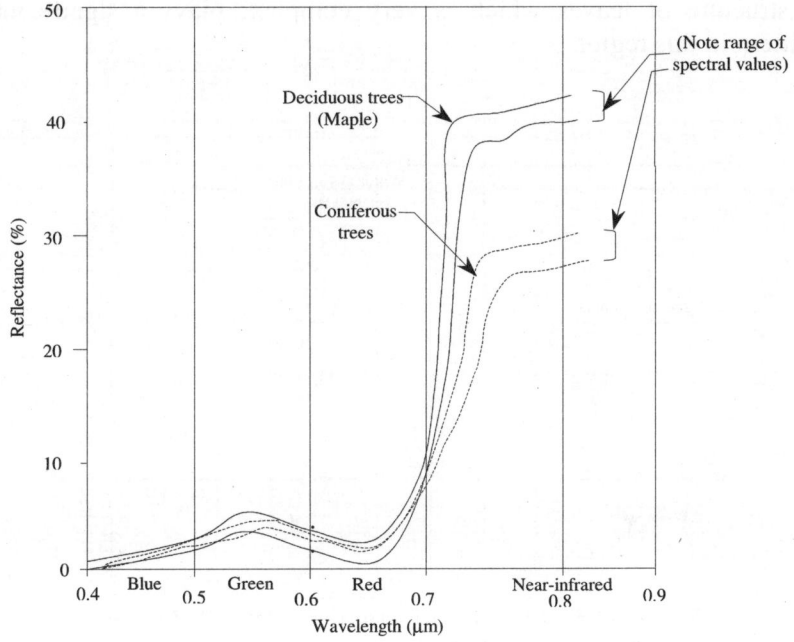

Fig. 2.8 Spectral reflectance curve for trees (*Source: Lilliesand and Kiefer, 1998*)

Vegetation

The spectral reflectance curve of green vegetation is distinctive and quite variable with wavelength (Fig. 2.9). In this we see there is low reflectance in the blue and red regions of the visible spectrum. This low reflectance corresponds to two chlorophyll absorptions band, because chlorophyll present in the green leaf, absorbs most of the incident energy. These absorption band are centered around 0.45 µm, hence a peak occurs due to absence of chlorophyll absorption band. Thus, when a plant is under stress, the chlorophyll content is less, and thus reflectance is high in this part of the spectrum, particularly in the red band, and appears yellow or chlorotic.

There are other pigments also *carotheses* and *xanthophylls* (yellow pigment), and *anthocyanins* (red). Carotheses and xanthophylls are frequently present in leaves, but have absorption band in blue region (0.45 µm) where chlorophyll is dominant; hence they are masked out. In the absence of chlorophyll, these pigments assume prominence. Some trees produce anthocyanins in large quantities, and hence appear red. If the effect of various pigments on the spectral reflectance is studied, it is found that they are large differences in spectral pattern in the visible portion of the spectrum, but there is no significant difference in the near infrared and middle infrared.

In the near infrared portion of the spectrum, there is a marked increase in reflectance beyond 0.7 µm as it passes from visible to infrared. In this region, green vegetation is characterized by high reflectance, high transmittance, and low absorption. It is found that reflectance and transmittance are of the order of 45-50% for each and absorption is about 5%. The internal structure of leaves which is very complex, plays a significant role in the reflectance pattern in this region.

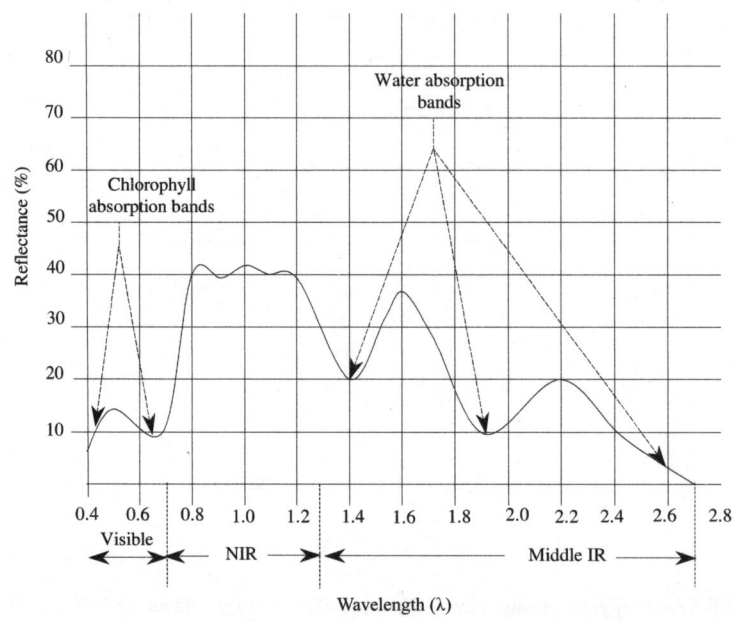

Fig. 2.9 Spectral reflectance curve of green vegetation

It has also been noticed that reflectance from multi-leaf layers as compared to single-leaf layer has high reflectance, and the difference between the two can be of the order of 85%. This is due to additive reflectance energy transmitted to the second layer through the first layer, and the reflected energy from second layer is partially transmitted back through the first layer. This effect has significant impact when data is taken at the centre of the field where there are many layers while at the edge of the field due to single layer, the reflectance is less.

In the middle infrared potion, the spectral response shows the effect on the reflectance due to the presence of dominant water absorption bands is centered at 1.4 µm, 1.9 µm and 2.7 µm. There are also other water absorption bands centered at 0.90 µm and 1.1 µm, which are very weak, and have no impact on the spectral reflectance.

The reflectance peaks in middle infrared occur at 1.6 μm and 2.2 μm. Furthermore, it is observed that leaf-moisture content has large effect. With decrease in moisture content the reflectance increases.

From the above, certain distinct spectral characteristics of green vegetation can be enumerated as below:

(*i*) Distinct differences in the reflectance are found amongst visible, near infrared, and middle infrared regions.
(*ii*) In the visible potion, pigmentation of leaves is the dominant factor.
(*iii*) The internal leaf structure controls the of leaves reflectance in the near infrared where nearly half is transmitted and half is reflected.
(*iv*) In the middle infrared, the total moisture content of the vegetation controls the reflectance where much of the incident energy is absorbed by the leaves.

Bare Soil

Spectral reflectance curves of most soils are not very complex one in appearance (Fig. 2.10). One of the most outstanding reflectance characteristics of dry soils is, generally, an increasing level of reflectance with increase in wavelength, particular in the visible and near infrared regions. The energy-matter interactions are perhaps less complex, as the whole incident energy is either absorbed or reflected, none being transmitted through the material. However, soil is a complex mixture of materials having various physical and chemical properties that can affect the absorptance and reflectance characteristics of soils. Although the shapes of the curves are similar, a number of interrelated soil properties must be considered when discussing the reasons for the difference in the amplitude of the curve. The moisture content, the amount of organic matter, the amount of iron oxide, and the relative percentages of dry sand, silt, and the roughness characteristics of the soil surface all significantly influence the spectral reflectance of soil.

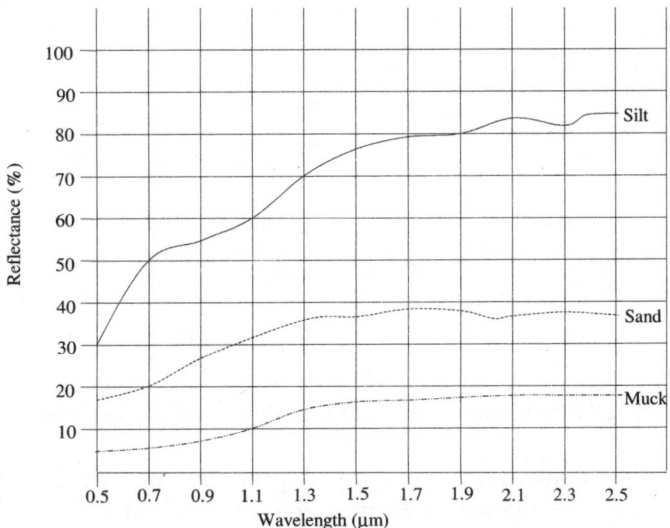

Fig. 2.10 Spectral reflectance of some soil types

The first characteristic of soil to be examined should be soil texture that refers to the proportions of sand, silt and clay. Clay has particles of size less than 0.002 mm in diameter; silt has 0.002 to 0.005 mm diameter particles, and sand 0.05 to 2.0 mm diameter particles. Due to the particle size, large number of particles is present in clay compared to sand. When moisture is present, each particle will be covered by a very thin layer of water, which will occupy some space between the soil particles. Even though the film is very thin, millions of such particles hold a large amount of water. Thus, it is evident that the relation between soil particles and moisture content has a significant impact on soils.

Fig. 2.11 illustrates typical reflectance curves for a sandy soil having different levels of moisture content. The curves show that there is no significant decrease in reflectance in dry sands, but sands having significant moisture contents, have distinct decrease in the reflectance at 1.4, 1.9, and 2.3 µm. In the visible portion of the spectrum, there is also a distinct decrease in reflectance for moist soil as compared to dry soil. So the texture of soil affects the spectral reflectance of soil because of its moisture retention capacity and the size of soil particles. If all the factors remain the same, as the particle size decreases, the soil surface becomes smoother, and more of the incident energy is reflected. It has been found that as the particle size increases from 0.22 mm to 2.66 mm, the increase in absorption is about 14%.

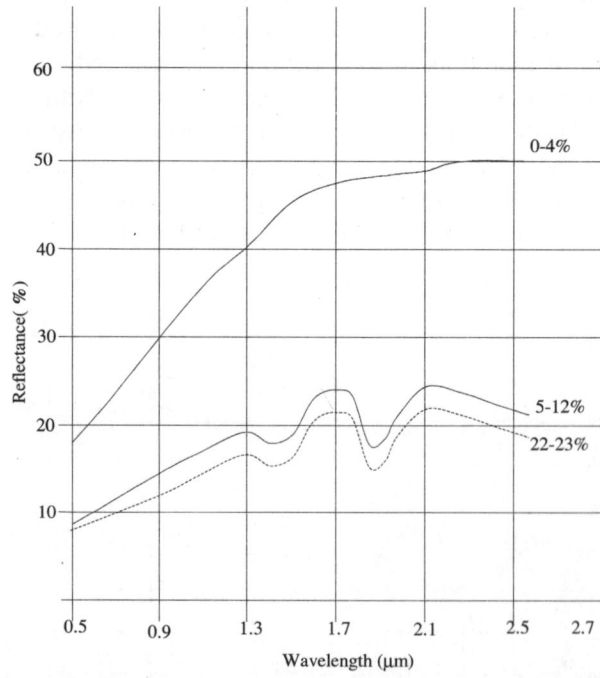

Fig. 2.11 Reflectance curves for a sandy soil at different levels of moisture

The organic matter content is another soil property that significantly influences the reflectance characteristics of the soil indicating the amount and form of nitrogen in the soil. Although it is found that in most temperate zones, the organic content varies from 0.5% to

5%. A soil with 5% organic matter usually appears dark brown or black in colour, where as soils with lower organic content are usually light brown in color or light shades of gray. The relationship between organic matter and reflectance within the visible range of the spectrum is curvilinear.

It has been observed that soils under different climatic conditions may not show the same relationship between colour and organic matter content. Under high temperature, well drained soils (coarse particles) have high organic matter since soils in cooler zone appears brown instead of black. Thus the climatic region and the drainage conditions must be taken into account when considering the relationship between organic matter and spectral reflectance.

Iron oxide present in soils, has significant influence on the spectral reflectance characteristics of soil as it can cause a significant decrease in reflectance at least in the visible bands. Fig. 2.12 shows an excellent inverse relationship between reflectance in the visible region against percentage iron oxide. It is found that removal of iron oxide causes a marked increase in reflectance from 0.5 to 1.1 µm, and the reflectance is insignificant beyond 1.1 µm. It is also seen that removal of organic matter causes a similar marked increase in reflectance over the same ranges of wavelengths (Fig. 2.13).

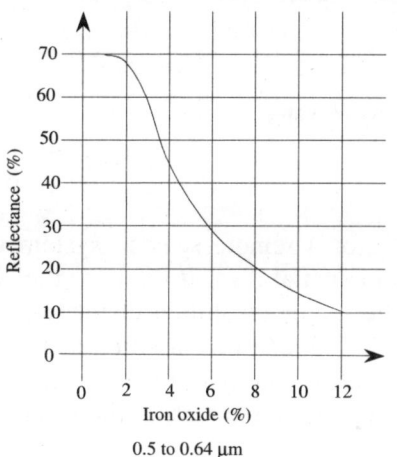

Fig. 2.12 Reflectance of iron oxide

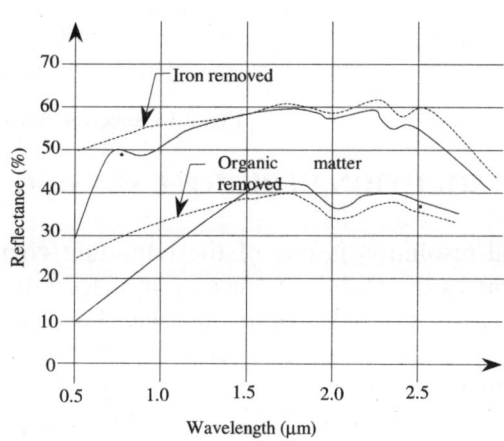

Fig. 2.13 Influence of iron oxide and organic material

On the basis of above, following conclusions can be drawn regarding the spectral characteristic of soil.

(*i*) Increase in moisture content causes a decrease in reflectance throughout the reflective portion of the spectrum.
(*ii*) Texture of soil increases the reflectance characteristics with decrease in particle size.
(*iii*) A decrease in surface roughness causes an increase level of reflectance.
(*iv*) Increase in organic matter causes a decrease in reflectance.
(*v*) Increase in iron oxide may cause a decrease in reflectance.

Water

Compared to vegetation and soils, water has the low reflectance. Vegetation may reflect up to 50%, soils up to 30-40% while water reflects at the most 10% of the incoming radiation. Water reflects electromagnetic energy in the visible to the near-infrared range. Beyond 1200 nm, all energy is absorbed. Some curves of different types of water are given in Fig. 2.14. High reflectance is given by turbid (silt loaded) water whereas water containing plants with chlorophyll II has a reflection peak in the green wavelength.

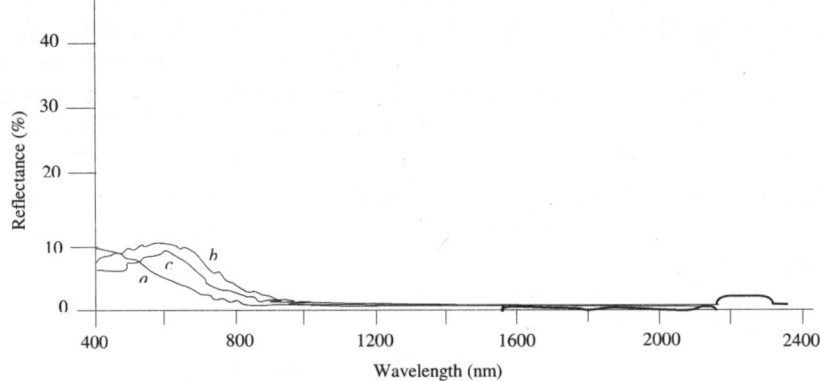

Fig. 2.14 Spectral reflectance curve of water

2.5 RESOLUTION IN REMOTE SENSING

Spatial resolution is one of the important characteristic of a remote sensing system which determines the ability of a sensor in discerning details in spatial data. It refers to the size of the smallest possible feature that can be identified. The major resolution characteristics of remote sensing imaging systems, whether analogue or digital, can be classified as (*i*) spatial resolution, (*ii*) spectral resolution, (*iii*) radiometric resolution, and (*iv*) temporal resolution. The success of data collection in remote sensing technology requires a clear understanding of the resolution characteristics of an imaging system.

Spatial Resolution

Spatial resolution is analogous to the sharpness of the image in conventional photography. In digital images produced by electro-optical scanning system, spatial resolution is described as Instantaneous Field of View (IFOV) or the solid angle, and it is as a linear measure of angles in milliradians. Factors affecting IFOV are height of the imaging platform, size of the detector element, and focal length of optical system. Spatial resolution can be determined using the following formula (Curran, 1985).

$$A = HB \qquad \ldots (2.6)$$

where
A = the ground dimension of the detector element in metres,
H = the flying height of the platform in metres, and
B = the IFOV in milliradians.

The area viewed on the ground is determined by resolution cell, and determines the maximum spatial resolution of a sensor. Images where only large features are visible are said to have *coarse* or *low resolution*. Fine or *high resolution* images make possible detection of small objects.

Spectral Resolution

Spectral resolution refers to the electromagnetic radiation wavelengths to which a remote sensing system is sensitive. It describes the ability of a sensor to define fine wavelength intervals. Finer the spectral resolution, the narrower the wavelength range for a particular band or channel. The two components which are considered in spectral resolution are (*i*) the number of wavelength bands or channels used and (*ii*) the width of each wave band. A higher spectral resolution is achieved from a larger number of bands and narrow band width of each band. It is important to select the correct spectral resolution for the type of information to be extracted by deciding the wave bands to be used.

The spectral resolution of black and white film is fairly coarse, as the various wavelengths of the visible spectrum extending over the visible portion of the electromagnetic spectrum are not individually distinguishable, and the overall reflectance in the entire visible portion is recorded. On the other hand, colour film has higher spectral resolution being sensitive individually to the reflected energy at the blue, green, and red wavelengths of the spectrum.

Many remote sensing systems record energy over several separate wavelength ranges having various spectral resolutions.

Radiometric Resolution

The smallest difference in radiant energy that can be detected is expressed as radiometric resolution. Radiometric characteristics of an image describe the actual information contain in the image. It is the sensitivity of a sensor to the magnitude of the electromagnetic energy. It is applicable to both photographs and digital images. In the case of photographs, radiometric resolution is the ability to resolve subtle changes in gray tones whereas in the case of digital images it refers to the number of discrete levels into which a signal may be divided during the analogue-to-digital conversion, also known as the *quantization levels*.

Temporal Resolution

Temporal resolution refers to the frequency of data collection. In remote sensing, to capture changes in environmental phenomenon occurring, the data may have to be collected either daily, monthly, seasonally or yearly. The ability to collect imagery of the same area of the Earth's surface at different periods of time is one of the most important applications of remote sensing. Change detection using imageries of the same area is only possible with good temporal resolution as the data collection should match the frequency of change.

The *absolute temporal resolution* of a remote sensing system to capture the same area at the same viewing angle a second time is equal to the period a satellite revisit frequency that may be several days. Special characteristics of features may change over time, and these changes can be detected by collecting and comparing multi-temporal imageries.

2.6 PIXEL AND MIXED PIXEL

Remote sensing images, raster in nature, are composed of a matrix of picture elements, called the *pixels*. A pixel represents a smallest unit of ground area in an image. It may be noted that the spatial resolution refers to a sensor used in a remote sensing system whereas the pixel refers to the image collected by the sensor. The pixel size can be changed in studying the ground surface feature but the resolution of sensor remains unchanged. For example, if a sensor has a spatial resolution of 10 m, a pixel at full resolution will represent an area of 10×10 m on the ground. Now reducing the image by four times, the pixel size will become 40×40 m, yet the spatial resolution will remain unchanged.

A pixel may be called as '*pure*' or '*mixed*'. A *pure pixel* occupies a completely single, homogeneous class of information (feature). In situations where a pixel occupies more than one class of information, is called as a *mixed pixel*. In a pure pixel, the average brightness over the entire pixel area represents the brightness of different parts of a non-homogeneous single class of feature. This does not happen in the case of mixed pixels, and the single digital value that represents the pixel may not accurately represent any of the classes present. The fact is that the pure spectral responses of specific features are mixed together with the pure responses of other features. In some cases, the digital values from mixed pixels may not resemble any of the several categories in the scene resulting in error and confusion.

Studies have shown that the number of mixed pixels increases as spatial resolution decreases. Hence, a fine spatial resolution would offer many practical advantages, including capture of fine details. On the other, the substantial increase in number of pixels to achieve this advantage, results in increased cost.

SENSORS AND PLATFORMS 3

3.1 INTRODUCTION

Electromagnetic energy from the Sun in form of reflected energy from the Earth, and the energy emitted energy by the Earth are measured and recorded to derive information for identification of surface features and their characteristics. There is a variation in the measured values of energy as it is dependent on the physical, chemical and biological characteristic of the features. Sensors placed on either static or moving platform make the measurement of electromagnetic radiation. The sensor-platform combination is employed to obtain the characteristics of the resulting image data. Different types of sensors are available for different applications. Aircraft and satellites are used as platforms to carry one or more sensors.

3.2 BROAD CLASSIFICATIONS OF SENSORS AND PLATFORM

Sensors can be classified on the basis of source of energy. The sensors which depend on external source of energy, usually the Sun, are known as *passive sensors*, while the sensors which have their own source of energy are known as *active sensors*. A normal photographic camera is one of the oldest sensors and under different operating conditions act as either as a passive or active sensor. Under good illumination operating condition when flash is not used, the camera behaves as a passive sensor. However, when the camera operates under poor illumination condition using a flash, it becomes an active sensor.

In order for a sensor to collect and record reflected energy or emitted energy from Earth surface, it is placed on a stable platform away from the surface being observed. The platform may be ground based, airborne, or spaceborne.

Sensors placed on ground-based platforms are employed for recording detailed information about the surface. The collected information may be used as reference data for subsequent analysis. Aircrafts or helicopters, which are airborne platforms, are used to collect detailed images over virtually any part of the Earth. Spaceborne-platforms are satellites launched for remote sensing purposes. Since the satellites provide repetitive coverage of an area under study, satellite data products have wide application in various fields due their multi-characteristics, as discussed in Sec. 1.4, and hence in the foregoing sections various platforms and sensors have been discussed in detail.

Satellites are placed into orbits tailored to match the capabilities of the sensors they carry and objectives of each satellite mission. Orbit selection depends upon the altitude of the satellite, and its orientation and rotation relative to the Earth.

If a satellite is positioned in the equatorial plane at an altitude of approximately 36,000 km moving in same direction as the Earth, it will have the same time period as the Earth, and hence appears to be stationary with respect to the Earth's surface. Such satellites are known as geostationary satellites. Geostationary satellites are ideal for meteorological or communication purposes.

If a satellite is positioned in a near north-south orbital plane, and made Sun-synchronous, it is called a *Sun synchronous satellite*. Such satellites have the capability of revisiting the same area under uniform illumination conditions at same local time in different seasons every year. This is an important factor, which helps in observing and analyzing the changes in the appearances of the features within each scene under the same conditions of observation, and does not require corrections for different illumination conditions.

Sun-synchronous satellites are useful for mapping of earth resources. Since they also provide a synoptic view of large area with fine detail and systematic repetitive coverage of land area, they are well suited to monitor many global environmental problems.

3.3 LAND OBSERVATION SATELLITES AND SENSORS

Worldwide three major land observation satellite missions launched are LANDSAT series by U.S.A., SPOT by France, and IRS by India. These satellites are low-altitude satellites having an altitude of less than 1000 km above the Earth's surface and sensors having low spatial resolution and spectral resolution of less than $0.1\mu m$.

LANDSAT Satellites

The first satellite LANDSAT-1 designed to monitor the Earth's surface, initially named as ERTS-1 (Earth Resources Technology Satellite), was launched by NASA, U.S.A, in July 23, 1972. The LANDSAT mission was transferred to NOAA (National Oceanographic and Atmospheric Administration) in 1983, and subsequently commercialized in 1985 for providing data to user community.

All LANDSAT satellites have been placed in a near polar Sun-synchronous orbit. Landsat-1 to -3 satellites was placed at an altitude of nearly 900 km with a revisit period of 18 days. The satellites launched later have an altitude of nearly 700 km with revisit period of 16 days. In order to optimize the illumination conditions, all LANDSAT satellites are desired to have same equatorial crossing time. The special features of these LANDSAT satellites are a combination of sensors with spectral bands tailored to Earth observation, functional spatial resolution, and a good aerial coverage.

A number of sensors have been carried by various LANDSAT satellites, which include Return Beam Vidicon (RBV) camera system, Multi-Spectral Scanner (MSS) system, Thematic Mapper (TM), Enhanced Thematic Mapper (ETM), and Enhanced Thematic Mapper Plus (ETM+). The aerial coverage of each LANDSAT scene is 185 km × 185 km. Various LANDSAT missions are presented in Table 3.1 and sensors used in Table 3.2. A sample of LANDSAT imagery is shown in Fig. 3.1. The operational period of various LANDSAT missions launched uptil now are shown in Fig. 3.2.

SPOT Satellites

SPOT (Systém Pour L' Observation de la Terre) is a series of imaging satellites designed and launched by CNES (Centre National d' Études Spatiales) of France in collaboration with Sweden and Belgium to provide Earth observation data. SPOT satellites ushered a new era in

Table 3.1 Various LANDSAT missions

Satellite	Date of launch	Altitude (km)	Orbit type	Orbital inclination	Orbital period (min)	Time of EQ crossing	Repeat cycle (days)	Sensor	Swath width (km)	Resolution (m)	No. of bands
LANDSAT-1	23.7.72	900	Sun-synchronous	91°	103	9:42 A.M.	18	MSS RBV	185 185	80 80	4 3
LANDSAT-2	22.1.75	900	Sun-synchronous	91°	103	9:42 A.M.	18	MSS RBV	185 185	80 80	4 3
LANDSAT-3	5.3.78	900	Sun-synchronous	91°	103	9:42 A.M.	18	MSS RBV	185 185	80 120 (Thermal band) 40	5 1
LANDSAT-4	16.7.82	705	Sun-synchronous	98.2°	99	9:45 A.M.	16	MSS TM	185 185	75 30 (120 for Band 6)	4 7
LANDSAT-5	1.1.84	705	Sun-synchronous	98.2°	98.9	9:45 A.M.	16	MSS TM	185 185	75 30 (120 for Band 6)	4 7
LANDSAT-6	5.10.93	705	Sun-synchronous	98.2°	98.9	10:00 A.M.	16	PAN TM (ETM)	- 185	13 × 15 30 (120 for Band 6)	1 7
LANDSAT-7	15.4.99	705	Sun-synchronous	98.2°	98	10:00 A.M.	16	ETM + TM Thermal PAN	- 185 -	30 m 60 m 15	6 1 1

Table 3.2 Sensors on board various Landsat mission

Sensor	Band	Spectral range (μm)	Application
RBV	1	0.475 - 0.575	
	2	0.580 - 0.680	
	3	0.090 - 0.830	
	4	0.505 - 0.750	
MSS	1	0.5 - 0.6	Emphasizes sediment-laden water and delineates areas of shallow water
	2	0.6 - 0.7	Emphasizes cultural details
	3	0.7 - 0.8	Emphasizes vegetation boundary between land and water and land forms
	4	0.8 - 1.1	Penetrates atmospheric haze best,
	5	10.4 - 12.6	Hydrothermal Mapping
TM	1	0.45 - 0.52	Coastal water mapping; soil/vegetation and coniferous/deciduous differentiations
	2	0.52 - 0.60	Green reflection from healthy vegetation
	3	0.63 - 0.69	Chlorophyll II absorption for plant species differentiation
	4	0.76 - 0.90	Biomass surveys; water body delineation
	5	1.55 - 1.75	Vegetation and moisture
	6	10.4 - 12.5	Hydro thermal mapping
	7	2.08 - 2.35	Plant heat stress, thermal mapping
ETM	All TM Bands		Same as for TM sensor
	PAN	0.50 - 0.90	
ETM+	Same as ETM with Thermal Band resolution of 60 m		Same as for ETM

remote sensing by introducing linear array sensors having pushbroom scanning facility. The SPOT sensors have pointable optics that enables site-to-site off-nadir view capabilities permitting full scene stereoscopic imaging of the same area. It is of tremendous value for terrain interpretation, mapping, and visual terrain simulation from two different satellite orbital paths. Another advantage of pointable optics is the steerability of the sensor by 27° on either side of the nadir. This allows the sensor to view an area a number of times within the revisit period of 26 days, and to monitor specific locations having dynamic phenomenon. It also increases the chance of obtaining cloud free scenes. Within the equatorial region, the revisit frequency reduces to a time interval of 3 days making possible viewing of an area 7 times during the revisit period of the satellite. If the latitudes close to 40° to 45°, an area can be imaged 11 times during the same revisit period. All satellites are in Sun-synchronous having near-polar orbits at an altitude of 830 km above the Earth, which results in orbit repetition every 26 days. They have equator crossing times around 10.30 AM local solar time. Uptil now five SPOT satellites have been launched.

The first three SPOT satellites *i.e.*, SPOT-1, -2 and -3 were identical, and their payload consisted of two identical HRV (Visible High Resolution) optical instruments, data recorders (on magnetic tapes), and a system for transmitting the images to the ground based receiving stations. Table 3.3 gives details of the orbital parameters and the various sensors carried on board by SPOT-1, -2, and -3 satellites.

Each HRV sensor can acquire the image either in panchromatic mode (P mode), a single wide band in the visible portion of the spectrum, or in multispectral mode (XS mode) in the green, red, and blue portions of the spectrum. The two HRV sensors can function independently or in tandem in either XS or P mode. In the P mode, the image has a spatial resolution of 10 m while in XS mode it is 20 m.

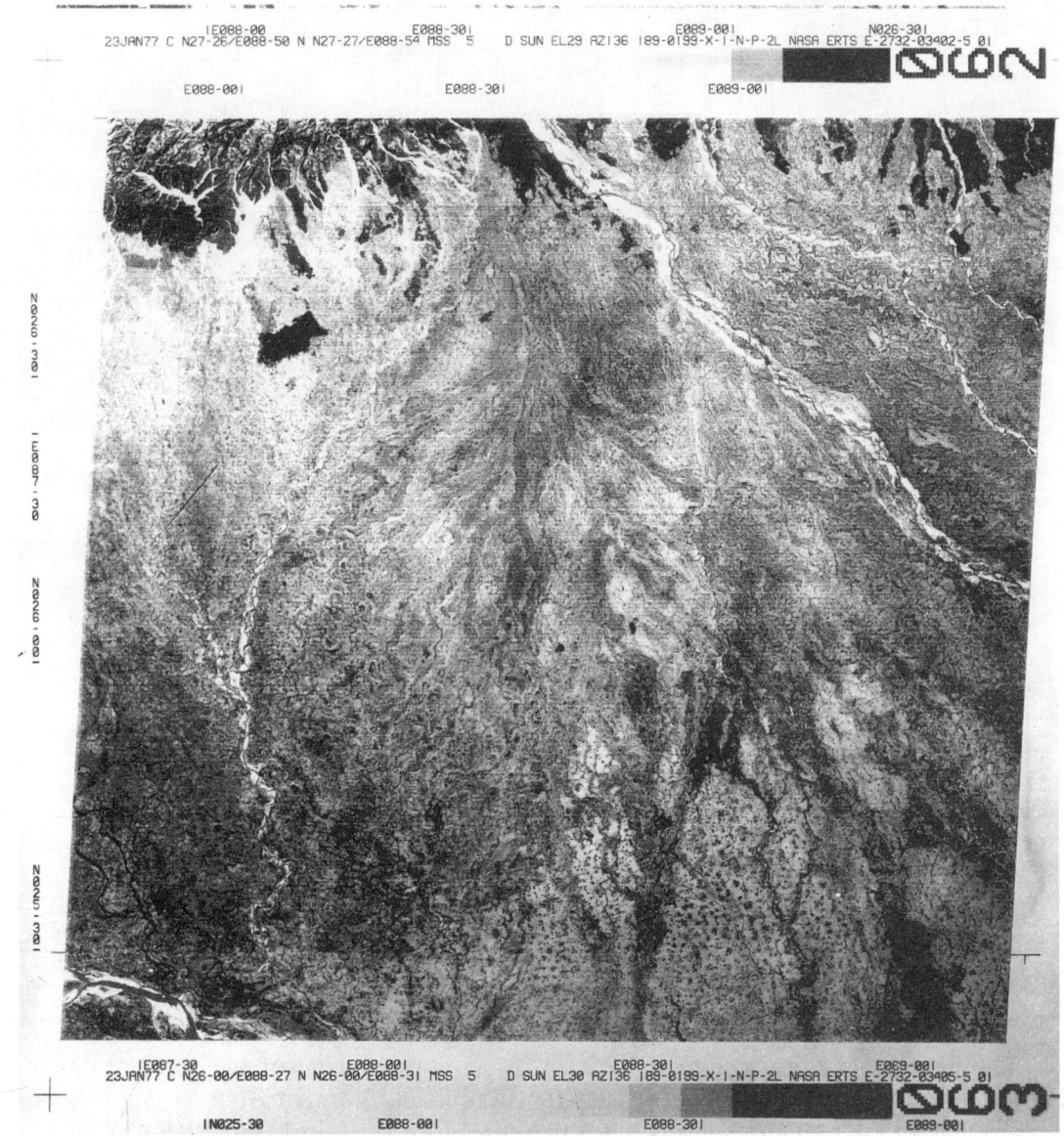

Fig. 3.1 LANDSAT imagery

The first element in the optical system for each HRV is a plane mirror that can be rotated to either side though an angle of ± 27° (in 45 steps of 0.6° each) making possible each instrument to image any point within a strip 475 km on either side of the satellite ground track. The ground swath covered varies with the pointing angle. For example, at the maximum value of 27°, the swath width for each instrument is 60 km. When the two instruments are pointed so as to cover adjacent image fields at nadir, the total swath width is 117 km with an overlap of 3 km between the two swaths.

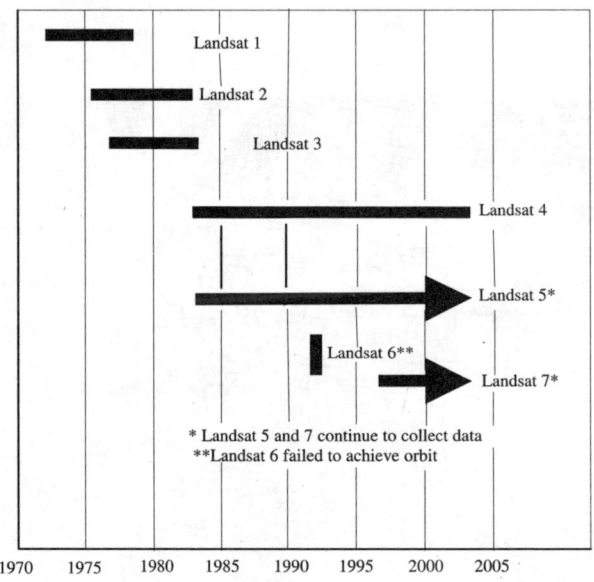

Fig. 3.2 Operational period of LANDSAT missions (*Source: USGS, 2004*)

Table 3.3 SPOT-1, 2 and 3 satellites (*Source: www.telsat.belspo.be*)

Orbital characteristics		Sensor characteristics			
Orbit	Near polar sun synchronous	HRV (Visible High-Resolution)			
Altitude	822 km				
Inclination	98°	Mode	Band	Spectral range (µm)	Resolution
Orbital period	101 minutes	XS - Multispectral	XS1	0.50 - 0.59	20 m
Equator crossing time	10:30 A.M. LST		XS2	0.61 - 0.68	20 m
Repeat cycle	26 days		XS3	0.79 - 0.89	20 m
Swath width	117 km (60 km per HRV, 3 km overlap)	P	PAN	0.51 - 0.73	10 m

The next generation satellite is SPOT-4, is an enhancement of the earlier satellites. It carries several sensors, two identical HRVIR (Visible and Infrared High-Resolution) optical sensors and VEGETATION sensors. HRVIR sensors are very similar to HRV sensors having the same spatial resolution, and observing in the off-nadir mode. However, in HRVIR, an additional spectral band in the middle-infrared band (1.98 - 1.73 µm) has been provided. The band improves the vegetation monitoring, mineral discrimination, and soil moisture mapping capabilities. Further, the panchromatic band (0.51 - 0.73 µm) has been replaced by B2 band

(0.61 - 0.68 μm) which can function both in 10 m (PAN) and 20 m (multispectral) spatial resolution mode. Table 3.4 gives the details of orbital parameters and the various sensors carried by SPOT-4.

A new sensor called VEGETATION capable of providing accurate measurements of the main characteristics of Earth's plant cover. This instrument is a part of VEGETATION programme co-financed by the European Union, Belgium, France, Italy, and Sweden, and is being conducted under the supervision of CNES. It provides a daily global coverage at a resolution of 1 km, making it an ideal sensor for observing long term regional and global environmental changes. VEGETATION works independently from HRVIR. It includes a wide angle radiometric camera operating in four spectral bands *i.e.*, blue, red, near-infrared, and middle-infrared (Table 3.4). Since it has a swath width of 2250 km, it is able to cover almost the Earth's entire land portion in a single day.

Table 3.4 SPOT-4 orbital and sensor characteristics (*Source: www.telsat.belspo.be*)

Orbital characteristics		Sensor characteristics			
Orbit	Near polar sun synchronous	HRVIS			
Altitude	830 km				
Inclination	98°	Mode	Band	Spectral range (μm)	Resolution
Orbital period	101 mins	Multispectral	B1	0.50 - 0.59	20 m
Equator crossing time	10:30 A.M. LST		B2	0.61 - 0.68	20 m
			B3	0.79 - 0.89	20 m
Repeat cycle	26 days		MIR	1.98 - 1.75	10 m
Swath width	117 km (80 km per HRV, 3 km overlap)	Monospectral	PAN	0.61-0.68	10 m
		VEGETATION			
		Band		Spectral range (μm)	Resolution
		B0		0.43 - 0.47	1165 m
		B2		0.61 - 0.68	1165 m
		B3		0.79 - 0.89	1165 m
		MIR		1.58 - 1.75	1165 m

In SPOT-5, the HRVIR sensors of SPOT-4 have been replaced by two High Resolution Geometric (HRG) sensors. The spectral bands of HRG sensors are same as HRVIR except that the spatial resolutions of B1, B2 and B3 have increased from 20 m to 10 m (Table 3.5). Further, the PAN band has been re-allocated the old spectral range of the PAN band of SPOT-1, -2 and -3, with improved spatial resolution of 5 m or 2.5 m. Thus HRG sensors are capable of generating data at four resolution levels for the same 60 km swath as given below:

(*i*) Images in SWIR band at 20 m resolution.
(*ii*) Multispectral images (green, red and near infrared) at 10 m resolution.
(*iii*) Panchromatic images at 5 m resolution.
(*iv*) Super mode panchromatic images at 2.5 m resolution.

Further a new sensor called High Resolution Stereoscopic (HRS) has been introduced. It has the ability to acquire stereopair images simultaneously both in fore-and-aft mode for generating high quality digital elevation model (DEM) at a resolution of 10 m. The instrument has a swath width of 120 km and a scene length of 600 km. The sensors have an off-nadir view angle of ± 20° (Table 3.5).The VEGETATION sensor has remained same as used in SPOT-4.

Table 3.5 SPOT-5 Orbital and Sensor Characteristics (*Source: www.telsat.belspo.be*)

Orbital characteristics		Sensor characteristics			
Orbit	Near polar sun synchronous	HRG (High-Resolution Geometric)			
Altitude	832 km				
Inclination	98°	Mode	Band	Spectral range (µm)	Resolution
Orbital period	101 mins	Multispectral	B1	0.50 - 0.59	10 m
Equator crossing time	10:30 A.M. LST		B2	0.61 - 0.68	10 m
			B3	0.79 - 0.89	10 m
Repeat cycle	26 days		SWIR	1.58 - 1.75	20 m
Swath width	117 km (80 km per HRV, 3 km)	Mono-spectral	PAN	0.51 - 0.73	5 m, 2.5 m
		HRS (High Resolution Stereoscopic)			
		Mono-spectral	PAN	0.51 - 0.73	10 m
		VEGETATION same as in SPOT 4			

IRS Satellites

The IRS Satellite Systems are under the umbrella of National Resources Management System (NNRMS), and coordinated at the national level by the Planning Committee of NNRMS (PC-NNRMS). The launch of the IRS-1A on 17 March, 1988, which is India's first civil remote sensing satellite, marked the beginning of a successful journey in the International Space Programme. The two LISS (Linear Imaging Self-scanning) sensors on board IRS-1A, have aided its capabilities in large-scale applications. Subsequently, the launch on IRS-1B on August 29, 1991 with same sensors provided better repetitive coverage. The introduction of PAN and WiFS (Wide Field Sensor) sensors on IRS-1C launched on December 28, 1995, and IRS-1D in September, 1997, further strengthened the scope of remote sensing application in the areas like resource survey and management, urban planning, forestry studies, and disaster monitoring and environmental studies. Table 3.6 gives details of the various IRS missions launched along with their some salient characteristics, while Table 3.7 gives details on the sensors carried onboard. Fig 3.3 shows the major satellite missions launched to date by India.

Table 3.6 Various IRS mission launched

Characteristics	IRS 1A & 1B	IRS 1C	IRS 1D
Orbit	Near polar	Near polar	
	Sun-synchronous	Sun-synchronous	Sun-synchronous
Altitude	914 km	817 km	821 km
Inclination	99.028°	98.69°	98.62°
Orbital period	103.192 minutes	101.35 minutes	100.56 minutes
Equatorial crossing time	1025 hours (local)	1030 hours (Local)	1040 hours (Local)
Repeat cycle	26 days	24 days	24 days, 5 days for PAN & WiFS
Swath width	148 km (LISS I)	141 km (LISS III VIS)	141 km (LISS III VIS)
	74 m (LISS II)	148 km (LISS III SWIR)	148 km (LISS III SWIR)
		70 km (PAN)	70 km (PAN)
		810 km (WiFS)	812 km (WiFS)
Spatial resolution	72.50 m (LISS I)	23.5 m (LISS III V & NIR)	23.5 m (LISS III V & NIR))
	36.25 m (LISS II)	70.0 m (LISS III MIR)	70.0 m (LISS III MIR)
		70.0 m (LISS III SWIR	70.0 m (LISS III SWIR
		5.8 m (PAN)	5.8 m (PAN)
		188.3 m (WiFS)	188 m (WiFS)

Table 3.7 Sensors onboard various IRS mission

Sensor	Band	IRS 1A & 1B Spectral range (µm)	IRS 1C Spectral range (µm)	IRS 1D Spectral range (µm)	Applications
LISS	1	0.45 - 0.52			Coastal environment; chlorophyll absorption
	2	0.52 - 0.59	0.52 - 0.59	0.52 - 0.59	Green vegetation; soil/rock discrimination
	3	0.62 - 0.68	0.62 - 0.68	0.62 - 0.68	Chlorophyll absorption for plant species
	4	0.77 - 0.86	0.77 - 0.86	0.77 - 0.86	Delineation of land and water
	5		1.55 - 1.70	1.55 - 1.70	Vegetation mapping
PAN			0.50 - 0.75	0.50 - 0.75	Resource identification
WiFS	1		0.62 - 0.68	0.62 - 0.68	Global vegetation mapping
	2		0.77 - 0.86	0.77 - 0.86	Global vegetation mapping
	3			1.55 - 1.75	Global vegetation mapping

An important aspect to be noted is than the PAN sensor carried by IRS satellite has the capability to view off-nadir, similar to SPOT system, by an angle of ±26° (398 km) from nadir. This feature allows the satellite to revisit any area within 5 days.

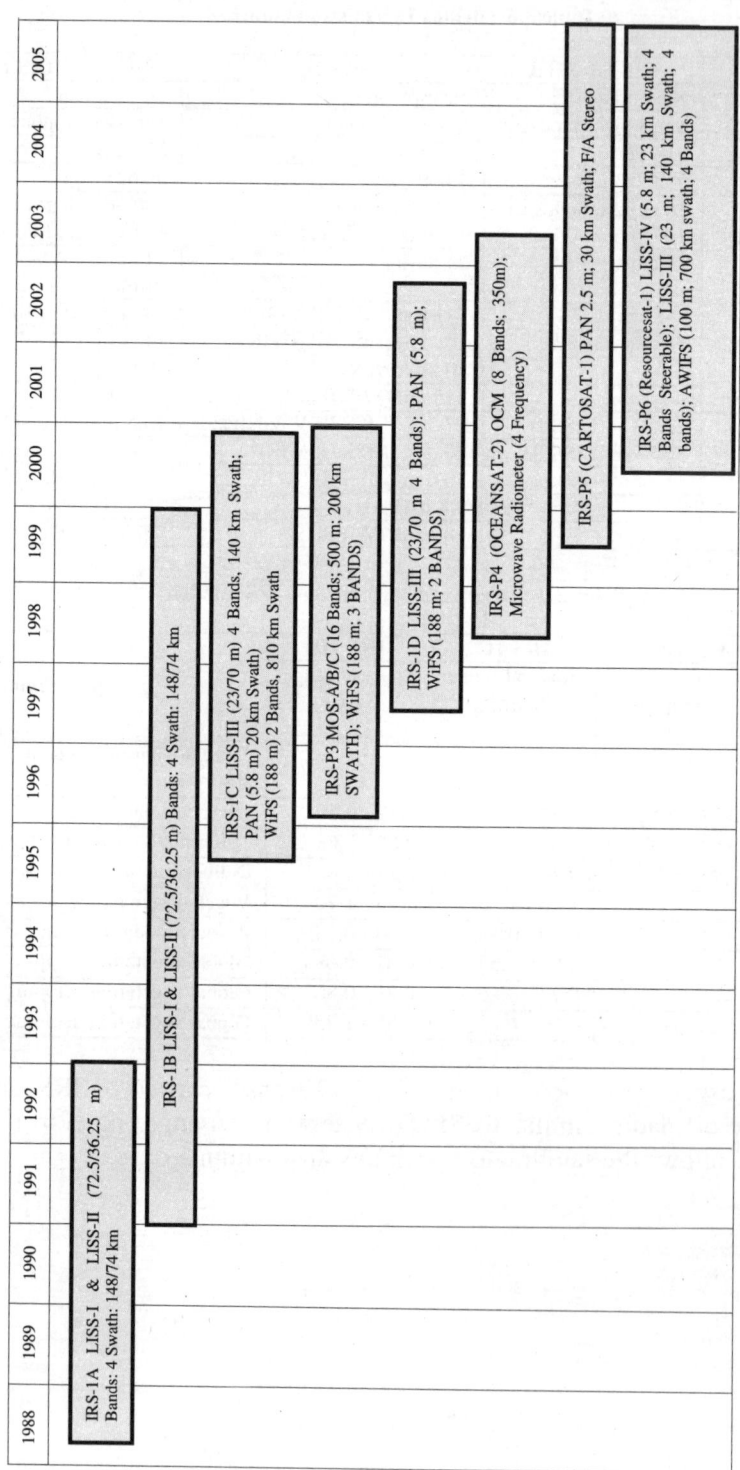

Fig. 3.3 Indian Earth Observation Systems: 1988 – 2005

In order to strengthen India's own capability to launch space vehicles, a series of satellites known as IRS-P was initiated. IRS-P3 and IRS-P4 was launched from Sriharikota, India, using a Polar Satellite Launched Vehicle (PSLV). IRS-P3 satellite was placed in a near polar, Sun-synchronous orbit at an altitude of 817 km with an equatorial crossing time of 10:30 A.M. It carried an X-ray astronomy payload and two remote sensing sensors, namely WiFS and MOS (Modular Opto-electronics Scanner). The IRS-P3 WiFS is similar to IRS-1C WiFS except for the inclusion of an additional band in Middle Infra-Red (MIR) to study dynamic vegetation studies, while MOS caters to oceanography applications.(Source NRSA, a). Table 3.8 gives the details of the characteristics of the sensors onboard IRS-P3.

Table 3.8 Sensor characteristics of IRS-P3 (*Source: NRSA, b*)

Characteristics	MOS-A	MOS-B	MOS-C	WiFS
Resolution (m)	1598 × 1395	523 × 523	523 × 644	188 × 188 (B3 & B4) 188 × 246 (B5)
Swath (km)	195	200	192	770
Repetivity (day)	24	24	24	24
Spectral band (μm)	0.755 - 0.768	0.408 – 1.01	1.5 - 1.7	0.62 - 0.68 (B3) 0.77 - 0.68 (B4) 1.55 - 1.69 (B5)

The next satellite in this series IRS-P4, also known as Oceansat, was launched on May 26, 1999 at 11:52 A.M. from Sriharikota using a PSLV vehicle. It carried two payloads, Ocean Colour Monitor (OCM) and Multi-frequency Scanning Microwave Radiometer (MSMR). The specifications of OCM and MSMR are given in Table 3.9.

Table 3.9 Specification of OCM and MSMR payloads (*Source: NRSA, c*)

OCM		MSMR
Band	Spectral range (nm)	Frequency (GHz)
1	402 - 422	6.60
2	433 - 453	10.65
3	480 - 500	18.00
4	500 - 520	21.00
5	545 - 565	
6	660 - 680	
7	745 - 785	
8	845 - 888	
Spatial Resolution (m)	360 × 236	Polarization V & H for all frequencies, spatial resolution of 120, 80, 40 and 40 Km respectively
Swath	1420 km	1360 km
Radiometric quantization	12 bits	12 bits

IRS-P6, also known as ResourceSat-1, was the next satellite in this series, which was launched on October 17, 2003 from Sriharikota, India on a PSLV satellite. The main objective of ResourceSat-1 is not only to provide continued remote sensing data for integrated land and water management, and agricultural and its related applications, but also to provide additional capabilities such as real time availability of data to ground stations anywhere in the World with its advanced onboard solid state recorder. Even though ResourceSat-1 carries three sensors similar to IRS-1C and IRS-1D, but with the following additional features (Source: NRSA, d):

(i) A high resolution Linear Imaging Self Scanner (LISS-4) operating in three spectral bands in the Visible and Near Infra Red (VNIR) regions with 5.8 m spatial resolution and steerable up to ± 26° across the track to obtain stereoscopic imagery and achieve 5-day revisit capability.

(ii) A medium resolution LISS-3 operating in three spectral bands in VNIR and one in Short-Wave Infrared (SWIR) band with 23.5 m spatial resolution.

(iii) An Advanced Wide Field Sensor (AWiFS) operating in three spectral bands in VNIR and one band in SWIR with 56 m spatial resolution.

The LISS-IV sensor is identical to LISS-III flown in IRS 1C/1P except that the spatial resolution of SWIR Band (B5) in Table 3.6 has been improved to 23.5 m. LISS-IV sensor is a high resolution multi-spectral sensor operating in three spectral bands (B2, B3, and B4). It provides a ground resolution of 5.8 m (at nadir), and can be operated in either of the two modes as below:

(i) Multi-spectral mode (Mx) having a swath width of 23 km selectable out of the total swath of 70 km in all the three bands.

(ii) Mono mode having a swath width of 70 km in any one of single band as defined by ground control commands.

In multi-spectral mode, the data corresponding to any 4000 pixels of each spectral band is transmitted selected anywhere out of the 12000 pixels available. Hence in this mode it is not a path based scheme, the user has to specify his area of interest in terms of latitude and longitude ranges. The LISS-IV can be tilted up to ± 26° in the across track direction thereby providing a revisit period of 5 days. Table 3.10 gives details of LISS-IV sensor characteristics.

Table 3.10 Details of LISS-IV sensor characteristics (*Source: NRSA, e*)

Sensor parameter	Specifications
Spectral bands	B2 0.52 - 0.59 µm B3 0.62 - 0.68 µm B4 0.77 - 0.86 µm
Spatial resolution	5.8 m (at nadir)
Radiometric resolution	10 bits

AWiFS sensor is an improved version of WiFS sensor carried onboard IRS-1C/1D. AWiFS operates in four spectral bands identical to LISS-III with a spatial resolution of 56 m and covers a swath of 70 km. In order to cover such a wide swath, the AWiFS camera is split into two separate electro-optic modules, AWiFS-A and AWiFS-B, which are tilted by 11.94° with respect to each other. AWiFS sensor characteristics are presented in Table 3.11.

Table 3.11 Details of AWiFS sensor characteristics (*Source: NRSA, f*)

Sensor Parameter	Specifications
Spectral bands	B2 0.52 - 0.59 µm B3 0.62 - 0.68 µm B4 0.77 - 0.86 µm B5 1.55 - 1.70 µm
Spatial resolution	370 km for each module 740 km combined
Radiometric resolution	10 bits

On the basis of the details of the various sensors used in Landsat, SPOT and IRS missions, a detailed insight in to the usage of various portions of EMR being made by different spectral bands can be highlighted (Table 3.12).

Cartosat

With the availability of IRS-1C/1D data, it was possible to prepare cartographic and town planning applications up to 1:10,000 scale. Also the stereo pairs of the imageries can provide height information to an accuracy of 10 m approximately. This provided the necessary impetus to further develop high resolution sensors dedicated to cartographic and mapping application. Cartosat-1 with two fore-and-aft PAN cameras having a spatial resolution of 2.5 m. The PAN cameras are mounted in such a manner that one camera is working at + 26° with respect to nadir and the other at − 5° with respect to nadir. The two cameras combinedly provide stereoscopic image pairs in the same pass. Table 3.13 gives details of the salient features of Cartosat-1 payload (Krishnaswamy & Kalyanaraman, 2002).

Cartosat-1, launched on May 5, 2005, is placed in a Polar Sun Synchronous orbit of altitude 618 km with an inclination of 97.87° and an equatorial crossing time of 10:30 hour (local time). It has repetitive cycle of 116 days, can be reduced to 5 days with the help of 26° off-nadir cross track.

Since the primary aim of Cartosat-1 is to provide data for cartographic purpose, various types of data products planned for using Cartosat-1 image are Image Data Products, Image Map Data products, and DEM data products. The level of Image data products, defined on the basis of their intended use with desired accuracy and turn-around time, covering both stereo and mono modes of operation, are given in Table 3.14.

Image Map Data Products (IMDP) contains co-registered ortho corrected Cartosat-PAN raster images with one or more layers of cartographic vector information (available apriori or derived from Cartosat image) including a layer containing ASCII text strings as labels of vector elements, with necessary additional ancillary information shall be generated and supplied to the users. The following types of products are planned:
  (*i*) 2-D Satellite Image Map Products in conventional map projection or with user-defined projection parameters.
  (*ii*) 2.5-D Satellite Image Map Products, representing the terrain elevation for one or more fixed, standard perspective view angles, possibly with artificially exaggerated scaling effects incorporated, to show the terrain relief.

(*iii*) 3-D Satellite Image Map Products, optionally including the 3-D viewing software as a part of the product.

Table 3.12 Used of different EMR region by various sensors (*Source: www.infoterra-global.com*)

Spectral Name	Type of Sensor	Applications
Visual Blue	TM Band 1	Designed for water penetration, useful for coastal water and lake bathymetry and sediment load mapping. Useful for differentiation of soil from vegetation, and deciduous from coniferous flora. Well fragmented and granular rocks (shales, phosphates, evaporites) scatter blue light and result in a high band 1 (and sometimes 2).
Visible Green	TM Band 2 MSS Band1 SPOT XS/X1 Band 1 IRS LISS Band 1	Designed to measure visible green reflectance peak of vegetation for vigour assessment. Also used to map sediment concentration in turbid waters, and is higher for ferrous iron rich rock compared to ferric iron.
Visible Red	TM Band 3 MSS Band 2 to 3 SPOT XS/X1 Band 2 IRS LISS Band 2	A chlorophyll absorption band important for vegetation discrimination. It is higher for rocks and soils rich in iron, especially ferric iron.
Near Infrared	TM Band 4 MSS Band 3 to 4 SPOT XS/X1 Band 3 IRS LISS Band 3	Useful for determining healthy vegetation and for delineation of water bodies. Peaks strongly for chlorophyll in healthy vegetation, resulting in a characteristic 'red-edge' between bands 3 and 4. In the absence of vegetation, ratios of bands 1 through 5 show ferric/ferrous iron differences in rocks and minerals.
Short wave Infrared	TM Band 5 SPOT X1 Band 4 IRS LISS Band 4	Indicative of vegetation moisture content and soil moisture. Dry material results in relatively higher values. Also useful for discriminating snow and clouds (low for snow, high for clouds). In vegetation free areas, band 5 varies according to the type of iron oxide present in rocks and soils, and is generally high for all minerals.
	TM Band 7	Useful for discriminating rocks, hydrothermal altered zones, and for mineral exploration. Hydroxyl (OH) molecular bonds in minerals stretch and the resultant electronic vibration causes absorption of energy around 2.2 μm, resulting in marked low values in band 7 for clay-rich minerals. Carbonate rich materials can also cause the same effect. Silica rich materials, dust in the air and bare soil are often relatively high in band 7.
Thermal Infrared	TM Band 6	Used for thermal mapping. Useful for heat intensity, vegetation and crop stress analysis and locating thermal pollution. It is usually higher for darker coloured rocks.

Table 3.13 Salient features of Cartosat-1 payload (*Source: Krishnaswamy & Kalyanaraman, 2002*)

Parameter	Specification
Spatial resolution (m) (Across-track × along track)	2.5 × 2.78 (Fore camera) 2.2 × 2.23 (Aft camera)
Spectral resolution a) No. of bands b) Band width	1 Panchromatic 500 nm to 850 nm
Radiometric resolution	10 bits
Swath (km)	30 (stereo) 26.855 (Fore+Aft combined in mono mode)

Table 3.14 Levels of data products (*Source: Krishnaswamy & Kalyanaraman, 2002*)

Level of product	Radiometric correction	Geometric correction	Intended end-use
Level 'OA' RAW	Not applied	Not applied	For internal use.
Level 'OB' RAD	Applied	Not applied	For stereo processing at user end
Level '1' SYS	Applied	System knowledge	Quick turn-around-time digital products, with acceptable quality for flat terrain or mono mode data acquisition.
Level '2' GCP	Applied	System knowledge + GCPs	For better location accuracy for cases when DEM is not available, for data acquired in mono mode or for flat terrain.
Level '3A' DEMA	Applied	System knowledge + GCPs + DEM (external input)	Orthoimage products primarily for IGS use, where GCPs and DEM are externally available.
Level '3B' DEMB	Applied	System knowledge + GCPs+DEM (internal)	For Ortho image product generation for data acquired in stereo mode.
Level '3C' DEMC	Applied	System knowledge + GCPs +DEM (external/internal, interactively edited for density and surface discontinuities)	For precision Ortho image/Ortho image map generation/map updating.

All the types of image map products are corrected Cartosat-PAN images, either on stand alone basis or desirably fused with available multi-spectral images with comparable spatial resolution as the base raster image.

DEM Data Products

The following types of DEM data products are planned to be generated:

Type I: As computed originally
- As randomly distributed point heights as computed originally.
- As Triangulated Irregular Network (TIN) retaining all the originally computed points.
- As progressively sampled rectangular grids, retaining all the originally computed points.

Type II: A completed originally and incorporating break-lines
- As a set of irregular point heights and break-lines showing the surface discontinuities
- As a TIN model retaining all the originally computed points and in addition having break- lines, either as part of the TIN edges (referred as soft break-lines) or as add-on specification (referred as hard break-lines) indicating abrupt surface changes.
- As progressively sampled rectangular grids, retaining all the originally computed points, including the break-lines manually identified.

Type III: As interpolated, mostly regular
- As a rectangular grid, generated by a suitable interpolation algorithm from the initial set of irregular points and break lines.

- As contours, connecting points of equal height at varying intervals.
- As a set of parallel vertical profiles, in any user desired direction.

With the launch of Cartosat-1, it is expected to meet the immediate demands for terrain visualization, updation of topographic maps, generation of National Topographic Database and other attitudes planning.

3.4 HIGH RESOLUTION SENSORS

The stereo ability of SPOT and IRS-1C/1D PAN data provided a new dimension to cartographic mapping applications up to 1:10,000 scale. The intense research has lead to significant improvement in spatial resolution close to 1 m. Some of the sensors capable of acquiring such high spatial resolution are IKONOS, Quick Bird, and OrbView. A brief description of these satellites and their applicability is discussed below.

IKONOS

Space Imaging Inc. made history on September 24, 1999, when it launched into space using the Athena II rocket, the IKONOS, the world's first commercial high-resolution imaging satellite from Vandenberg Air Force Base in California. Spacecraft is manufactured by Lockheed Martin, and the investors in the IKONOS system include Lockheed Martin Corporation, Raytheon Company, Japan's Mitsubishi Corporation, Singapore's Van Der Horst Ltd., Korea's Hyundai Space & Aircraft, Europe's Remote Sensing Affiliates, Swedish Space Corporation, and Thailand's Loxley Public Company Ltd (SISEA, a).

IKONOS is derived from the Greek word for image. Driven by an increasing market need for detailed and accurate satellite imagery for a multitude of important applications, such as mapping, agriculture monitoring, and urban planning, the IKONOS satellite is the first of its kind with the capability to simultaneously collect 1-m resolution panchromatic (B&W) images and 4-m resolution multispectral (color) images. Designed to take digital images of the Earth from an altitude of 680 km and moving at a speed of about 7 km/s, the satellite camera can distinguish objects on the Earth's surface as small as one square meter. The IKONOS satellite revolves around the Earth in a Sun-synchronous orbit once every 98 minutes in 14 revolutions, and passes a given longitude at about the same local time of 10:30 A.M. daily, and can produce 1-meter imagery of the same geography every three days. Image collection can be taken in a swath of 700 km at the minimum collection area of 100 km^2, and a maximum of 10,000 km^2 per pass. The satellite is expected to have an operational life of more than seven years. Table 3.15 gives some details of IKONOS. IKONOS images cover a surface area of 11 km × 11 km.

The user will define the necessary area of interest of any shape but with a minimum surface area, which can be made up of one or more parts of scene. IKONOS images will be available with different levels of geometric processing which are:

Table 3.15 Salient features of IKONOS (*Source: SISEA, b*)

IKONOS Characteristics	
Launch date	September 24, 1999 at 11:21:08 A.M. PDT
Launch vehicle	Athena II
Manufacturer	Lockheed Martin
Launch location	Vandenberg Air Force Base, California, U.S.A.
Viewing angle	Agile spacecraft-in-track and cross-track pointing
Weight	725 kg.
Orbit	
Altitude	423 miles or 681 km.
Inclination	98.1°
Speed	4 miles/sec. (17500 miles/hr) or 7 km/sec (26000 km/hr)
Descending nodal crossing time	10:30 A.M.
Orbit time	98 minutes
Orbit type	Near-polar, sun-synchronous
Sensors	Panchromatic and Multispectral
Resolution	Ground resolution of each band: 1-m panchromatic (nominal at <26° off nadir) 4-m multispectral (nominal at <26° off nadir)
Imagery spectral response	Panchromatic: 0.45 - 0.90 μm Multispectral: (same as Landsat 4 and 5 TM Bands 1-4)
Swath widths	Nominal swath width: 11 km.
Revisit frequency	2.9 days at 1 m resolution (at 40° latitude) 1.5 days at 1.5 m resolution (at 40° latitude)
Dynamic range	11-bit data or 8-bit data
Scene sizes	A nominal single image at 11 km × 11 km Strips of 11 km × 100 km Image mosaics of up to 10, 000 km^2 Up to two 10, 000 km^2 continuous areas in a single pass within a region In-track stereo imagery capability of 22 km. × 130 km. perpendicular to the ground track

(*i*) Geo-products corrected according to a specific ellipsoid and projection chosen by the user,

(*ii*) Ortho-products corrected for distortion due to the relief by means of a DEM based on IKONOS data. Only precision and precision plus products use ground control.

Table 3.16 provides information regarding the different types of data product available and their intended application.

The next generation IKONOS satellite is IKONOS-2. IKONOS-2 is capable of generating 1-m panchromatic, 4 m multispectral, and 1 m color-enhanced Earth-imagery products. Spatial resolution is related primarily to the ground-sample distance (GSD). IKONOS-2 panchromatic imagery yields a GSD of 0.82 m at nadir. The multi-spectral resolution at nadir is 3.2 m. The available revisit frequency for areas near the equator for an image collected, at a GSD of 1 m in the panchromatic and 4 m in multi-spectral bands, is 3.9 days. At latitude of 30°, the revisit interval is improved to 3.35 days.

Table 3.16 Type of products available and their application (*Source: SISEA, b*)

Product	Horizontal (accuracy)	Compatible scale	Examples of applications
Geo	± 50 m	N/A	Photo-interpretation and analyses which do not require a high degree of location accuracy.
Reference	± 25.4 m	1:50 000	Mapping of large areas and GIS applications for the media, insurance companies and other commercial markets.
Map	± 12 m	1:24 000	Urban planning, impact studies, for local and regional authorities.
Pro	± 10 m	1:12 000	Environmental impact studies, transport and urban planning, for local and regional authorities, agriculture, telecommunications and services for users.
Precision	± 4 m	1:4 800	Urban mapping, GIS applications requiring high degree of location accuracy.
Precision Plus	± 2 m	1:2 400	Urban mapping, GIS applications requiring a high degree of location accuracy.

The IKONOS-2 satellite operations are supported by the Primary Operations Center in Thornton, CO, and a second terminal in Norman, OK. Polar visibility is provided by a Space Imaging facility outside of Fairbanks, AK, and a shared receiving station capability provided by the Swedish Space Corporation, Kiruna, Sweden. IKONOS-2 has the following basic characteristics (Table 3.17).

Table 3.17 IKONOS 2 characteristics (*Source: www.tec.army.mil*)

Characteristics	
Launch vehicle	Athena II
Resolution:	10 m panchromatic, 4 m multispectral
Swath width:	13 km at nadir
Scene sizes:	nominal single image-13 km × 13 km Strips-11 km ×100 km up to 11 km × 1000 km Image mosaics-up to 12, 000 sq. km
Revisit frequency:	3 days at 1 m resolution 1.5 days at 1.5 m resolution 1 day at 2 m resolution
Orbit type:	sun synchronous

Quick Bird

The Quick Bird satellite launched in October 2002, by Digital Globe Inc., acquired black and white images with a resolution of 61 cm and colour images (4 bands) with a resolution of 2.44 m covering a surface area of 16.5 km × 16.5 km (www.spotimage.fr). The characteristics of Quick Bird are given in Table 3.18. Quick Bird provides panchromatic, multi-spectral and colour image products that enable superior image classification and analysis based on 4 spectral bands at 11-bit dynamic range. The imagery products are available at different processing levels (basic, standard, and ortho) serving the needs of different user groups.

Table 3.18 Characteristics of Quick Bird (*Source: www.imagery-central.com*)

Sensor Specifications	
Orbital type	Sun-synchronous
Orbital altitude	450 km
Swath width	22 km
Bits per pixel	11
Image area	16.5 × 16.5 sq km
Revisit time	1 - 3.5 days
Launch date	18 Oct. 2001
Sensor type	PAN
Resolution	0.61m
Channels	1
Spectral range	0.45 - 0.90 µm
Sensor type	MSI
Resolution	2.44 m
Channels	4
Band 1	Blue: 0.45 - 0.52 µm
Band 2	Green: 0.52 - 0.60 µm
Band 3	Red: 0.63 - 0.69 µm
Band 4	Near IR: 0.76 - 0.90 µm

Panchromatic imagery is collected in 11-bit format (2048 gray levels) and delivered in 16-bit format for superior image interpretation (shadow detail) or 8-bit format (256 gray level) supported by desktop GIS and mapping application packages. The multi-spectral imagery consists of blue, green, red and near infrared bands, delivered in 16-bit and 8-bit. Pan-sharpened imagery is available as 70 cm natural colour (red, green, blue bands) and 70 cm false colour infrared (red, green, infrared) imagery, in 8-bit.

QuickBird products are offered at three levels of processing:

(*i*) Basic imagery
(*ii*) Standard imagery
(*iii*) Ortho-rectified imagery.

Basic Imagery Products

Basic imagery products are the least processed Quick Bird imagery products. They are designed for customers with advanced image processing capabilities on their own. Basic products are supplied with attitude, ephemeris and camera model information required for use in advanced photogrammetric processing (*e.g.,* ortho-rectification). Basic products are radiometrically and sensor corrected, but not geometrically corrected nor mapped to any cartographic projection and ellipsoid. The actual image resolution varies between 61 cm at nadir and 72 cm at 25° off-nadir look angle for panchromatic images and 2.44 m (at nadir) and 2.88 m (25° off-nadir look angle) for multi-spectral imagery. The accuracies of the basic imagery products are essentially "geometrically raw". However, when the data are processed with the supplied Image Support Data (ISD), a horizontal geo-locational accuracy of 14 m RMSE (Root Mean Square Error), excluding any topographic displacement, may be achieved.

Standard Imagery Products

Standard imagery products are suitable for users requiring modest absolute geometric accuracy and/or large area coverage. In addition to the corrections of the basic imagery products (radiometric and sensor corrections), these products are also geometrically corrected and mapped to a cartographic projection. All standard imagery products have a uniform pixel size across the entire product. The geometric corrections of the imagery account for the spacecraft orbit position, earth rotation, and curvature as well as panoramic distortions. Standard imagery comes in two varieties as given below.

Standard imagery: It has a coarse (DEM) applied to it, which is used to normalize for topographic relief with respect to the reference ellipsoid. The geometric corrections are done without the use of ground control points, based entirely on the satellite attitude and ephemeris information. The degree of normalization is relatively small, so while this product has terrain corrections, it is not considered ortho-rectified.

Ortho Ready Standard imagery: This imagery has no topographic corrections, making it suitable for ortho-correction. This product is ideal for those customers who have a high quality DEM available and want to do the ortho-rectification by themselves. If ortho-rectification is not the goal, then Standard imagery is recommended.

Ortho-rectified Imagery Products

Ortho-rectified products are ready for use in a GIS and ideally suited as image base map for creating and/or revising mapping and GIS databases. The ortho-rectified imagery products provide the highest degree of geometric accuracy available. Ortho-rectified products require a Digital Elevation Model DEM and Ground Control Points (GCPs). In some cases these may need to be provided by the customer.

Basic, Standard and Ortho ready Standard imagery all have different attributes, making them appropriate for different uses. Basic imagery allows ortho-rectification using the QuickBird Rigourous Sensor Model or RPCs. Standard imagery is most useful for applications requiring geo-referenced imagery that are not terrain corrected. Standard imagery has a coarse DEM applied to it making it unsuitable for ortho-rectification. Ortho ready Standard imagery is a geo-referenced product, designed for ortho-rectification; using supplied RPCs. Achievable accuracies will be comparable to those attainable using Basic imagery with RPCs. If ortho-rectification is not the goal, then Standard imagery is recommended. Ortho Ready Standard imagery is projected to a base elevation of zero, it could be off by several hundred meters, especially in area of high elevation, if the user does not apply terrain corrections. Table 3.19 provides an overview of the important image product characteristics.

Table 3.19 Overview of the important image product characteristics of Quick Bird (*Source: www.nl.gim.be, 2002*)

	Basic imagery	Standard imagery	Ortho imagery
Image resolution			
Panchromatic	61 cm (at nadir) to 72 cm (at 25°off nadir)	0.6 m or 0.7 m	0.6 m or 0.7 m
Multi-spectral	2.44 m (at nadir) to 2.88 m (at 25°off nadir)	2.4 m or 2.8 m	2.4 m or 2.8 m
Pan-sharpening	N/A	0.6 m or 0.7 m	0.6 m or 0.7 m
Radiometric corrections	Yes		Yes
Sensor corrections	Yes	Yes	Yes
Geometric corrections	No	Yes	Yes
Ortho-rectification	No	No	Yes
Horizontal accuracy	No geometric correction provided with product	14 m RMSE excluding any topographic displacement	7.7 m RMSE to meet 1:25.000 scale
If order polygon intersects with several image scenes	No mosaic No spectral balance	No mosaic No spectral balance	

The next generation of Quick Bird is Quick Bird 1. Table 3.20 shows the salient features of Quick Bird 1.

Table 3.20 Salient features of Quick Bird 1 (*Source: www.imagery-central.com, 2001*)

Sensor Specifications	
Orbital type	66° medium inclination non Sun-synchronous
Orbital altitude	Variable
Swath width	22 km
Bits per pixel	11
Image area	22 × 22 sq km
Revisit time	1-5 days (latitude dependent)
Local nodal equatorial crossing time	Variable
Launch date	N/A
Sensor type	PAN
Resolution	1m
Channels	1
Spectral range	0.45 - 0.90 µm
Sensor type	MSI
Resolution	4 m
Channels	4
Band 1	Blue: 0.45 - 0.52 µm
Band 2	Green: 0.52 - 0.60 µm
Band 3	Red: 0.63 - 0.69 µm
Band 4	Near IR: 0.76 - 0.90 µm

OrbView-1

Successfully launched in April 1995 and now completed, ORBIMAGE's OrbView-1 satellite contained two atmospheric instruments that improved weather forecasting capabilities around the world. The satellite's miniaturized camera provided daily several weather images and global lightning information during day and night operations. Its atmospheric monitoring instrument provided global meteorological data useful for improving long-term weather forecasts.

OrbView-1 provided the world's first broad-area cloud-to-cloud lighting imagery. The lightning imaging instrument was a single channel imaging device with a center wavelength of 777nm and a spatial resolution of 10 km. The lighting imagery provided nearly global coverage every two days with a swath width of 1,300 km and an orbital inclination of 70° relative to the equator. OrbView-1 also offered meteorological research customers data for atmospheric monitoring and weather applications. The lightning and atmospheric data were stored onboard the spacecraft and downlinked twice a day to ORBIMAGE's central U.S. ground station located in West Virginia. ORBIMAGE delivered both lightning and atmospheric data directly to customers in near real-time using the Internet. Table 3.21 gives the specification of OrbView-1.

Table 3.21 OrbView-1 specifications (*Source: ORBIMAGE, 2004 a*)

Parameter	Specification
Imaging mode	Monochromatic
Spatial resolution	10 km
Imaging channels	1 channel
Spectral range	777 nm
Swath width	1,300 km
Image area	Continuous
Revisit time	Less than 2 days
Orbital altitude	740 km
System life	5 years (completed)

OrbView-2

The OrbView-2 satellite, launched successfully in August 1997, provides unprecedented multispectral imagery of the Earth's land and ocean surfaces every day. By detecting subtle color changes on the Earth's surface, OrbView-2's imagery is valuable for monitoring plankton and sedimentation levels in oceans and assessing the health of land-based vegetation on a global basis. OrbView-2 provides the world's first daily color imagery of the Earth. The satellite's imaging instrument has eight channels, six in the visible and two in the near infrared spectrum, with a spatial resolution of 1.1 km.

OrbView-2 provides daily coverage of the Earth with a swath width of 2,800 km from a polar orbital path. OrbView-2 imagery is continuously down linked in real-time and can be acquired with a standard High Resolution Picture Transmission (HRPT) ground station. To access OrbView-2 imagery, customers can either directly downlink the data with their own ground station by purchasing a downlink license or purchase individual images from ORBIMAGE's online service. OrbView-2 is useful for a variety of applications such as fishing, agriculture, naval operations, scientific research, and environmental monitoring. Table 3.22 gives details of OrbView-2 specification.

Table 3.22 OrbView-2 specifications (*Source: ORBIMAGE, 2004 b*)

Parameter	Specification
Imaging mode	Multispectral
Spatial resolution	1 km
Imaging channels	8 channel
Spectral range	402 - 422 nm
	433 - 453 nm
	480 - 500 nm
	500 - 520 nm
	545 - 565 nm
	660 - 680 nm
	745 - 785 nm
	845 - 885 nm
Swath width	2,800 km
Image area	Continuous
Maximum data rate	2 Mbps
Revisit time	1 day
Orbital altitude	705 km
Nodal crossing	12:00 P.M.
System life	7.5 years

OrbView-3

ORBIMAGE's OrbView-3 satellite is among the world's first commercial satellites to provide high-resolution imagery from space. OrbView-3 produces 1-m resolution panchromatic and 4-m resolution multispectral imagery. 1-m imagery enables the viewing of houses, automobiles and aircraft, and makes it possible to create highly precise digital maps and three-dimensional fly-through scenes. 4-m multispectral imagery provides color and infrared information to further characterize cities, rural areas, and undeveloped land from space.

OrbView-3's imaging instrument provides panchromatic imagery and multispectral imagery with a swath width of 8 km. The satellite revisits each location on the Earth in less than three days with an ability to turn from side-to-side up to 45°. OrbView-3 imagery can be down linked in real-time to ground stations located around the world or stored onboard the spacecraft and down linked to ORBIMAGE's master U.S. ground stations.

OrbView-3 provides images useful for a variety of applications such as telecommunications and utilities, oil and gas, mapping and surveying, agriculture and forestry, and national. Table 3.23 gives details of OrbView-3 specifications.

Table 3.23 OrbView-3 specifications (*Source: ORBIMAGE, 2004 c*)

Parameter	Specification	
Imaging mode	Panchromatic	Multispectral
Spatial resolution	1 meter	4 meter
Imaging channels	1 channel	4 channel
Spectral range	450-900 nm	450 - 520 nm 520 - 600 nm 625 - 695 nm 760 - 900 nm
Swath width	8 km	
Image area	Use defined	
Revisit time	Less than 3 days	
Orbital altitude	470 km	
Nodal crossing	10:30 A.M.	
System life	Minimum 5 years	

3.5 EARTH OBSERVING-1(EO-1)

In 1996, NASA started a New Millennium Program (NMP), designed to identify, develop, and flight-validate key instrument and spacecraft technologies that can enable new or more cost-effective approaches to conducting scientific missions in the 21st century. The first of these New Millenium Program Earth-orbiting missions is Earth Observing-1 (EO-1), an advanced land-imaging mission that will demonstrate new instruments and spacecraft systems. EO-1 will validate technologies contributing to the significant reduction in cost of follow-on Landsat missions (www. spaceflightnow.com).

EO-1 was launched on a Delta 7320 from Vandenberg Air Force Base on November 21, 2000. It was inserted into a 705 km circular, sun-synchronous orbit at an inclination of 98.7° such that it flies in formation 1 minute behind Landsat 7 in the same ground track and maintaining the separation within 2 seconds. This close separation has enabled EO-1 to observe the same ground location through the same atmospheric region so that paired scene comparisons between the two satellites can be made [www.eo1.gsfc.nasa.gov (a)].

Both Landsat 7 and EO-1 image the same ground areas. All three of the EO-1 land imaging instruments views all or subsegments of the Landsat 7 swath. Reflected light from the ground is imaged onto the focal plane of each instrument. Each of the imaging instruments has unique filtering methods for passing light in only specific spectral bands. Bands are selected to provide best look for specific surface features or land characteristics based on scientific or commercial applications.

There are three imagers in the EO-1 instrument suite: the Advanced Land Imager (ALI), Hyperion, and the Linear Etalon Imaging Spectral Array (LEISA) Atmospheric Corrector (LAC). The latter two are spectral imaging (also known as hyperspectral) instruments with narrow bandwidth, contiguous bands providing high-resolution spectral measurements as contrasted to measurements of the wider bandwidth, discrete bands of traditional multi-spectral imagers. The essential spatial and spectral characteristics of the EO-1 instrument suite in comparison to Landsat-7 Enhanced Thematic Mapper Plus (ETM+) are summarized in Table 3.24. Each of the instruments operates in a "pushbroom" mode where the forward motion of the satellite is responsible for sweeping out a two-dimensional (2-D) image (Ungar, et al, 2003).

Table 3.24 EO-1 Instrument overviews (*Source: Ungar, et al, 2003*)

Parameter	Landsat 7	EO-1		
	ETM+	ALI	HYPERION	AC
Spectral range	0.4-2.4 µm*	0.4-2.4 µm	0.4-2.5 µm	0.9-1.6 µm
Spatial resolution	30 m	30 m	30 m	250 m
Swath width	185 km	37 km	7.7 km	185 km
Spectral resolution	Variable	Variable	10 nm	3-9 nm**
Spectral coverage	Discrete	Discrete	Continuous	Continuous
Pan band resolution	15 m	10 m	N/A	N/A
Number of bands	7	10	220	256

*Excludes Thermal channel   ** Constant resolution 35/55 cm

Advanced Land Imager (ALI)

The EO-1 Advanced Land Imager is a technology verification instrument under the New Millennium Program (NMP). The focal plane for this instrument is partially populated with four sensor chip assemblies (SCA) and covers 3° × 1.625°. Operating in a pushbroom fashion at an orbit of 705 km, the ALI provides Landsat type panchromatic and multi-spectral bands. These bands have been designed to mimic six Landsat bands with three additional bands covering 0.433-0.453, 0.845-0.890, and 1.20-1.30 µm (Table 3.25). The ALI also contains wide-angle optics designed to provide a continuous 15° × 1.625° field of view for a fully populated focal plane with 30m resolution for the multi-spectral pixels and 10m resolution for the panchromatic pixels.

Use of the ALI technologies has the potential for reducing the cost and size of future Landsat-type instruments by a factor of 4 to 5. Table 3.26 shows a comparative evaluation of instrument carried by EO-1 and Landsat mission.

Table 3.25 EO-1 ALI Spectral Coverage (*Source: www.eo1.gsfc.nasa.gov (b)*)

Band	Wavelength (µm)	Ground resolution (m)
Pan	0.480 - 0.690	10
MS-1*	0.433 - 0.453	30
MS-1	0.450 - 0.515	30
MS-2	0.525 - 0.605	30
MS-3	0.630 - 0.690	30
MS-4	0.775 - 0.805	30
MS-4*	0.845 - 0.890	30
MS-5*	1.200 - 1.300	30
MS-6	1.550 - 1.750	30
MS-7	2.080 - 2.350	30

*New spectral bands

Table 3.26 EO-1 Landsat instrument comparison (*Source: www.eo1.gsfc.nasa.gov (b)*)

Parameter	ALI	Enhanced Thematic Mapper (ETM+)
Mass (kg)	100	425
Power (W)	100	545
Size (m^3)	0.2	1.4
VNIR/SWIR bands	10	7
Detectors per band	6200	16
Thermal bands	None	1
Data rate (Mbps)	300	150
Pan resolution (m)	10	15
Relative SNR	4x	1x

Hyperion Instrument

The Hyperion instrument provides a new class of Earth observation data for improved Earth surface characterization. The Hyperion provides a science grade instrument with quality calibration based on heritage from the LEWIS Hyperspectral Imaging Instrument (HIS). The Hyperion capabilities provide resolution of surface properties into hundreds of spectral bands *versus* the ten multispectral bands flown on traditional Landsat imaging missions. Through this large number of spectral bands, complex land eco-systems shall be imaged and accurately classified (www.gsfc.nasa.gov).

The Hyperion provides a high resolution hyperspectral imager capable of resolving 220 spectral bands (from 0.4 to 2.5 µm) with a 30 meter spatial resolution. The instrument images a 7.5 km × 100 km land area per image and provides detailed spectral mapping across all 220 channels with high radiometric accuracy. The major components of the instrument include the following:

(*i*) System fore-optics design based on the KOMPSAT EOC mission. The telescope provides for two separate grating image spectrometers to improve signal-to-noise ration (SNR).

(*ii*) A focal plane array which provides separate short wave infrared (SWIR) and visible/near infrared (VNIR) detectors based on spare hardware from the LEWIS HIS program.

(*iii*) A cryo-cooler identical to that fabricated for the LEWIS HSI mission for cooling of the SWIR focal plane.

Table 3.27 lists the characteristics of the Hyperion sensor. Hyperspectral imaging has wide ranging applications in mining, geology, forestry, agriculture, and environmental management. Detailed classification of land assets through the Hyperion will enable more accurate remote mineral exploration, better predictions of crop yield, and assessments, and better containment mapping.

Table 3.27 Characteristics of the Hyperion sensor (*Source: Pearlman, et al, 2003*)

Parameter	Hyperion
Volume (L × W × H, cm)	75 × 39 × 65
Weight (kg)	49
Average power (W)	51
Aperture (cm)	12
IFOV (mrad)	0.043
Cross-track FOV (deg)	0.63
Wavelength range (nm)	400 - 2500
Spectral resolution (nm)	10
Number of spectral bands	242
Digitization (bits)	12
Frame rate (Hz)	223.4

Atmospheric Corrector

Earth imagery is degraded by atmospheric absorption and scattering. The New Millennium Program's EO-1 is providing the first space-based test of an Atmospheric Corrector (AC) for increasing the accuracy of surface reflectance estimates [www.eo1.gsfc.nasa.gov (c)].

The Atmospheric Corrector (AC) provides the following capabilities via a compact and simple bolt on design for future Earth Science and land imaging missions (www.gsfc.nasa.gov):

(*i*) High spectral, moderate spatial resolution hyperspectral imager using a wedge filter technology.
(*ii*) Spectral coverage of 0.89 - 1.58 µm. Bands is selected for optimal correction of high spatial resolution images.
(*iii*) Correction of surface imagery for atmospheric variability (primarily water vapor).

The Atmospheric Corrector is applicable to any scientific or commercial Earth remote sensing mission where atmospheric absorption due to water vapor or aerosols degrades surface reflectance measurements. Using the Atmospheric Corrector, instrument measurements of actual rather than modeled absorption values enables more precise predictive models to be constructed for remote sensing applications. The algorithms developed will enable more accurate measurement and classification of land resources, and better models for land management in the future.

3.6 WEATHER SATELLITES/SENSORS

Weather monitoring and forecasting were one of the first civilian applications of satellite remote sensing, dating back to the first true weather satellite, TIROS-1 (Television and Infrared Observation Satellite -1) launched in 1960 by the United States. Several other weather satellites were launched over the next five years in near-polar orbits providing repetitive coverage of global weather patterns. In 1966, NASA (the U.S. National Aeronautics and Space Administration) launched the geostationary Applications Technology Satellite (ATS-1), which provided hemispheric images of the Earth's surface and cloud cover every half hour. For the first time, the development and movement of weather systems could be routinely monitored. Today, several countries operate weather, or meteorological satellites to monitor weather conditions around the globe. Generally speaking, these satellites use sensors, which have fairly coarse spatial resolution (when compared to systems for observing land), and provide large aerial coverage. Their temporal resolutions are generally quite high, providing frequent observations of the Earth's surface, atmospheric moisture, and cloud cover, which allows for near continuous monitoring of global weather conditions, and hence forecasting. Here a review of some of the representative satellites sensors used for meteorological applications is given.

Geostationary Operational Environmental Satellite (GOES)
The GOES System is the follow-up to the ATS series. They were designed by NASA for the National Oceanic and Atmospheric Administration (NOAA) to provide the United States National Weather Service with frequent, small-scale imaging of the Earth's surface and cloud cover. Meteorologists for weather monitoring and forecasting for over 30 years have used the GOES series of satellites extensively. These satellites are part of a global network of meteorological satellites spaced at approximately 70° longitude intervals around the Earth in order to provide near-global coverage. Two GOES satellites, placed in geostationary orbits 36000 km above the equator, each view approximately one-third of the Earth. One is situated at 75° W longitude and monitors North and South America, and most of the Atlantic Ocean. The other is situated at 135° W longitude and monitors North America and the Pacific Ocean basin. Together they cover from 20° W to 165° E longitudes.

Two generations of GOES satellites have been launched, each measuring emitted and reflected radiation from which atmospheric temperature, winds, moisture, and cloud cover can be derived. The first generation of satellites consisted of GOES-1 (launched in 1975) through GOES-7 (launched in 1992). Due to their design, these satellites were capable of viewing the Earth only a small percentage of the time (approximately five per cent). The second generation of satellites began with the launching of GOES-8 in 1994, and has numerous technological improvements over the first series. They provide near-continuous observation of the Earth allowing more frequent imaging (as often as every 15 minutes). This increase in temporal resolution coupled with improvements in the spatial and radiometric resolution of the sensors provides timely information and improved data quality for forecasting meteorological conditions.

The second generation GOES satellites have separate *imaging* and *sounding* instruments. The imager has five channels sensing visible and infrared reflected and emitted solar radiation. The infrared capability allows for day and night imaging. Sensor pointing and scan selection capability enable imaging of an entire hemisphere, or small-scale imaging of selected areas. The latter allows meteorologists to monitor specific weather trouble spots to assist in improved short-term forecasting. The imager data are 10-bit radiometric resolution, and can be transmitted directly to local user terminals on the Earth's surface. Table 3.28 describes the individual bands, their spatial resolution, and their meteorological applications of GOES.

Table 3.28 GOES Bands (*Source: CCRS, 1998*)

Band	Wavelength range (μm)	Spatial resolution	Application
1	0.52 - 0.72 (visible)	1 km	Cloud, pollution, and haze detection; severe storm identification
2	3.78 - 4.03 (shortwave IR)	4 km	Identification of fog at night; discriminating water clouds and snow or ice clouds during daytime; detecting fires and volcanoes; nighttime determination of sea surface temperatures
3	6.47 - 7.02 (upper level water vapour)	4 km	Estimating regions of mid-level moisture content and advection; tracking mid-level atmospheric motion
4	10.2 - 11.2 (longwave IR)	4 km	Identifying cloud-drift winds, severe storms, and heavy rainfall
5	11.5 - 12.5 (IR window sensitive to water vapour)	4 km	Identification of low-level moisture; determination of sea surface temperature; detection of airborne dust and volcanic ash

The 19 channel sounder measures emitted radiation in 18 thermal infrared bands and reflected radiation in one visible band. These data have a spatial resolution of 8 km and 13-bit radiometric resolution. Sounder data are used for surface and cloud-top temperatures, multi-level moisture profiling in the atmosphere, and ozone distribution analysis.

NOAA AVHRR

NOAA is also responsible for another series of satellites that are useful for meteorological, as well as other application. These satellites, in Sun-synchronous, near-polar orbits (830-870 km above the Earth), are a part of the Advanced TIROS series and provide complementary information to the geostationary meteorological satellites (such as GOES). Two satellites, each providing global coverage, work together to ensure that data for any region of the Earth, are not more than six hours old. One satellite crosses the equator in the early morning from north-to-south while the other crosses in the afternoon.

The primary sensor onboard the NOAA satellites, used for both meteorology and small-scale Earth observation and reconnaissance, is the Advanced Very High Resolution Radiometer (AHVRR). The AVHRR sensor detects radiation in the visible, near and mid infrared, and thermal infrared portions of the electromagnetic spectrum over a swath width of 3000 km. Table 3.29 outlines the AVHRR bands, their wavelengths and spatial resolution (at swath nadir), and general applications of each.

AVHRR data can be acquired and formatted in four operational modes, differing in resolution and method of transmission. Data can be transmitted directly to the ground, and viewed as data are collected, or recorded on board the satellite for later transmission and processing. Table 3.30 describes the various data formats and their characteristics.

Table 3.29 NOAA AVHRR Bands (*Source: CCRS, 1998*)

Band	Wavelength range (μm)	Spatial resolution	Application
1	0.58 - 0.68 (red)	1.1 km	Cloud, snow, and ice monitoring
2	0.725 - 1.1 (near IR)	1.1 km	Water, vegetation, and agriculture surveys
3	3.55 - 3.93 (mid IR)	1.1 km	Sea surface temperature, volcanoes, and forest fire activity
4	10.3 - 11.3 (Thermal IR)	1.1 km	Sea surface temperature, soil moisture
5	11.5 - 12.5 (Thermal IR)	1.1 km	Sea surface temperature, soil moisture

Table 3.30 AVHRR data formats (*Source: CCRS, 1998*)

Format	Spatial resolution	Transmission and processing
APT (Automatic Picture Transmission)	4 km	Low-resolution direct transmission and display
HRPT (High Resolution Picture Transmission)	1.1 km	Full-resolution direct transmission and display
GAC (Global Area Coverage)	4 km	Low-resolution coverage from recorded data
LAC (Local Area Coverage)	1.1 km	Selected full-resolution local area data from recorded data

Although AVHRR data are widely used for weather system forecasting and analysis, the sensor is also well suited to observation and monitoring of land features. AVHRR has much coarser spatial resolution than other typical land observations sensors but is used extensively for monitoring regional and small-scale phenomena including mapping of sea surface temperature, and natural vegetation and crop conditions. Mosaics covering large areas can be created from several AVHRR data sets allowing small-scale analysis and mapping of broad vegetation cover.

INSAT

INSAT is an operational multipurpose satellite system catering to the requirements of three different services, viz Television & Radio Broadcasting, Communications, and Meteorology. The INSAT project is a joint venture of the Department of Telecommunications (DOT), the Indian Meteorological Department (IMD), Doordarshan, and All India Radio (AIR). The responsibility for overall management and coordination of the INSAT system among the user agencies rests with the INSAT Co-ordination Committee (ICC) (Eumetsat, 2003).

The first satellite (INSAT-1A) of INSAT-1 series was launched in April 1982 and it ceased to function on September 6, 1982 as a result of major anomaly in the satellite. The second satellite INSAT-1B was launched on August 30, 1983, and it became operational on October 15, 1983. It was the main operational satellite throughout the 1980s, and provided very good services during its entire mission life. It was de-orbited in July 1993 and the third satellite of the series INSAT-1C was launched on July 22, 1988. Due to some technical problem it lost control on November 22, 1989, after which it was not available for operational services. The last satellite of INSAT-1 series INSAT-1D was launched on June 12, 1990, and became operational on July 17, 1990. After providing very useful services for almost 12 years, the VHRR payload of this satellite was switched off from May 14, 2002, due to non-availability of fuel for attitude control.

Table 3.31 INSAT-1 Geostationary Satellite Series (*Source: Eumetsat, 2003*)

Satellite	Launch date	Met. payload with wavelength bands	Major applications
INSAT-1A	April 10, 1982	Very High Resolution Radiometer (VHRR)	Monitoring cyclones & monsoon, CMV Winds, OLR, Rainfall estimation
INSAT-1B	August 30, 1983		
INSAT-1C	July 22, 1988		
INSAT-1D	June 12, 1990	Vis: 0.55-0.75 µm (2.75 km resolution) IR: 10.5-12.5 km (11km resolution)	

The second generation of INSAT satellites (INSAT-2 series) was started from July 1992 with the successful launch of the first satellite of the series (INSAT-2A) on July 10, 1992. The second satellite of INSAT-2 programme i.e. (INSAT-2B) was launched successfully on July 22, 1993. All INSAT satellites are three-axis body stabilized spacecrafts. The last satellite of INSAT-2E was launched successfully on April 3, 1999 and was made operational from May 1999 (Table 3.32). It has a new payload, called Charged Coupled Device (CCD) camera capable of taking 1 km resolution images in 3 bands. The meteorological imaging capability in thermal IR band has also been upgraded on this satellite, as compared to its predecessors, by providing a water vapor channel with 8 km resolution in the VHRR, the imaging instrument of the satellite. However, VHRR onboard INSAT-3E is not working due to anomaly in the scan-mechanism. A dedicated Meteorological Satellite METSAT (Kalpana-1) has been launched by India in September 2002, for earth imaging with three channels Very High Resolution Radiometer (VHRR) (Table 3.33). A data Relay Transponder (DRT) for collection of meteorological and hydrological data from automatic weather stations has also been provided on this satellite. METSAT is operational from September 24, 2002.

The first satellite INSAT-3A with meteorological payloads, of third generation of INSAT satellites (INSAT-3 series) was launched on April 10, 2003. Its meteorological payloads are identical to those of INSAT-2E, *i.e.* 3-channel VHRR and a 3-channel CCD. INSAT-3A has also a data Relay Transponder. The satellite has been declared operational from May 2003 (Table 3.33).

Table 3.32 INSAT-2 Geostationary Satellite Series (*Source: Eumetsat, 2003*)

Satellite	Launch date	Met. payload with wavelength bands	Major applications
INSAT-2A	July 10, 1992	Very High Resolution Radiometer (VHRR) Bands: 0.55 - 0.75 µm 10.5 – 12.5 µm	Monitoring cyclones & monsoon CMV winds, OLR, Rainfall estimation, Meso-scale features, Flood/intense precipitation advisory, Snow detection
INSAT-2B	July 23, 1993	Very High Resolution Radiometer (VHRR) Bands: 0.55 - 0.75 µm 10.5 – 12.5 µm	
INSAT-2E	April, 1999	1. VHRR: as above + WV Band: 5.0 - 7.1 µm 2. CCD Payload Bands: 0.63 – 0.79µm 0.77 - 0.86 µm 1.55 - 1.70 µm	

Table 3.33 INSAT-3 Geostationary Satellite Series – PRESENT (*Source: Eumetsat, 2003*)

Satellite	Launch date	Met. payload with wavelength bands	Major applications
METSAT (Kalpana-1)	September, 2002	VHRR: similar to INSAT-2E/3A	Monitoring cyclones & monsoon CMV Winds, OLR, Rainfall estimation
INSAT-3A (similar to INSAT-2E)	April, 2003	1. VHRR: as above + WV Band: 5.0 - 7.1 µm 2. CCD Payload Bands: 0.63 - 0.79 µm 0.77 - 0.86 µm 1.55 - 1.70 µm	Monitoring cyclones & monsoon, CMV winds, OLR, Rainfall estimation, Mesoscale features, Flood/intense precipitation advisory Snow detection, Crop discrimination Aerosols studies, Temperature/humidity profile (with INSAT-3D)

The imaging mission is working satisfactorily with METSAT (Kalpana-1) satellite and INSAT-3A, and they continue to be used operationally. High resolution (1 km) images in 3-channels are also available operationally from INSAT-3A and INSAT-2E CCD cameras. The activities like image processing, derivation of meteorological products, data archival and dissemination of products to field stations for operational use, are being done on routing basis.

VHRR images are normally received at three-hourly intervals. More frequent images are taken for monitoring the development of special weather phenomena as and when the situation demands. CCD images from INSAT-3A are also being taken every three hours for operational use during daytime. More frequent images can be taken if situation demands.

3.7 OTHER WEATHER SATELLITES

The United States operates the DMSP (Defense Meteorological Satellite Program) series of satellites which are also used for weather monitoring. These are near-polar orbiting satellites whose Operational Linescan System (OLS) sensor provides twice daily coverage with a swath width of 3000 km at a spatial resolution of 2.7 km. It has two fairly broad wavelength bands: a visible and near infrared band (0.4 to 1.1 µm), and a thermal infrared band (10.0 to 13.4 µm). An interesting feature of the sensor is its ability to acquire visible band nighttime imagery under very low illumination conditions. With this sensor, it is possible to collect striking images of the Earth showing typically the nighttime lights of large urban centres.

There are several other meteorological satellites in orbit, launched and operated by other countries, or groups of countries. These include Japan, with the GMS satellite series, and the consortium of European communities, with the Meteosat satellites. Both are geostationary satellites situated above the equator over Japan and Europe, respectively. Both provide half-hourly imaging of the Earth similar to GOES. GMS has two bands: 0.5 to 0.75 µm (1.25 km resolution), and 10.5 to 12.5 µm (5 km resolution). Meteosat has three bands: visible band (0.4 to 1.1 µm); 2.5 km resolution), mid-IR (5.7 to 7.1 µm; 5 km resolution), and IR (10.5 to 12.5 µm; 5 km resolution).

3.8 MARINE OBSERVATION SATELLITES/SENSORS

The oceans cover more than two-thirds of the Earth's surface, and play an important role in the global climate system. They also contain an abundance of living organisms and natural resources, which are susceptible to pollution and other man-induced hazards. The meteorological and land observations satellites/sensors discussed earlier can be used for monitoring the oceans of the planet, but there are other satellite/sensor systems, that have been designed specifically for this purpose.

The Nimbus-7 satellite, launched in 1978, carried the first sensor, the Coastal Zone Colour Scanner (CZCS), specifically intended for monitoring the oceans and water bodies. The primary objective of this sensor was to observe ocean colour and temperature, particularly in coastal zones with sufficient spatial and spectral resolution to detect pollutants in the upper levels of the ocean, and to determine the nature of materials suspended in the water column. The Nimbus satellite was placed in a Sun-synchronous, near-polar orbit at an altitude of 955 km. Equator crossing times were local noon for ascending passes and local midnight for descending passes. The repeat cycle of the satellite allowed for global coverage every six days, or every 83 orbits. The CZCS sensor consisted of six spectral bands in the visible, near-IR and thermal portions of the spectrum, each collecting data at a spatial resolution of 825 m at nadir over a 1566 km swath width. Table 3.34 outlines the spectral ranges of each band and the primary parameter measured by each.

The first four bands of the CZCS sensor are very narrow. They were optimized to allow detailed discrimination of differences in water reflectance due to phytoplankton concentrations and other suspended particulates in the water. In addition to detecting surface vegetation on the water, band 5 was used to discriminate water from land prior to processing the other bands of information. The CZCS sensor ceased operation in 1986.

Marine Observation Satellite (MOS)

The first Marine Observation Satellite (MOS-1) was launched by Japan in February 1987, and was followed by its successor MOS-1b, in February of 1990. These satellites carry three different sensors; a 4-channel Multispectral Electronic Self-Scanning Radiometer (MESSR), a 4-channel Visible and Thermal Infrared Radiometer (VTIR), and a 2-channel Microwave Scanning Radiometer (MSR), in the microwave portion of the spectrum. The characteristics of the two sensors in the visible/infrared are described in the accompanying Table 3.35.

Table 3.34 CZCS Spectral bands (*Source: CCRS, 1998*)

Channel	Wavelength range (μm)	Primary measured parameter
1	0.43 - 0.45	Chlorophyll absorption
2	0.51 - 0.53	Chlorophyll absorption
3	0.54 - 0.56	Gelbstoffe (yellow substance)
4	0.66 - 0.68	Chlorophyll concentration
5	0.70 - 0.80	Surface vegetation
6	10.5 - 12.50	Surface temperature

Table 3.35 MOS visible/infrared instruments (*Source: CCRS, 1998*)

Sensor	Wavelength ranges (μm)	Spatial resolution (m)	Swath width (km)
MESSR	0.51 - 0.59	50	100
	0.61 - 0.69	50	100
	0.72 - 0.80	50	100
	0.80 - 1.10	50	100
VTIR	0.50 - 0.70	900	1500
	6.0 - 7.0	2700	1500
	10.5 - 11.5	2700	1500
	11.5 - 12.5	2700	1500

The MESSR bands are quite similar in spectral range to the Landsat MSS sensor,, and are thus useful for land applications in addition to observations of marine environments. The MOS system orbit at an altitude around 900 km, and have revisit periods of 17 days.

3.25 SeaWiFS

The SeaWiFS (Sea-viewing Wide-Field-of View Sensor) onboard the SeaStar spacecraft is an advanced sensor designed for ocean monitoring. It consists of eight spectral bands of very narrow wavelength ranges (Table 3.36) tailored for very specific detection and monitoring of various ocean phenomena including: ocean primary production and phytoplankton processes, ocean influences on climate processes (heat storage and aerosol formation), and monitoring of the cycles of carbon, sulfur, and nitrogen. The orbit altitude is 705 km with a local equatorial crossing time of 12 mid night. Two combinations of spatial resolution and swath width are available for each band: a higher resolution mode of 1.1 km (at nadir) over a swath of 2800 km, and a lower resolution mode of 4.5 km (at nadir) over a swath of 1500 km.

Table 3.36 SeaWiFS spectral bands (*Source: CCRS, 1998*)

Channel	Wavelength ranges (μm)
1	0.402 - 0.422
2	0.433 - 0.453
3	0.480 - 0.500
4	0.500 - 0.520
5	0.545 - 0.565
6	0.660 - 0.680
7	0.745 - 0.785
8	0.845 - 0.885

These ocean-observing satellite systems are important for global and regional scale monitoring of ocean pollution and health, and assist scientists in understanding the influence and impact of the oceans on the global climate system.

SATELLITE DATA PRODUCTS 4

4.1 INTRODUCTION

The data collected by different types of sensors, are to be disseminated to the user community for analysis. It is important that the user should know regarding the manner in which the data is stored, *i.e.*, the type of format, different types of data products and their characteristics. This chapter provides, details regarding the format, data products and their characteristics with special emphasis to IRS data products, both photographic and digital.

4.2 DATA RECEPTION, TRANSMISSION, AND PROCESSING

There are three main options for transmitting the data acquired by satellites to the surface. The data can be directly transmitted to Earth as shown in Fig. 4.1, if a Ground Receiving Station (GRS) is in the line of sight of the satellite S_1. If this is not the case, the data can be recorded onboard the satellite using some recording device for transmission to a GRS at a later time. Data can also be relayed to the GRS through the Tracking and Data Relay Satellite System (TDRSS) S_2, which consists of a series of communication satellites in geosynchronous orbit. The data are transmitted from one satellite to another until they reach the appropriate GRS.

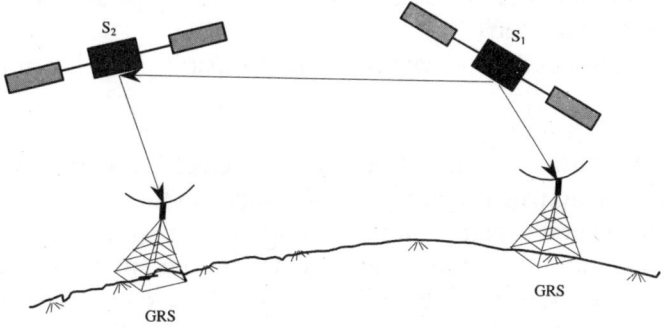

Fig. 4.1 Satellite data transmission

The data received at GRS is in a raw digital format, which may then, if required, be processed to correct for systematic, geometric, and atmospheric distortions present in the imagery, and be transformed into a standardized format. The data are written on CCT, disk or CD. The data are typically archived at most receiving and processing stations, and whole libraries of data are managed by government agencies such as NRSA in India or commercial companies, such as EOSAT in USA.

For many sensors, it is possible to provide to the users with quick-turnaround imagery when data are needed immediately after collection. Near real-time processing systems are used to produce low-resolution imagery in hard copy or soft copy (digital) format within hours of data acquisition. Such imagery can then be transmitted in the form of hard copy or soft copy to the end users. One of the applications of this type of fast data processing is to provide imagery to ships sailing in the Arctic to allow them to assess current ice conditions in order to make safe navigation. Real-time processing of imagery in airborne systems has been used, for example, to pass thermal infrared imagery to forest fire fighters right at the scene.

Low resolution quick-look imagery is used to preview archived imagery prior to purchase. The spatial and radiometric quality of these types of data products is degraded, but they are useful for ensuring that the overall quality, coverage, and cloud cover of the data are appropriate.

4.3 REMOTE SENSING DATA

In remote sensing, the terms 'image' and 'photograph' are often used in a confusing manner, though they have a precise meaning. A *photograph* is a representation of an object or scene that has been recorded on a photographic film using an optical lens camera. On the contrary, an *image* is a representation of an object recorded by a scanner system. A scanner generally operates within small and particular ranges of the electromagnetic spectrum, and records information as digital number (DN) that relates to some property of the object, such as its reflectance characteristics.

In remote sensing, in general, data acquisition is through scanner based devices, hence the primary data collected at the sensor level is digital in nature. However, this data when distributed to user community is made available both in digital and photographic form as per choice or preference of the user.

Hence, a detailed discussion on data in digital format is provided in the next section.

Digital Data

As a sensor scans the earth's surface, it generates an electrical current that varies in intensity depending on the variation in brightness of the land surface. If the sensor detects several spectral bands, then several separate electrical currents are generated. Each electrical current is, of course, a continuously varying signal that must be subdivided into distinct units to create the discrete values necessary for digital analysis. The conversion from the continuously varying analog signal to the discrete digital values is accomplished by sampling the electric current at uniform intervals (Fig. 4.2). All signal values within this interval are represented by an average value and all variation within this interval is lost. Thus, the choice of sampling interval forms one dimension to the resolution of the sensor. In addition, digital values are usually scaled in such a way that they portray relative, rather than absolute brightness, *i.e.*, the digital values do not represent true radiometric values from the scene.

Another limit upon image detail is the manner in which digital values are quantized from the analog signal. Each digital value is recorded as a series of binary digits or bits. Each bit records an exponent of a power of 2, with the value of the exponent determined by the position of the bit in the sequence. As an example, let us consider a system designed to record seven bits for each digital value. This means that seven binary places are available to record the brightness sensed for each band of the sensor. The seven values are recorded in sequence of successive powers of 2. A binary value (either 0 or 1) denotes whether or not that specific value is to be added to the total value for a given pixel. A "1" signifies that a specific power of 2 (determined by its position within the sequence) is to be evoked and a "0" indicates a value of zero for that position. Thus, the seven-bit binary number "1111111" signifies $2^6 + 2^5 + 2^4 + 2^3 + 2^2 + 2^1 + 2^0 = 64 + 32 + 8 + 4 + 2 + 1 = 127$ and "1001011" signifies $2^6 + 0^5 + 0^4 + 2^3 + 0^2 + 2^1 + 2^0 = 64 + 0 + 0 + 8 + 0 + 2 + 1 = 75$.

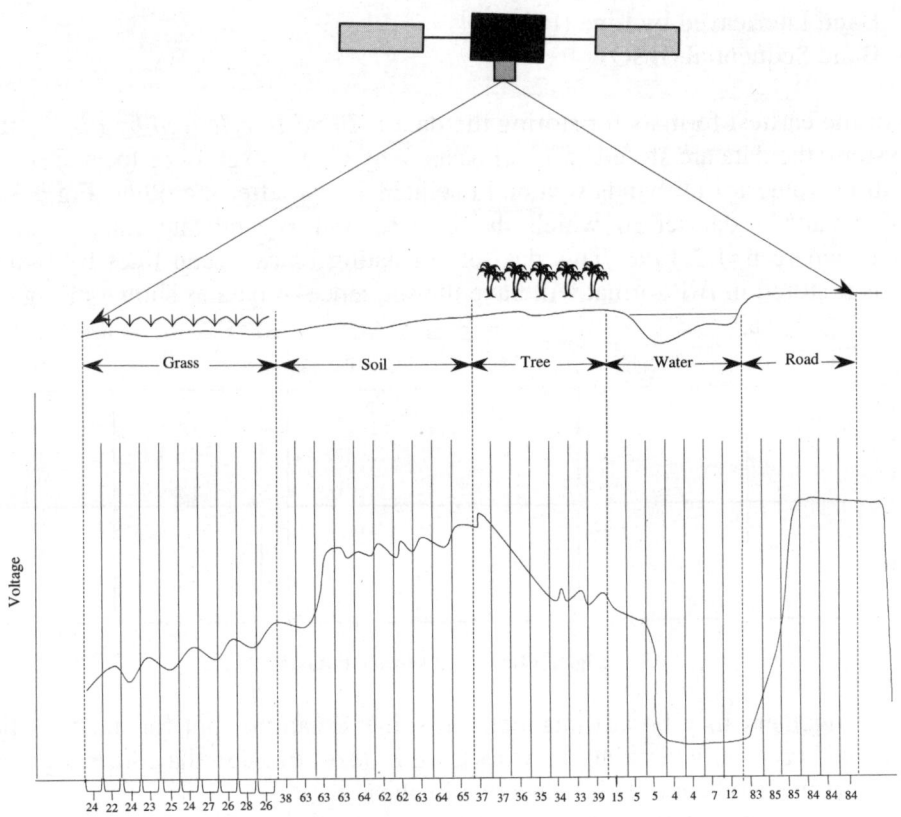

Fig. 4.2 Schematic representation of data collection as DN values

In this manner discrete digital values for each pixel are recorded in a form suitable for storage on tapes or disks, and for subsequent analysis by a digital computer. The values, as read from tape or disk, are popularly known as *digital numbers* (DN), *brightness values* (BV), or *digital counts*.

It may be noted that the number of brightness values within a digital image are determined by the number of bits available. The six-bit would allow for a maximum range of 64 possible values, *i.e.*, a pixel may have a value ranging from 0 to 63. Similarly, a 7-bit would mean that 128 range of brightness values, *i.e.*, from 0 to 127, and 8-bit would extend the range to 256, *i.e.*, from 0 to 255. Hence, the number of bits determines the radiometric resolution of a digital image.

Tape Formats
The digital image data must be stored in an organized for easy and fast retrieval for analysis. Digital remote sensing data can be stored in any one of the following three formats commonly used to store image data:

(*i*) Band Interleaved by Pixel (BIP)
(*ii*) Band Interleaved by Line (BIL)
(*iii*) Band Sequential (BSQ).

One of the earliest formats for storing the data is *Band Interleaved by Pixel* (BIP). In this format system, the data are stored on pixel basis. Any given pixel, once located on the tape, is found with its value for all bands written in sequence, one after the other. Fig 4.3 shows the layout of a sample dataset in which the L_n, P_n, and B_n indicate line, pixel and band respectively where n=1,2,3 *etc*. This dataset consisting of two scan lines by two pixels for three bands if stored in BIP format will have the sequence of data as shown in Fig 4.4.

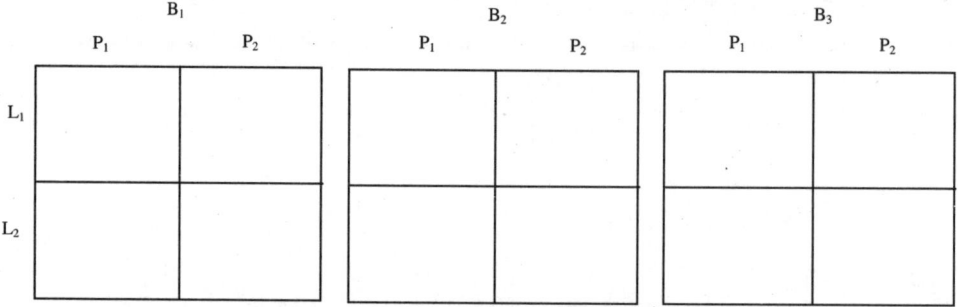

Fig. 4.3 Layout of sample dataset

This arrangement may be advantageous in some situations, but for most applications in which the data is typically very large it is awkward to sort through the entire sequence of data in order to sort the bands into their respective images.

Band Interleaved by Line (BIL) format treats each line of data as a separate unit. Each line is represented in all bands before the next line is encountered. A typical example of BIL data format is shown in Fig. 4.5.

A third format for recording remotely sensed data is the Band Sequential (BSQ) format. All data for band 1 are written in sequence, followed by all data for band 2, then bands 3 and 4 in sequence as shown in Fig. 4.6. Each band is treated as a separate unit. For many applications, this format is among the most practical as it presents data in the format that most closely resembles that used for the display and analysis. However, if areas smaller than the entire scene are to be examined, the analyst must read all four images before the sub area can be identified and extracted.

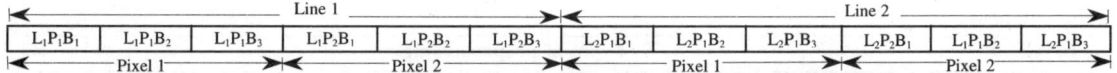

Fig. 4.4 Band Interleaved by Pixel (BIP) data format

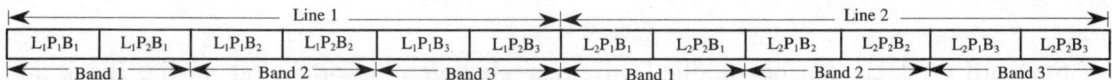

Fig. 4.5 Band Interleaved by Line (BIL) data format

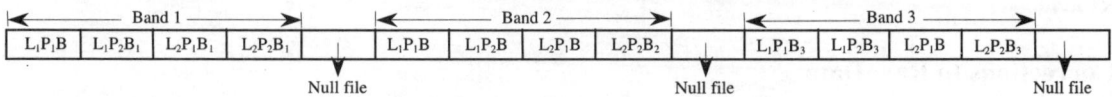

Fig. 4.6 Band Sequence (BSQ) data format

The best tape format depends upon the context of the study, and often upon the software and equipment available to a specific analyst. If all bands for an entire image are to be used, then the BSQ and BIL formats are useful, because they are convenient for reconstructing the entire scene in all bands. On the other hand, if the exact position on the image of the sub-area to be studied is known, then the BIP format is useful, because values for all bands are found together, and it is not necessary to read through the entire data set to find a specific region. In general, facilities should be available to read the data in the available format, and to convert them into the format most convenient for use.

4.4 DATA PRODUCTS

Primarily satellite data when received at the Ground Receiving Station, it is digital in nature. However to the user community, it may be available both in digital or photographic form. In the next section, the different types of data products as supplied by National Remote Sensing Agency (NRSA), India to the user committee.

Data Products Supplied by NRSA

The two categories in which data products are generated are digital and photographic. These may be standard and special products. Standard products are generated after applying radiometric and geometric corrections, whereas special products are generated after further processing of the standard products.

The raw data recorded at the Earth station are corrected to various levels of processing at the Data Processing System (DPS) as shown in Table 4.1.

Table 4.1 Various levels of processing (*Source: NRSA, 1997*)

Level	Types of preprocessing
0	Uncorrected (Raw Data)
1	Radiometrically corrected and geometrically corrected only for Earth station (for Browse Facility).
2	Radiometrically and geometrically corrected (Standard Product)
3	Special processing such as merging, enhancement on standard product or level 2 product

Standard products can be supplied on either photographic or digital media. Black and White (B/W) and False Colour Composite (FCC) photographic products are available in the form of films or paper prints. Digital products are supplied on various magnetic media that are currently popular, *viz.*, Computer Compatible Tape (CCT), 8 mm Exabyte tape, and CD-ROM.

Corrections to Raw Data

Raw data suffer from both geometric and radiometric distortions which have to be corrected. The various corrections applied are as follows:

Radiometric corrections

Radiometric distortions arise due to:

(*i*) non-uniform response of the detectors and detector elements,
(*ii*) detector element failure,
(*iii*) data losses during communication or archival/retrieval,
(*iv*) narrow dynamic range,
(*v*) image to image variations.

A radiometric correction Look-Up-Table (LUT) is prepared for normalizing the responses of all detector elements with respect to a desired common response. The least saturation radiance value realized over the whole array is used as the reference. The same value can be used for conversion of radiometrically corrected Digital Number (DN) values back to absolute units by the users of the data products. This can be done using the ground calibration data for all detectors.

The correction for scan line losses is done by averaging the neighbouring pixel values. If data losses occur in more than two consecutive scan lines, then these are replaced by a line consisting of minimum DN value. The failed detector pixel values are replaced by the average of the adjacent pixels on the same scan line.

Geometric corrections

Geometric distortions in the data may be due to following reasons:

(*i*) Scene related distortions are earth rotation effect and earth shape (geoid) induced distortions.

(*ii*) Sensor related distortions are caused by sensor focal plane detector geometry, alignment of optical axis with respect to spacecraft attitude reference, multi-band and multi-array misregistration and off-nadir pointing (for PAN) induced distortions.

(*iii*) Spacecraft related distortions are due to image orientation with respect to spacecraft heading, altitude and velocity variations affecting image scale, and attitude variations in roll, pitch and yaw directions.

(*iv*) Measurement/Calibration distortions are due to wrong estimation of spacecraft state vectors, altitude and pointing angles measurement calibration of various alignment angles involved, and synchronization of onboard and ground reception times.

Geometric corrections are performed through a dynamic model, which represents the imaging geometry. In this model, through a series of transformations, the image is converted from image coordinate system to ground coordinate system.

A grid of input co-ordinates (scan line and pixel) on the radiometrically corrected image, are selected and the corresponding output co-ordinates (latitude and longitude) are calculated for all the grid points. For a user area given in the output space, a grid is defined, and the input coordinates for these grid points are obtained through interpolation from the earlier computed points. The input co-ordinates for the intermediate points of output space are obtained by another interpolation, now in the output space only. The grey values for all the output points are obtained by resampling the input image.

Map projection and the image orientation (for geocoded products) are incorporated at the time of fixing the output grid. Finally, the data is formatted for generating the photographic or digital products, in the required format.

4.5 REFERENCING SCHEME

Referencing scheme is unique for each satellite mission and is a means of conveniently identifying the geographic location of points on the earth. This scheme is designated by *Path* and *Row*. Here the concept has been explained with respect to IRS satellite system.

Path

An orbit is the course of motion taken by a satellite in space, its ground trace on the Earth is called a path. IRS-1C/1D satellite has an orbital period of 101.35 minutes, thus it completes approximately 14 orbits per day. Since it requires 24 days to completely cover the whole Earth, it requires 341 orbits before the next repeat cycle begins. Though the number of paths and orbits are same, the designation of number of orbit and path in the referencing scheme is not same.

Path number one is assigned to that orbit which is at 29.7°W longitude. The next orbit is spaced apart by 1.055° on the westward side. So if on Day 1, the satellite on Orbit No. 1 passes over Path No. 1, then the next orbit as per referencing scheme will be on Path No. 318. Similarly, for orbit 3 the Path No. will be 294 and so on. Similarly, on Day 2, the first orbit will be Path 6 which will be east of Path 1 and is separated by 5 paths.

Row

Along a path, the continuous stream of data is segmented into a number of scenes of convenient size. While framing the scenes, the equator is taken as the reference line for segmentation. The scenes are framed in such a manner that the centre of one of the scene lies on the equator. In case of IRS-LISS-III scene consisting of 6000 lines, is framed such that, the centre of scene lies on the equator. The next scene is defined such that its centre lies exactly 5703 lines from equator. The centre of third scene is then defined 5703 lines northwards, and so on. This continued up to 81° North. The lines joining the corresponding scene centre of different paths are parallel to the equator and are called the rows.

The uniformly separated scene centres are such that same rows of different paths fall on the same latitude. The row number 1 falls around 81° N latitude, while row number 41 is near to 40° N and row number 75 lies on Equator. The Indian region is covered by Path Numbers 65 to 130 and Row numbers 30 to 90.

4.6 STANDARD PRODUCTS

The various kinds of standard products which are available are

  (*i*)   Path/Row products,
  (*ii*) Shift Along Track products,
  (*iii*) Quadrant products,
  (*iv*) Basic Stereo products, and
  (*v*)   Geocoded products.

Path/Row Based Products

These products are based on the referencing scheme of each sensor. The user has to specify the following:

  (*i*)   Path/Row number as per referencing scheme,
  (*ii*) Sensor Identification,
  (*iii*) Subscene Identification (for PAN),
  (*iv*) Date of Pass,
  (*v*)   Band number for B/W products, band combination for FCC products, and
  (*vi*) Product Code.

Shift along Track Products

If an area of interest is less than the dimensions of a full scene, and falls in between two successive rows of the same path, then the data can be supplied by sliding the scene in the forward direction (along the path) accommodating the required area in a single product. These products are called *Shift Along Track* (SAT) products.

In the case of SAT products, the percentage of shift has to be specified, in addition to the inputs specified by the user for Path/Row based products. The percentage of shift along the path has to be specified between 10% and 90%, in multiples of 10%. Fig. 4.7 depicts the concept of a scene which has been shifted along track.

Satellite Data Products 67

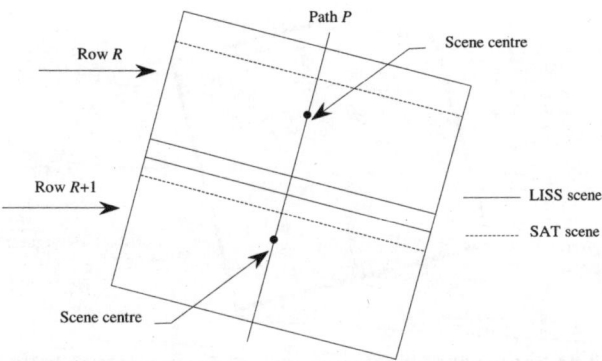

Fig. 4.7 Scene shifting along a track (*Source: NRSA, 1997*)

Quadrant Products
In the case of LISS-III, the full scene is divided into four nominal and eight derived quadrants as shown in Fig. 4.8. Quadrant number 1, 2, 3, 4 are nominal quadrants and the remaining eight quadrants are obtained by sliding quadrants 1, 2, 3, and 4 by 25% along and across the scene and within the path. LISS-III quadrant products are generated on 1:125,000 scale.

In the case of PAN, the full scene is divided into four quadrants as shown in Fig. 4.9. Here, each quadrant corresponds to one and a half array data.

Stereo Products
The oblique viewing capability of PAN sensor can be used to acquire two images of the same area, obtained on different dates and from different angles. Such images are called the *stereopairs*, and therefore, PAN sensors can only generate stereopairs. Stereo products are supplied with only radiometric correction. Stereopairs are widely used in photointerpretation for relief perception, and also in photogrammetric studies for deriving DTM models.

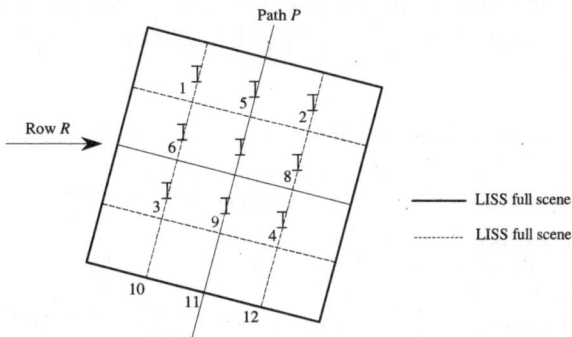

Fig. 4.8 LISS-III scene divided into quadrants (*Source: NRSA, 1997*)

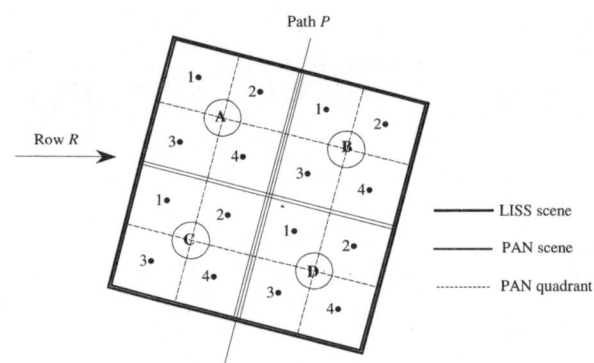

Fig. 4.9 PAN scene divided into four quadrants (*Source: NRSA, 1997*)

Geocoded Products

Geocoding is done to correct the imagery to a source independent format, whereby multidate and multi-satellite data can be handled with ease. Geocoded products are generated after applying radiometric and geometric corrections, orienting the image to true north, and generating the products with an output resolution appropriate to the map scale. The advantages of a geocoded product are that it can be overlaid on a Survey of India (SOI) toposheet map.

The generation of geocoded products is based upon the SOI toposheet scales, *e.g.*, PAN at 1:25,000 scale and LISS-III at 1:50,000 scale. The inputs required to be specified by the user is the SOI toposheet number in addition to those provided in the case of Path/Row based products. Table 4.2 gives the details of geocoded products.

In addition to the map sheet based geocoded products, PAN 5' × 5' geocoded products are also supplied. Here, an area corresponding to 5' × 5' within a scene is extracted around a user specified point, and aligned to true North after applying standard corrections. The inputs to be specified by the user are the coordinates of the point around which the 5' × 5' data are required, in addition to the other details as in the case of Path/Row based products. The main advantage of this product over the map sheet based geocoded products is that it is on 1:12,500 scale.

Table-4.2 IRS Geocoded product (*Source: NRSA, 1997*)

Product type	Aerial extent	Output resolution	Scale
Map sheet based products LISS-II geocoded products in visible band resolution (B/W and FCC)	15' × 15'	12.5 m	1:50,000
PAN Geocoded products	7.5' × 7.5'	6.25 m	1:25,000
Special geocoded products User specified PAN point geocoded products	5' × 5'	3.125 m	1:12,500

Special Products

Value added products are generated after further processing of the standard products, by extracting a specific area, mosaicing, merging, and enhancing the data. These products are generated as per user requirements. In addition to the above, the following products can be made available.

(*i*) PAN + LISS-III merged products, and
(*ii*) Orthoimage.

PAN + LISS-III merged products: In order to exploit the dual advantage of the spectral resolution of LISS-III and the spatial resolution of PAN, merged products of PAN+LISS-III at PAN resolution, can be made available to the user. The criteria that will be considered while selecting the PAN and LISS-III scenes are:

(*i*) PAN tilt is near nadir and the scene fits into a LISS-III scene.
(*ii*) Day of passing is not separated by more than a few days.

These products are available at 1:25,000 scale (7½' × 7½' map sheet based product) and 1:12,500 (floating geocoded product with an area coverage of 5' × 5') scale as FCC products.

Orthoimage: One of the important special products from IRS-1C/1D data is the orthoimage generated from PAN stereo pairs. In the case of LISS-III, orthoimages can be generated using the DEM provided by the user and on the availability of at least four GCP's. The product is corrected for all types of errors present in the raw data including the distortions due to terrain relief and camera tilt. These products are made available on photographic and digital media. The scale of the photographic product is 1:50,000 and 1:25,000 for LISS-III and PAN, respectively.

4.7 DIGITAL DATA PRODUCTS

The data for all the sensors of IRS-1C/1D are supplied on digital media like Computer Compatible Tapes (CCT'), CD-ROM, Cartridge, and 8 mm Exabyte tape based on user request. The file formats and structures in User CCT (UCCT) are the same for all levels of processing. The two formats in which digital data are supplied are Fast format and LGSOWG Superstructure format. All digital data in super structure format are provided in BIL or BSQ modes. However, in Fast format, data is supplied in BSQ format only.

Computer Compatible Tapes

CCTs are supplied with 6250 BPI density. Data on CCTs are supplied in the following formats:

Fast format: The two files in the UCCT for Fast Format are header file and image file. The Header file is the first file on each volume contains header data in ASCII format. It contains map projection, resampling options, and tick mark locations. The image files contain only video data. There is no prefix and suffix data with the individual image record.

LGSOWG format: In LGSOWG format, in addition to the video data for a scene, each product contains scene identification, location information, sensor, platform, and processing related information. In the LGSOWG format, these are following five files:

- (i) Logical volume file,
- (ii) Volume director file,
- (iii) Leader file,
- (iv) Image data file,
- (v) Trailer file, and
- (vi) Null Volume file.

Logical volume: A logical volume is a logical collection of one or more files, recorded consecutively. A logical volume contains one or more band data of a scene. All logical volumes have a volume directory as the first file and conclude with a null volume directory. When a logical volume is split between physical volumes, the volume directory is repeated in the continuation tape.

Volume directory: The volume directory file is the first file of every logical volume. It is composed of volume descriptor record, a series of file pointer records, and a text record. The volume descriptor record identifies the logical volume and the number of files that it contains. A text record follows the volume descriptor record, and identifies the data contained in the logical volume. There is a file pointer record for each type of data in the logical volume that indicates each file class, format, and attributes.

Leader File: The leader file is composed of a file descriptor record and two types of data records. The data records are header and ancillary. Header record contains information related to mission, sensor, calibration coefficients, and processing parameters. Ancillary records contain information pertaining to ephemeris, attitude, map projection, GCPs for image correction and image location, and annotation.

Image file: Image file consists of file descriptor record and actual image record. Image data contains actual video data in BIL or BSQ format. In addition to the image or video data, it also contains pixel counts, scan line identification starting and ending of actual data in the line.

Trailer file: The trailer file shows the calibration data file and ancillary information file. This is composed of a file descriptor record and one trailer record for each band.

Null volume directory file: The file which terminates a logical volume is null volume directory file. The file is referred to as 'NULL', because it identifies a non existent logical volume. This file consists of a volume descriptor record only.

CD-ROM Products
Compact Disk (CD) have the advantage of being compact, reliable, immune to magnetic fields, rugged and cost effective, with high memory capacity and random access. CD-ROM is a Write Once Read Many Times (WROM) device, with a memory of 650 MB. It can be read on any system with a CD-ROM drive, in compliance with the ISO 9660.

CD-ROM Logical Description
A CD-ROM may contain a full scene, quadrant or sub-scene. Apart from image (video) data, ancillary data and format document are copied onto the "PRODUCT" directory. The file CDINFO contains the identification of the product and request number. A general purpose display program (DISPLAY.EXE) is also available on each CD-ROM that will display B&W data, only on an EVGA monitor.

Format of products on CD: Data is supplied in two formats namely LGSOWG and Fast format. The contents of the files are exactly the same as that on a CCT.

File naming convention: The following conventions of naming files are in use:

LGSOWG format
1. VOLUME.PAN/L3/WIF   Volume directory file
2. LEADER.PAN/L3/WIF   Leader File
3. IMAGERY.PAN/l3 WIF   Imagery file
4. TRAILER.PAN/L3/WIF   Trailer file
5. NULL.PAN/L3/WIF   Null volume file

Fast format
1. HEADER.PAN/L3/WIF   Header file
2. BANDx.PAN/L3/WIF   Imagery file of x where x = 1,2,3,4 or 5.

IMAGE INTERPRETATION 5

5.1 INTRODUCTION

An image is a detailed photographic documentation or record of features on the ground at the time of data acquisition. Such acquired images are examined by an analyst in a systematic manner with the help of some supporting information collected from maps, field visit reports, or previously interpreted images of the same area. The interpretation of the information is carried out on the basis of certain physical characteristics of the object(s) and phenomena appearing in the image. The success of an image interpretation is dependent upon the experience of the analyst, the type of object or phenomena being interpreted, and the quality of the image.

5.2 INTERPRETATION PROCEDURE

Photographic/Image Interpretation is defined as the act of examining photographic images for the purpose of identifying objects and judging their significance. The image interpretation procedure is a complex task, and requires several tasks to be conducted in a well-defined routine consisting of the process of classification, enumeration, mensuration, and delineation (Campbell, 1987).

Classification is the first task to be performed by an interpreter, where based on the appearance of an object or feature, the analyst assigns a class or informational group. At this stage, the analyst first carries out the determination of presence or absence of an object or phenomena through the process of *detection*. This is followed by *recognition* where an object or phenomena is assigned an identity to a class or category. This generally requires a higher level of knowledge of feature. Finally, the feature is identified with a certain degree of confidence to a specific class. This process is known as *identification*.

Enumeration is the next step, and it relates to listing and counting of objects or phenomena that are visible on an image. However, it is dependent upon the ability of the analyst to classify items accurately.

Mensuration is the process of measurement, wherein an analyst makes measurements of objects in terms of length, area, volume or height. Another form of measurement could be in terms of image brightness characteristics known as *densitometry*.

Delineation which is the final task to be performed is outlining the regions of homogenous objects or areas. These are characterized by specific tones and textures. When sharp boundaries between objects occur, the delineation process becomes simpler in comparison to those areas where there is a gradual variation, *e.g.*, sandy area in the close proximity of water bodies.

In order to carry out the above processes, it is essential to understand the basic characteristics of images which govern the appearance of objects or phenomena on an image, discussed in the following section.

5.3 ELEMENTS OF PHOTOINTERPRETATION

An object or phenomena on an image can be identified, recognized, and delineated on the basis of some of the typical characteristics of the image. These characteristics, which allow for a systematic and logical approach in carrying out image interpretation, are known as the elements of photointerpretation. These image characteristics are described below.

Size of an object is a function of scale. Generally, relative sizes of objects must be taken into consideration within a same image. Individual houses would be smaller in size in comparison to commercial buildings.

Shape normally refers to the general form or outline of individual features. It is a very distinctive clue for interpretation. Normally, man-made features tend to have defined edges leading to regular shape, while natural objects will be having irregular shape. Roads, canals, buildings are man-made objects, and have regular shape while forest areas; water bodies tend to have irregular shape.

Tone of an object or phenomena refers to the relative brightness or colour in an image. It is one of the fundamental elements for distinguishing between different objects or phenomena, and is a qualitative measure. Generally, tone is designated as dark, medium or light.

Pattern refers to the spatial arrangement of visibly discernable objects. Typically, the repetition of similar tones and texture produces a distinctive and recognizable pattern. For example, houses laid out in orderly manner in urban area or the trees in an orchard create different patterns.

Texture is referred to as the frequency of tonal changes in particular areas of an image. Texture is a qualitative characteristic, and is normally categorized as rough or smooth. An area of dry sand will appear on an image having smooth texture, since the variation of tone for long stretches is not present. Similarly, an area covered with forest having variety of tree species with varying canopy size, shape, and density will appear with a rough texture, as tone will be changing very rapidly.

Shadow is an important characteristic of image interpretation. It gives an idea of the profile and relative height of an object, hence making identification easier. In mountainous areas, shadow is an important characteristic of images as it enhances the ground topography, and hence helpful for identifying variations in geological landforms.

Association is another important characteristic as it considers the interrelation with the objects within the close proximity of an object or phenomena. For example, white irregular patches adjacent to river indicate presence of dry sand banks of the river. A regular grid of lines having small regular shaped box like objects is an indication of urban areas.

Site refers to the vocational characteristics of objects such as topography, soil, vegetation, and cultural features.

Fig. 5.1 shows the primary ordering of each element in the interpretation procedure with respect to degree of complexity.

Image Interpretation 75

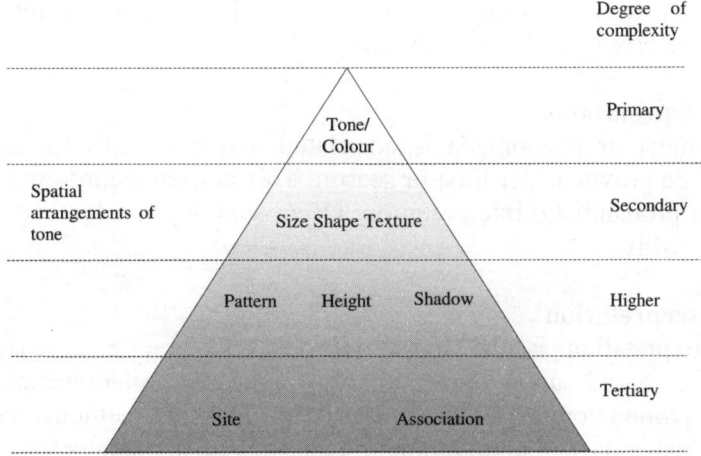

Fig. 5.1 Order of element of photointerpretation (*Source: Lo and Yeung, 2002*)

5.4 IMAGE INTERPRETATION STRATEGIES

An image interpretation strategy can be defined as a disciplined procedure that enables the interpreter to relate geographic patterns on the ground to their appearance on the image. Campbell (1978) has defined five categories of image interpretation strategies discussed below.

Field Observation
Many a times, the ground conditions as depicted on an image is complex, and an interpreter on the basis of his knowledge and experience, is unable to correlate the relationships between ground and image. Hence, the analyst is forced to visit the ground to make proper identification. Field observation is an important part of any interpretation task in order to assess the accuracy of identification.

Direct Recognition
Direct recognition is the application of an interpreter's experience, skill, and judgment to associate the image patterns with informational classes. This process is essentially a qualitative and subjective analysis of the image using the elements of image interpretation as visual and logical clues. Generally, direct recognition is an intuitive process, and hence, it requires very careful and systematic examination of the image.

Interpretation by Inference
In interpretation by inference, the analyst identifies information on the basis of the presence of some other information to which it is closely related to. Such inference information is known as *surrogate* or *proxy*. For example, as soil distributions are closely related to patterns of landforms and vegetation that are recorded on the image, landforms and vegetation form surrogates for the soil pattern, and hence the interpreter infers the invisible soil distribution

from patterns that are visible. Application of this strategy requires a complete knowledge of the link between the proxy and the distribution.

Probabilistic Interpretations

Sometimes, an object or phenomena is correlated to certain specific characteristics *e.g.*, certain crops will be grown as per time or season. This non-image information or knowledge can be utilized in probabilistic interpretation. Often such knowledge can be expressed as a statement of probability.

Deterministic Interpretation

Deterministic interpretation is the most rigorous and precise approach. Deterministic interpretations are based upon quantitatively expressed relationships that tie image characteristics to ground conditions. In contrast with the other methods, most information is derived from the image itself. Photogrammetric analysis of stereopairs for terrain information is a good example of deterministic interpretation. A scene is imaged from two separate positions along a flight path, and the photogrammetrist measures the apparent displacement in the position of an object. Based upon knowledge of the geometry of the photographic system, a topographic model of the landscape can be reconstructed. The result is, therefore, the derivation of precise information about the landscape using only the image itself and knowledge of its geometric relationship with the landscape. Relative to the other methods, very little non-mage information is required.

Image interpreters, of course, may apply a mixture of several strategies in a given situation. Interpretation of soil patterns, for example, may require direct recognition to identify specific classes of vegetation, than application of interpretation by proxy to relate the vegetation pattern to the underlying soil pattern.

5.5 PHOTOMORPHIC ANALYSIS

Another approach to interpretation of complex patterns is to identify areas of uniform appearance on the image, *i.e.,* search for photomorphic regions. Here the interpreter does not attempt to resolve the individual components within the landscape, but looks for their combined influence on image pattern. Photomorphic regions then are simply image regions of relatively uniform tone and texture.

In the first step, the interpreter delineates regions of uniform image appearance using tone, texture, shadow, and the other elements of image interpretation as a means of separating regions, and then tries to match photomorphic regions to useful classes of interest. This step obviously requires field observations or collateral information, because regions cannot be identified by image information alone. As the interpretation is refined, the interpreter may find that it is necessary to combine or divide photo regions to produce an acceptable interpretation.

Delineation of photomorphic regions is a powerful interpretation tool, but it must be applied with caution. Photomorphic regions do not always correspond to the categories of interest to the interpreter. The appearance of one region may be dominated by factors related to geology and topography, whereas that of another region in the same image may be controlled by the vegetation pattern.

5.6 IMAGE INTERPRETATION KEYS

Image interpretation keys are valuable aids for summarizing complex information. Such keys serve either or both of the two purposes:

(i) A means of training inexperienced personnel in the interpretation of complex or unfamiliar topics.
(ii) A reference aid for experienced interpreters to organize information and examples pertaining to specific topics.

An *image interpretation key* is simply a reference material designed to permit rapid and accurate identification of objects or features represented on aerial images. A key usually consists of two parts. The first part is a collection of annotated or captioned images or stereograms, while the second part is a graphic and/or word description. These materials are organized in a systematic manner that permits retrieval of desired images by, for example, data, season, region, or subject.

The key is a means of organizing the essential characteristics of an object or phenomena in an orderly manner. It must be noted that scientific keys of all forms require a basic familiarity with the subject matter. A key, then, is not a substitute for experience and knowledge of an interpreter but a means of systematically ordered information to help in the interpretation process so that a user can learn quickly.

The interpretation key forms an effective way of organizing and presenting the expert knowledge of a few individuals. The construction of a key tends to sharpen one's interpretation skills and encourages the interpreter to think more clearly about the interpretation process.

5.7 EQUIPMENT FOR IMAGE INTERPRETATION

Image interpretation can be conducted with relatively simple and inexpensive equipment, although some of the optional items can be expensive. An image interpretation laboratory must have sufficient storage and handling facility for images, both as paper prints and as film transparencies. In order to carry out the interpretation process, the following equipment may be required.

Light tables having a translucent surface and illuminated from behind, permit convenient viewing of film transparencies. If roll film is to be used, light tables must be equipped with special brackets to hold the film spools and rollers at the edges to permit the film to move freely without damage.
Rulers used in interpretation are an engineer's scales or rulers with accurate graduations. Ordinary household rulers are not satisfactory for photo interpretation.

Stereoscopes are devices that facilitate stereoscopic viewing of aerial photographs. The simple and most common is the *pocket stereoscope*. Its compact size and inexpensive cost makes it one of the most widely used instruments in remote sensing for visual interpretation.

Magnifiers are used to magnify the objects or features appearing in the image for detailed study. Image interpreters almost always wish to examine images using magnifiers, although the exact magnification depends upon individual preference and the nature of the task at hand.

Densitometer is an instrument that measures image density by directing a light of known brightness through a small portion of the image, and then measuring its brightness as altered by the film. Such instruments find densities for selected regions within an image so that an interpreter can make quantitative measurements of image tone.

Parallax bar is an instrument designed for use with a stereoscope. It permits estimation of topographic elevation or heights of features form stereo aerial photographs. A parallax bar has two glass plates; one under each lens of the stereoscope is attached to the base of the stereoscope. Both glass plates are marked with a small black dot, but one glass plate is fixed in position, whereas the other can be moved from side to side along a scale that measures its movement left to right parallel to the bar. This movable dot appears to float till it fuses with the fixed dot and appears to rest on the ground. At this position, the reading of the parallax bar is noted and used for computing height of that point.

Zoom transferscope is an instrument manufactured by Bausch and Lomb Corporation, is tailored for the visual matching of images. Separate maps or images at different scales can be manipulated optically so that they register to one another. The operator views both images through binocular eyepieces, and has control over magnification and orientation of one of the images. This control is used to bring the two images into registration. The operator can then trace detail from one image onto an overlay that registers to the second.

5.8 AUTOMATED APPROACH TO IMAGE CLASSIFICATION

The fundamentals of visual image interpretation were developed for interpretation of aerial photographs, and have been suitably adapted for remote sensing image interpretation. With the availability of remote sensing images in digital form and high speed computing facilities, digital image analysis, discussed in Chapter 6, provides an automated approach to image interpretation.

DIGITAL IMAGE PROCESSING

6.1 INTRODUCTION

Remote sensing data can be interpreted visually using a hard copy of the image or by digital method using digital image employing suitable software. Though the digital method cannot replace the visual interpretation discussed in Chapter 5, it has some advantages such as consistency in results, discrimination of more shades of gray tones, quantitative analysis, etc. Table 6.1 presents a broad comparison between visual and digital interpretation.

Digital image processing is an extremely broad subject, and often involves mathematically complex procedures. The objective of this chapter is to introduce the basic principles of digital image processing.

Table 6.1 Visual versus digital interpretation procedure

Visual	Digital
It is a traditional approach based on human intuition, and its success is experienced based.	It is a recent approach and requires specialized training.
It requires simple and inexpensive equipment.	It is complex, highly mathematical and requires expensive equipment.
It uses the brightness characteristics of the object, and accounts for the spatial content of the image.	It relies heavily upon the brightness and spectral content of the object and does not use the spatial content of the image.
Usually a single band of data is used for analysis. However, colour products generated from three bands of data, can be used for analysis.	Multiple bands of data are used for analysis.
The analysis process is subjective, qualitative, and dependent on analyst bias, however deductions are concrete.	The process is objective and quantitative in nature, yet abstract in nature.

6.2 DIGITAL IMAGE PROCESSING

Image processing is a vital part of most remote sensing operations. All digital images must be of use in the majority of applications. The digital image processing is the task of processing and analyzing the digital data using some image processing algorithm. The analysis relies solely upon multispectral characteristic of the feature represented in the form of tone and colour.

The digitally processed data are the results of computations for each pixel which forms a new digital image that may be displayed or recorded in pictorial format or it maybe further manipulated.

The digital image processing has the following broad operations:

(*i*) Image rectification and restoration,
(*ii*) Image enhancement
(*iii*) Image classification, and
(*iv*) Data merging and GIS integration.

Fig. 6.1 is a flow chart of an idealized sequence for digital image processing.

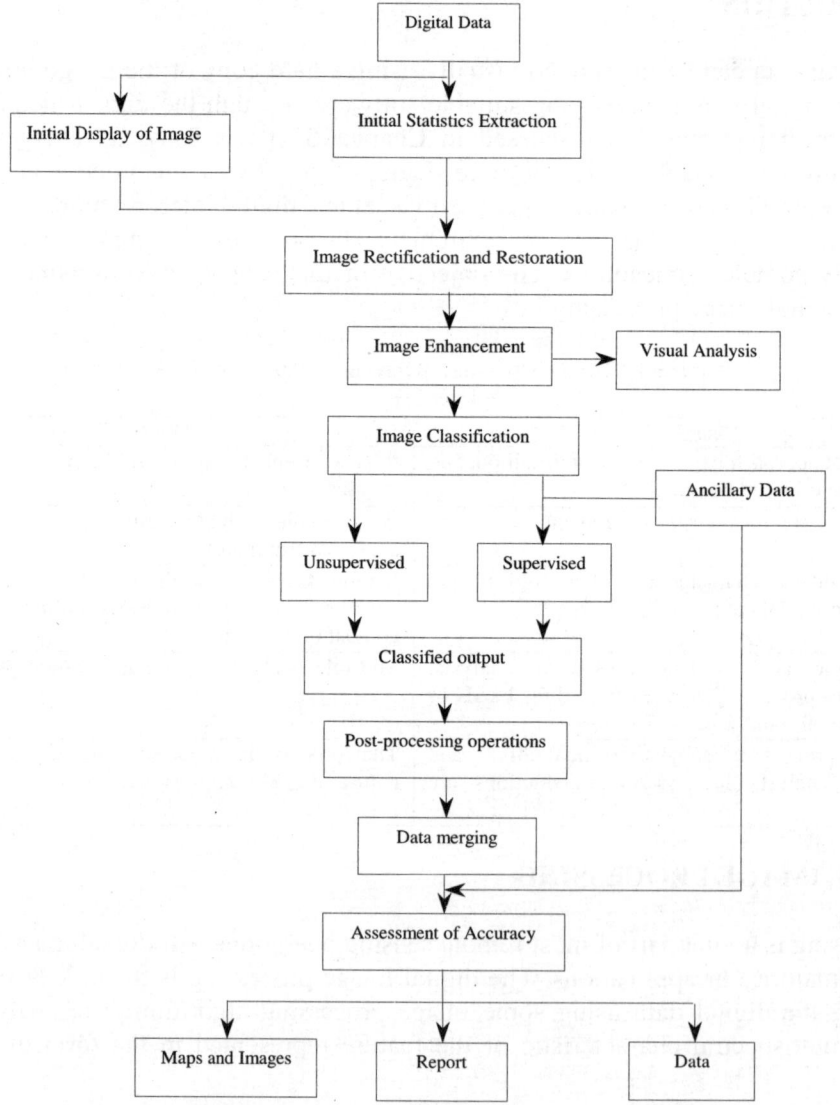

Fig. 6.1 Flowchart of digital image processing sequence.

6.3 OVERVIEW OF DIGITAL ANALYSIS STEPS

The common image processing steps employed as available in image analysis systems are discussed below.

Initial Data Statistics

Initial statistics provides an insight into the raw data as received from satellite. Statistical information such as the minimum and maximum values of the data set, mean, standard deviation, and variance for each band are calculated. Also inter-band relationships through variance-covariance and correlation matrixes are examined. Histogram and scattergrams provide a graphic view of the nature of data for different bands. These are important statistical information required for the next stage of preprocessing and image enhancement.

Image Rectification and Restoration

Image rectification and restoration are required to correct the distorted or degraded image to create a more faithful representation of the original image. The raw data are corrected for geometric distortions which are due to sensor, earth geometry variations, and geocoding and registration of image to real world coordinate system, calibrated radiometrically for sensors irregularities, and removal of noise. Thus, the characteristics of the sensor used to acquire the image data decide the nature of image restoration. The term preprocessing often used includes both image rectification and restoration.

Image Enhancement

The purpose of image enhancement is to improve the appearance of the imagery and to assist in subsequent visual interpretation and analysis. Normally, image enhancement involves techniques for increasing the visual distinctions between features by improving tonal distinction between various features in a scene using technique of contrast stretching, while the technique of spatial filtering enhances or suppresses specific spatial patterns in an image. The enhanced images can be displayed interactively on a monitor or recorded in a hardcopy format, either in black and white or colour.

Image Transformations

These are operations similar in concept to image enhancement. Generally, image enhancement operation is carried out on a single band of data, while image transformations are usually carried on multiple bands. Arithmetic operations such as subtraction, addition, multiplication, and division, are performed to combine and transform the original bands into new images which display better or highlight certain features in the scene. Further, in multiband data set, when the spectral range of bands are located very close to each other, repetitive information available leads to redundancy of data. Principal Component Analysis helps in reducing the number of bands for analysis and hence redundancy.

Image Classification

The objective of classification is to replace visual analysis of the image data with quantitative techniques for automating the identifications of features in a scene. Thus the operations involve digital identification, and classification of all pixels in a digital image into one of the several land cover classes or themes. The statistical based decision rules are applied for determining the land cover identity of each pixel in an image. When these decision rules utilize the spectral radiances for the classification, the classification process is called *spectral pattern recognition*. When the decisions are based on the geometric shapes, sizes, and pattern present in the image data, the classification procedure falls into the domain of *spatial pattern recognition*. The categorized data may be used to produce thematic maps and/or summary of statistics on the areas covered by each land cover type.

There are a variety of approaches to perform digital image classification; however, only two generic approaches most commonly used are supervised and unsupervised classification.

Data Merging and GIS Integration

Data merging techniques are applied to combine data for a given geographic area with other geographically referenced data set of the same area. The other data sets may be image data generated on other dates by the same sensor or by other remote sensing systems. The data merging technique is frequently used to combine remotely sensed data with other sources of information in the context of a Geographic Information System (GIS).

6.4 INITIAL STATISTICS EXTRACTION

Once the data has been extracted from the CCT, CD, or any other media on which the data is made available, some fundamental univariate and multivariate statistics of the multispectral data are extracted to help in undertaking some of the image processing operations to be carried out later.

Normally at this stage, the minimum and maximum values, mean, standard deviation, and variance of each band is computed along with the variance-covariance and correlation matrices. Frequency distribution of values for each band is represented by histogram, and interband relationship is depicted by 2D or 3D scatterogram.

Minimum and maximum values provide an idea of the range of brightness values within a brightness scale. Minimum value for each band is an indicator of atmospheric correction to be provided to each band of data, while minimum and maximum values are important for image contrast enhancement. Mean value of each band is also an indicator of the overall distribution of brightness of an image. If the mean value of a data set is close to the central range of the brightness scale, it is an indicator of the image contrast. On the other hand, if the mean is left of the central value of image, the initial display of the image will be darker, *i.e.* objects are generally having low reflectance, and *vice-versa*.

Variance-covariance and correlation matrices provide an insight to data redundancy, and are used in principal component analysis, feature selection, and classification. The structure of variance-covariance matrix is given in Table 6.2.

Table 6.2 Variance-covariance matrix

	Band 1	Band 2	Band 3	Band 4
Band 1	σ_1^2	Cov_{12}	Cov_{13}	Cov_{14}
Band 2	Cov_{21}	σ_2^2	Cov_{23}	Cov_{24}
Band 3	Cov_{31}	Cov_{32}	σ_3^2	Cov_{34}
Band 4	Cov_{41}	Cov_{42}	Cov_{43}	σ_4^2

where

σ_1^2 = the square of standard deviation or variance, and

$$Cov_{ij} = \frac{SP_{ij}}{n-1} = \frac{n \sum_{k=1}^{n} (x_{ki} \times x_{kj}) - \sum_{k=1}^{n} x_{ki} \sum_{k=1}^{n} x_{kj}}{n(n-1)} \qquad ...(6.1)$$

where

n = the total number of pixels in the image, and
x_{ki} = the brightness value of kth pixel in band i.

The correlation coefficient between two bands i and j can be expressed as

$$r_{ij} = \frac{Cov_{ij}}{\sigma_i \sigma_j} \qquad ...(6.2)$$

where σ_i & σ_j are standard deviation of band i and j.

Since correlation coefficient is a ratio, it is a dimensionless parameter and it ranges between – 1 to + 1. A correlation coefficient of – 1 means that information in one band is inversely related to the other, while + 1 is indicative of 1 : 1 correspondence. In remote sensing, two bands having high positive values of correlation coefficient is an indication of redundancy of data as, information is repetitive or highly similar.

Apart of the above statistical parameters, histogram plots are extremely useful to an analyst. It provides a graphical view of the data within a given brightness scale in terms of the spread of data, contrast, and the possible number of informational class. A small spread is indicative of low contrast, while the number of peaks in the histograms, *i.e.*, multimodal, gives an insight to the number of informational classes.

6.5 IMAGE RECTIFICATION AND RESTORATION

Image rectification and restoration refer to those operations that are preliminary to the main analysis. It produces a corrected image that is as close as possible, both radiometrically and geometrically, to the radiant energy characteristics of the original scene. In order to correct the image data internal and external errors must be determined. Internal errors are due to the sensor itself. These errors are systematic and constant in nature, and can be known from pre-launch or in-flight calibration measurements. External errors, which are unsystematic in nature, are due to perturbations of the platform and modulation of scene characteristics. These unsystematic errors can be determined by relating known ground points to their corresponding points in the image. Typical preprocessing operations include radiometric and geometric corrections.

Radiometric corrections may be due to variations in scene illumination and viewing geometry, atmospheric conditions, and sensor noise and response. Each of these will vary depending on the types of sensor and platform used to acquire the data, and the environmental conditions during data acquisition. Sometimes, it may be desirable to convert and/or calibrate the data to known (absolute) radiation or reflectance units to facilitate comparison between different dates. This is often required so as to be able to readily compare images collected by different sensors at different dates or times, or to mosaic multiple images from a single sensor while maintaining uniform illumination conditions from scene to scene.

Any sensor, which observes the Earth surface, will record a mixture of two parts of brightness. One part of the brightness is due to the reflected energy of the Earth surface, while the other part is due to scattered energy from the atmosphere. The second part of the brightness is an additional energy/brightness that has been received at the sensor, and hence is an error/noise in the image. This effect is commonly known as *atmospheric effect*.

Preprocessing operations to correct for atmospheric degradation can be categorized into three broad categories. The first category of correction methods is based on *physical modeling* of the behaviour of radiation as it passes through the atmosphere. These models permit observed reflectances to be adjusted close to the true value that might be observed under clear atmospheric conditions. These models provide a rigorous and accurate solution with flexibility to model a large variety of atmospheric conditions. However, these also have many significant disadvantages. Generally, these are complex models requiring the use of intricate computer programs. Further, these models require detailed meteorological information related to humidity and concentration of atmospheric gases and particles. These informations vary with time and altitude, hence it is difficult to obtain such detail informations. Therefore, these models are not routinely applied to remote sensing data.

The second category of correction methods is based upon the examination of histograms of data sets of all the band of a given image. Atmospheric effects are generally low or negligible in the infrared region and high in the visible region. When the histograms of the visible region are examined, the lowest value of the histogram amongst of the visible bands is subtracted from the brightness values of all the pixels in the visible bands. It is one of the simplest and direct method for correcting atmospheric degradation, and is called as *Histogram Minimum* Method (HMM). The advantages are its simplicity, directness, and universal applicability, as it uses the information present within the image itself. However, this method is an approximate one.

The third category of correction method, while retaining the concept of examining the brightness of object within each scene, also accounts for the interrelationship between separate bands. In this approach, a visible band is plotted against an infrared band, and a best-fit (least-squares) straight line is computed using standard regression methods. The offset a on the x-axis as shown in Fig. 6.2, represents an estimate of the atmospheric correction to be applied for that particular band. This method is known as the *regression method*.

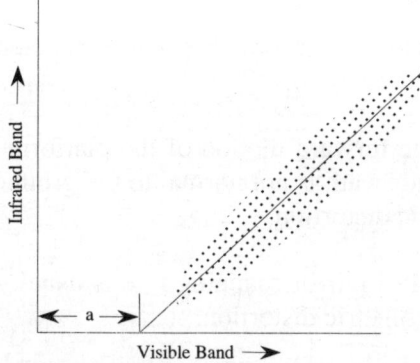

Fig. 6.2 Regression method for computing atmospheric correction (*Source: Mather, 1999*)

The advantage of using regression method is that it can be used for local areas also. An extension of this method proposed by Switzer, *et al*, (1981), known as *covariance matrix method* (CMM), which uses data from areas having homogeneous reflectance properties. Generally, in this method areas having low reflectance are selected, such as shadow areas or water bodies, in which the variation of brightness reveals the effect of atmospheric degradation. This procedure does not provide per pixel correction. Generally, the result is applied to all pixels in a given area of an image. It also does not take into account the variations in atmospheric conditions and illumination geometry.

Geometric Correction

All remote sensing imageries are inherently subject to geometric distortions. These distortions may be due to several factors such as

    (*i*) the perspective of the sensor optics,
   (*ii*) the motion of the scanning system,
  (*iii*) the motion of the platform,
  (*iv*) the platform altitude, attitude, and velocity,
   (*v*) the terrain relief, and
  (*vi*) the curvature and rotation of the Earth.

Geometric corrections are intended to compensate for those distortions which make the geometric representation of the imagery as close as possible to the real world. Many of these variations are *systematic*, or predictable in nature, and can be accounted for by accurate modeling of the sensor and platform motion, and the geometric relationship of the platform with the Earth. Other *non-systematic* or random errors cannot be modeled and corrected in this way. Therefore, geometric registration of the imagery to a known ground coordinate system must be performed. The systematic and non-systematic errors in remote sensing images for Landsat MSS and TM data are discussed below.

Systematic distortions
The different types of systematic distortions that may be present in a satellite are discussed below.

Scan skew: It is caused by the forward motion of the platform during the time required for each mirror sweep. The ground swath is not normal to the ground track but is slightly skewed, producing cross-scan geometric distortion.

Mirror scan velocity: The MSS mirror scanning rate is usually not constant across a given scan, producing along-scan geometric distortion.

Panoramic distortion: The ground area imaged is proportional to the tangent of the scan angle rather than to the angle itself. Since data are sampled at regular intervals, this produces along-scan distortion.

Platform velocity: If the speed of the platform changes the ground track covered by successive mirror scans changes producing along-track scale distortion.

Earth rotation: The Earth rotates as the MSS sensor scans the terrain. This results in a shift of the ground swath being scanned, causing along-scan distortion.

Perspective: For some applications it is desirable to have images represent the projection of points on the Earth upon a plane tangent to the Earth with all projection lines normal to the plane. This introduces along-scan distortion.

Nonsystematic distortions
The different types of non-systematic distortions that may be present in a satellite are discussed below.

Altitude: If the MSS platform departs from its normal altitude, changes in scale occur.

Attitude: One of the sensor system axes usually maintained normal to the earth's surface and introduces geometric distortion.

The errors discussed above can be corrected by using a suitable number of Ground Control Points (*GCP*) that are points on the Earth surface whose both image coordinates (measured in rows in pixel number) and map coordinates (measured in latitude and longitude or any cartesian coordinate system) are known.

To remove systematic errors, it requires an in-depth knowledge of a large number of parameters related to satellite ephemeris data and these errors are corrected at the Master Data Centre. Generally, non-systematic errors are corrected by the user himself by selecting suitable number of GCPs. There are two geometric correction procedures, *i.e.*, geometric rectification and geometric registration, which are generally applied (Jensen, 1986).

Geometric rectification is the process by which the geometry of an image is made planmetric. It requires the use of GCPs whose image coordinates in terms of rows and pixel number are known along with the map or ground coordinates. It is an accurate process since each pixel is referenced not only by its image coordinates but also by a ground coordinate system in a standard map projection, and hence known as *image to map* rectification. Geometric rectification has to be done when accurate linear or areal measurements are needed.

Many a times accurate geometric rectification is not needed especially when two or more images are used to find out the changes in an area or region. Here instead of carrying geometric rectification of all the images and then determining the changes, the images are registered with respect to each other. This is known as *image registration* or *image-to-image registration*.

The difference between rectification and registration is that in rectification the image is referenced to a map having a standard map projection while in registration two images are referenced to each other and not to any map having a standard map projection. As most of the remote sensing applications yield a map, hence image to map rectification has been discussed in details in the following section.

Geometric Rectification

In order to rectify a remote sensing image to a map coordinate system, two basic operations have to be performed, *i.e.*, spatial interpolation and intensity interpolation. In spatial interpolation, a large number of GCPs are identified on the image and ground. A polynomial is fit to the control points using least squares criteria to model the geometric error directly into the image without explicitly identifying the source. Generally, a six-parameter affine transformation is sufficient to rectify a remote sensing image as it can model for six kinds of distortions, *i.e.*, translation in x and y, scale changes in x and y direction, skew, and rotation. This can be expressed mathematically by the following equations:

$$x' = a_0 + a_1 x + a_2 y$$
$$y' = b_0 + b_1 x + b_2 y \qquad \ldots(6.3)$$

where x and y are the ground coordinate values and x' and y' are the image coordinates. It may be noted that minimum of 3 GCPs can yield a solution, however more than 3 GCPs are used to compute the coefficients $a_0, a_1, a_2, b_0, b_1,$ and b_2. This method involves the computation of root mean square error (RMS_{error}) for each GCP.

$$RMS_{error} = \sqrt{(x'-x_0)^2 + (y'-y_0)^2} \qquad \ldots(6.4)$$

where (x_0, y_0) and (x', y') are the original and computed image coordinates of the GCP respectively. By computing RMS_{error} for all the GCPs, it is possible to identify the GCP which exhibits the greatest error and also the sum total of RMS_{error} of all GCPs.

Generally, the analyst specifies a threshold value for total error. If the total error exceeds the threshold value, then the GCP having the highest individual RMS_{error} is deleted, and a new set of coefficients of $a_0, a_1, a_2, b_0, b_1,$ and b_2 are computed. This process is an iterative in nature and continues till total RMS_{error} of all the GCPs is within a user specified threshold, or the number of GCPs falls below an acceptable limit, usually 3. Once the acceptable RMS_{error} is achieved, the intensity interpolation of geometrically rectified image starts.

Intensity Interpolation

After the spatial interpolation is carried out, the next step is to extract or identify the brightness values of new image from the old image. This process is known as *resampling*.

The resampling process calculates the new pixel values from the original digital pixel values in the uncorrected image. There are there common methods for resampling; nearest neighbour, bilinear interpolation, and cubic convolution. *Nearest neighbour* resampling uses the digital value from the pixel in the original image which is nearest to the new pixel location in the corrected image. This is one of the simplest methods and does not alter the original values, but may result in some pixel values being duplicated while others are lost. This method also tends to result in a disjointed or blocky image appearance. *Bilinear interpolation* resampling takes a weighted average of four pixels (BV_{wt}) in the original image nearest to the new pixel location. The average process alters the original pixel values, and creates entirely new digital values in the output image.

$$BV_{wt} = \frac{\sum_{k=1}^{4}(z_k / D_k^2)}{\sum_{k=1}^{4}(1/ D_k^2)} \qquad \ldots(6.5)$$

where z_k are the surrounding four data point values, and D_k^2 are the distances from the point in question to the data points.

This may be undesirable if further processing and analysis, such as image classification is to be done. If this is the case, resampling may best be done after the classification process. *Cubic convolution* resampling goes even further to calculate a distance weighted average of a block of sixteen pixels from the original image which surrounds the new output pixel location. As in bilinear interpolation, this method also results in a completely new pixel value. However, both the methods produce images which are sharper in appearance and avoid the blocky appearance of the nearest neighbour method.

6.6 IMAGE ENHANCEMENT

Image enhancement is a process of making the visual interpretation and understanding of imagery easier. The advantage of digital imagery is that it allows manipulating the digital values of a pixel in an image. Although radiometric correction for illumination, atmospheric influences, and sensor characteristics may be done prior to distribution of data to the user, the image may still not be optimized for visual interpretation. With large variations in spectral response from a diverse range of targets (*e.g.*, forest, deserts, snowfields, water, etc.) no generic radiometric correction could optimally account for and display the optimum brightness range and contrast for all targets. Thus, for each application and each image, an adjustment of the range and distribution of brightness values are usually necessary.

In raw imagery, the useful data occupies only a small portion of the available range of digital values (commonly 8 bits or 256 levels). Contrast enhancement involves changing of the original values so that the full range of the brightness scale is used, thereby increasing the contrast between targets and their backgrounds. The key to understanding contrast enhancements is to understand the concept of an *image histogram*. An image histogram is a graphical representation of the brightness values that comprise an image. The brightness values (*i.e.*, 0-255) are displayed along the *x*-axis of the graph, and the frequency of occurrence of each of these values in the image is on the *y*-axis. By manipulating the range of digital values in an image, *i.e.*, graphically represented by its histogram, various enhancements can be applied to the data. There are many different techniques and methods of enhancing contrast and detail in an image; however these can be grouped under two broad categories, *i.e.*, linear and non-linear methods.

Linear Contrast Enhancement

The basic purpose of contrast enhancement is to expand the original brightness values to make use of the full range of the radiometric scale of the sensor. Normally, a brightness scale is a binary scale, and remote sensing data are recorded on an 8-bit binary scale on which a black and white image can be displayed by 256 levels of grey ranging from 0 to 255, where 0 corresponds to black and 255 represents white colour. When the histogram of the original image does not utilize the whole range of the brightness scale, it is found that the visual quality of the image is poor.

One of the simplest methods of improving the contrast is the Min-Max stretching. Here the lower and upper values of the histogram of the original image are made to fit between the full ranges of the brightness scale. The lower value of the original histogram is assigned a zero brightness value while the upper value of the original histogram is assigned a value of 255 (Fig 6.3). In between all other intermediate values of the histogram, are reassigned new brightness values by linear interpolation, and are expressed as:

$$x_{new} = \frac{x_{in} - x_{min}}{x_{max} - x_{min}} x_r \qquad \ldots(6.6)$$

where
x_{new} = the new brightness value after stretching,
x_{min} = the minimum brightness value of the original data,
x_{max} = the maximum brightness value of the original data,
x_{in} = the original input brightness value, and
x_r = the range of brightness scale (0-255 for a 8-bit data).

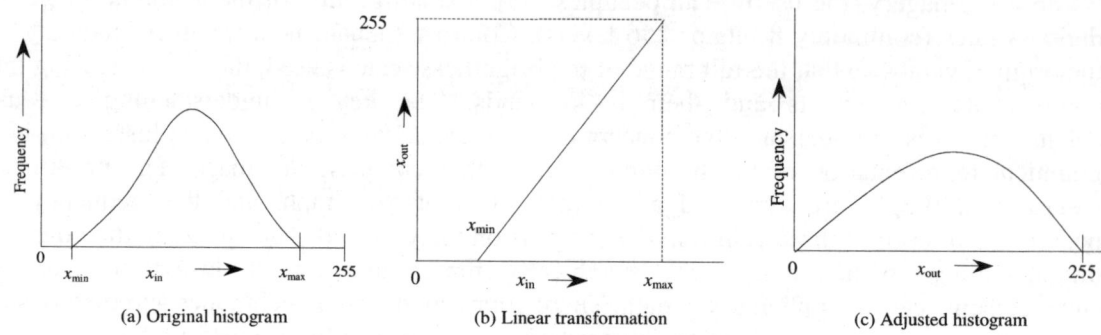

Fig. 6.3 Concept of Min-Max stretch

Sometimes, when Min-max stretching is performed, the tail ends of the histogram are long resulting in no improvement in image quality. Under such circumstances, it is better to clip a certain percentage, *i.e.*, 1%, 2%, 5% of the data from the two tail ends of the histogram of the original image. The method is known as *percentile stretching*. Fig. 6.4 illustrates the principle of percentile stretching in which the histogram of an image has long tail.

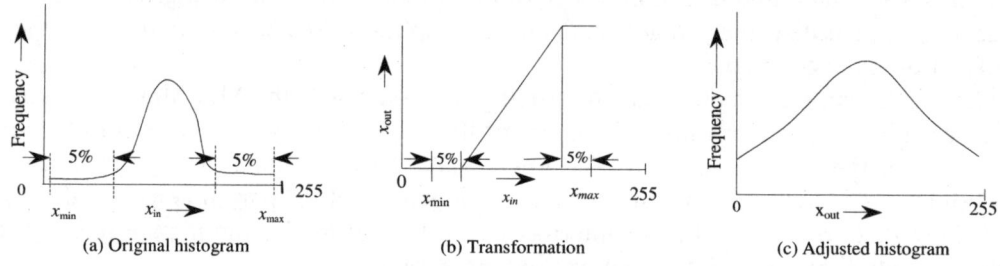

Fig. 6.4 Percentile stretching

A 5% cut off is applied to both the tail ends of the histogram, and the remaining portion of the histogram is linear stretched as in the case of Min-Max stretching. This increases the slope of the linear stretch, thereby enhancing the details of the image. The two methods are generally suitable when the histogram is Gaussian in nature, *i.e.*, it is unimodal. When the histogram is non- Gaussian in nature, then piecewise enhancement has to be adopted in which separate linear stretching is applied between each node of the histogram (Fig. 6.5).

Non-Linear Enhancement

Generally, it is observed that histogram of the input image does not show a uniform distribution. Under such situation, non linear enhancements are more suited. The most common or simplest is *the Histogram Equalization Method*. In this method, the basic idea is to re-distribute the original histogram so that each brightness level has approximately equal number of pixels. Further, the analyst can also change the number of brightness levels, *i.e.*, from 8-bit to 7-bit data, and *vice-versa*. Histogram equalization increases the contrast in the heavily populated range of the histogram while reducing the contrast at sparsely populated range of the histogram.

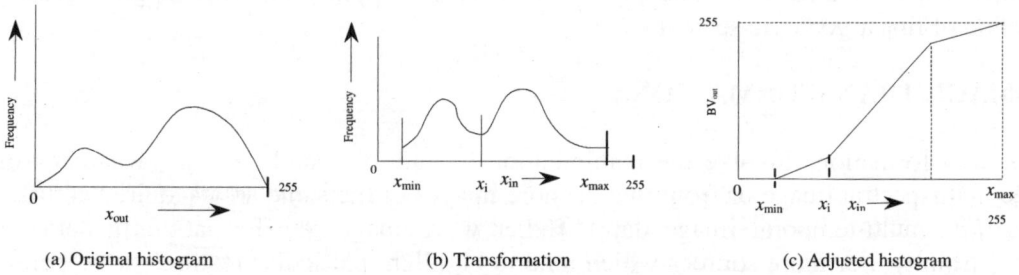

(a) Original histogram (b) Transformation (c) Adjusted histogram

Fig. 6.5 Piecewise linear contrast stretch

Another method known as the *Gaussian stretch*, involves the fitting of the original histogram to a Gaussian histogram within a normal distribution limit of ± 3 times the standard deviation. The Gaussian histogram is compared to the original histogram to redistribute the number of points for each brightness level. In general, Gaussian stretch increases contrast at the tail ends of the histogram. Instead of a Gaussian distribution model, different exponential and logarithmic models can also be used.

6.7 SPATIAL FILTERING

Spatial filtering encompasses another set of digital processing functions that are used to enhance the appearance of an image. Spatial filters are designed to highlight or suppress specific features in an image based on their spatial frequency. Spatial frequency is related to the concept of image texture. It refers to the frequency of the variations in tone that appear in an image. *Rough* textured areas of an image, where the changes in tone are abrupt over a small area, have high spatial frequencies, while *smooth* areas with little variation in tone over several pixels, have low spatial frequencies.

A common filtering procedure involves moving a window of a few pixels in dimension (*e.g.* 3×3, 5×5, etc.) over each pixel in the image, applying a mathematical technique using the pixel values under that window, and replacing the central pixel with the new value. The window moves both in the row and column directions, examining one pixel at a time till the entire image has been filtered and a new image has been generated. By varying the calculation performed and the weightings of the individual pixels in the filter window, filters can be designed to enhance or suppress different types of features.

A *low-pass filter* is designed to emphasize larger, homogeneous areas of similar tone, and reduce the smaller detail in an image. Thus, low-pass filters generally serve to smooth the appearance of an image. Average and median filters are examples of low-pass filters. *High-pass filters* do the opposite job, and serve to sharpen the appearance of fine details in an image. One implementation of a high-pass filter is to first apply a low-pass filter to an image, and then to subtract the result from the original, leaving behind only the high-spatial frequency information. This is known as a *high boost filter*.

Directional or *edge detection filters* are designed to highlight linear features, such as roads or field boundaries. These filters can also be designed to enhance features which are oriented in specific directions. These filters are useful in applications such as geology, for the detection of linear geologic structures.

6.8 IMAGE TRANSFORMATIONS

Image transformations involve the manipulation of multiple bands of data, whether from a single multispectral image or from two or more images of the same area acquired at different times (*i.e.*, multi-temporal image data). Either way, image transformations generate new images from two or more sources which tend to highlight particular features or properties of interest, better than the original input images.

Basic image transformations use simple arithmetic operations such as addition, subtraction, division, and multiplication to the image data. In *Image subtraction*, two geometrically registered images acquired at two different dates are subtracted pixel by pixel to yield a new image called as the *difference image*. In these new images, areas where there has been little or no change are represented in mid-grey tones, while those areas where significant change has taken place, and are shown brighter or darker tones depending on the direction of change in reflectance between the two images. This type of image transformation can be useful for mapping growth of cities, deforestation, crop acreage, etc. Hence image subtraction is also known as *change detection*.

Image addition is basically an averaging operation, in order to reduce the overall effect of noise. It is very commonly used in spatial filtering to enhance features. *Image multiplication* of two real images is rarely performed in remote sensing.

Image division or *spectral ratioing* is one of the most common transforms applied to image data. It serves to highlight subtle variations in the spectral responses of various surface covers. By ratioing the data from two different spectral bands, yields useful information regarding the objects. Healthy and green vegetation reflects strongly in the near-infrared portion of the spectrum while absorbs strongly in the visible red. Other types of surface, such as soil and water, show near equal reflectances in both the near-infrared and red portions. Thus, a ratio image of near-infrared band (0.8 to 1.1 µm) divided by visible Red band (0.6 to 0.7 µm) would result in ratios much greater than 1.0 for vegetation, and ratios around 1.0 for soil and water. Hence the discrimination of vegetation from other surface cover types is significantly enhanced. This ratio is commonly known as *Vegetation Index (VI)*.

Another advantage of ratioing is that, it yields relative values instead of absolute brightness values, hence variations in scene illumination which as a result of topographic effects are reduced. Thus, although the absolute reflectances for forest covered slopes may vary depending on their orientation relative to the sun's illumination, the ratio of their reflectances between the two bands should always be very similar. The further extension of ratioing is the formulation of Normalized Difference Vegetation Index (*NDVI*). It is based on the normalized difference between Near-Infra Red (NIR) and Visible Red (VR) (Rouse., *et al*, 1973), and is expressed as

$$NDVI = \frac{NIR - VR}{NIR + VR} \qquad ...(6.7)$$

The ranges of values for NDVI vary from - 1 to + 1. For vegetative areas, the value of NDVI will be greater than 1.0, while for soil areas it will be close to 1.0, and for water bodies it will be less than 1.0.

However, it is found that soil and vegetation tends to provide more mixed information due to varying cover and hence another index known as *Perpendicular Vegetation Index* (PVI) was proposed (Richardson and Wiegand, 1977). This index is a measure of plant development and a soil background. A plot of the red band on the *x*-axis and infrared band on the *y*-axis shows that bare soils pixel lie-along a line (known as the *soil line* at an angle of 45° with the axes), where all vegetation point lies perpendicular and left of it (Fig. 6.6). Larger the distance from the soil line is an indication of greater amount of vegetation present. Further, if a point lies on the soil line, then the position of the point on the line is an indication of the moisture content of the soil. If this point is on the lower side of the line, it indicates greater amount of moisture content. PVI was developed for Landsat MSS sensor and has been expressed as follows:

$$PVI\ 7 = \sqrt{(0.355\,MSS7 - 0.149\,MSS5\,)^2 + (0.355\,MSS5 - 0.852\,MSS7\,)^2}$$

$$PVI\ 6 = \sqrt{(-2.507 - 0.457\,MSS + 0.498\,MSS6\,)^2 + (2.734 + 0.498\,MSS5 - 0.543\,MSS6\,)^2}$$
$$...(6.8)$$

Tassel Cap Transformation

PVI considers spectral variations in two of the four Landsat MSS bands, and uses distance from a soil line in the 2D space as a measure of biomass or green leaf area index. Kauth and Thomas (1976) used similar concept using all the four bands of Landsat MSS. Here all the pixels representing soil, fall along an axis that is oblique with respect to each pair of four MSS axes. A triangular region in the 4D MSS space is occupied by pixels representing vegetation in various stage of growth. The shapes of triangular regions are in the form of tassel cap, and hence this transformation is known as the *Tassel Cap Transformation*. This transformation is based on *Gram–Schmidt Sequential Orthogonalization technique* that produces an orthogonal transformation of the original 4 band MSS data space into a new 4D space. The axes of this new coordinate system are termed as brightness, greenness, yellowness and non-such. The *brightness axis* is associated with variation in the soil background reflectance. The *greenness axis* is correlated with variations in the vigour of green vegetation, while *yellowness axis* is related to variations in the yellowness of senescent vegetation. The *non-such axis* is correlated to atmospheric conditions.

94 *Remote Sensing and Geographical Information System*

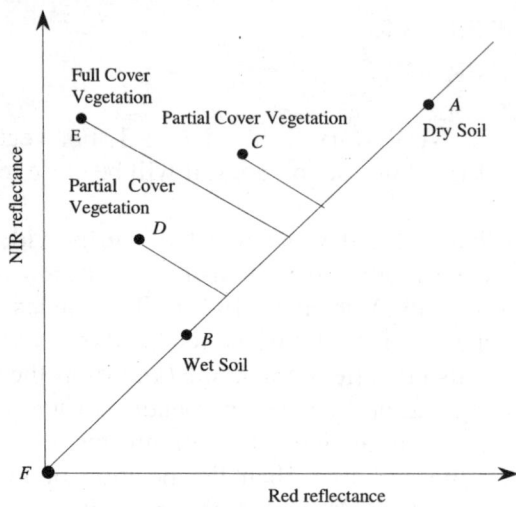

Fig. 6.6 Perpendicular vegetation index

Tassel Cap Transformation has been used primarily to monitor the growth of agricultural crops at different stages by using the brightness and greenness information. The advantage of using this transformation is that the axes provide a consistent, physically-based coordinate system as these are defined aprior, thus any variation in crop cover and stage of growth from image to image will not be affected.

Crist and Cicone (1984) explained this concept for 6 bands TM Data (excluding the thermal band). It was observed that TM bands contained significant information related to *wetness* in the third dimension Fig. 6.7 shows the TM Tasseled Cap functionality between brightness, greenness and wetness in two planes.

The brightness information is weighted sum of the entire 6 TM bands and is a measure of overall reflectance. It is helpful in differentiating light soils from dark soils. Greenness is the contrast between Near Infrared and Visible reflectance, and is a measure of presence and density of green vegetation, while wetness is the contrast between Shortwave Infra Red and Visible/Near Infrared reflectance. It is a measure of soil moisture content, vegetation density and other class characteristics. Fig. 6.8 shows the general locations of some of the important features in TM Tasseled Cap feature space.

The plane formed between Brightness and Greenness is known as the *Plane of Vegetation* while the plane formed between Brightness and Wetness is known as the *Plane of Soil*. By plotting the Brightness, Greenness and Wetness, certain specific and interesting information are revealed. When Greenness and wetness are plotted, it provides informations regarding the strong correlation to the percentage of vegetation cover (Fig. 6.9). The distinction between forest/natural vegetation and cultivated vegetation is enhanced in the Wetness dimension. In the Greenness/Brightness projection it is found that there is a distinct separation between cultivated vegetation and forest. The location of forest is for interest as it forms a *badge of Tree* in front of the cap [Fig. 6.8 (a)]. Table 6.3 lists the Landsat-5 TM Tasseled Cap coefficient.

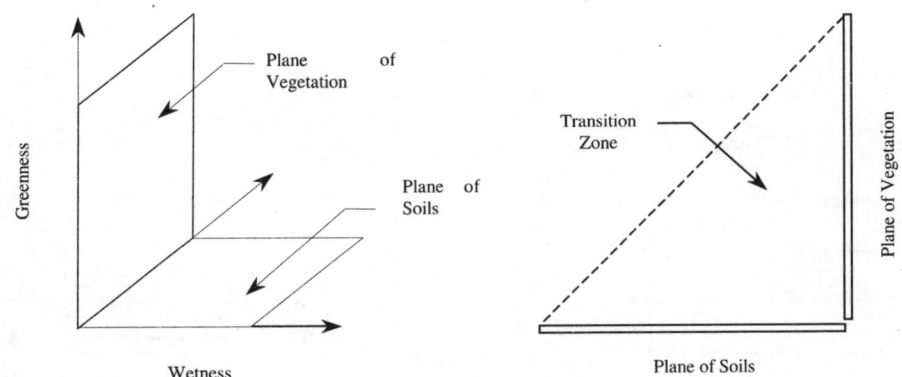

Fig. 6.7 – TM Tasseled Cap transformation axes system (*Source: Crist and Cicone, 1986*)

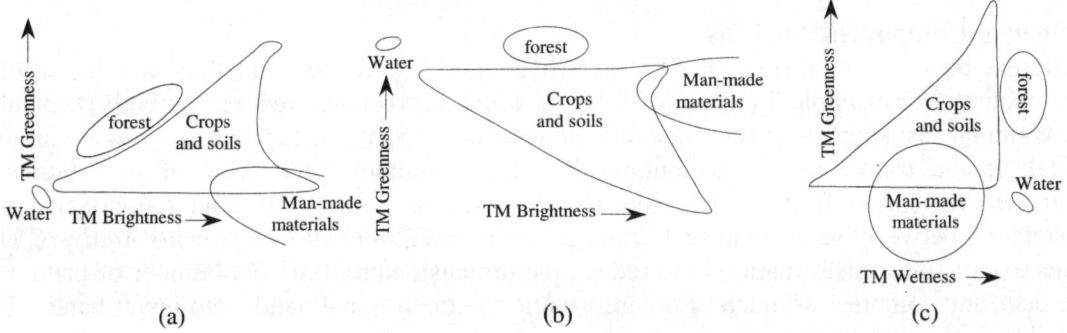

Fig. 6.8 Approximate locations of important classes in TM Tasseled Cap feature space
(*Source: Crist and Cicone, 1986*)

Table 6.3 Landsat-5 TM Tasseled Cap coefficient (*Source: Crist and Cicone, 1986*)

Feature	Coefficients						Additive term
	TM1	TM2	TM3	TM4	TM5	TM7	
Brightness	0.2909	0.2493	0.4806	0.5568	0.4438	0.1706	10.3695
Greenness	−0.2728	−0.2174	−0.5508	0.7221	0.0733	−0.1648	−0.7310
Wetness	0.1446	0.1761	0.3322	0.3396	−0.6210	−0.4186	−3.3828
Haze	0.8461	−0.0731	−0.4640	−0.0032	−0.0492	0.0119	0.7879
Fifth	0.0549	−0.0232	0.0339	−0.1937	0.4162	−0.7823	−2.4750
Sixth	0.1186	−0.8069	0.4094	0.571	−0.0228	0.0220	−0.0336

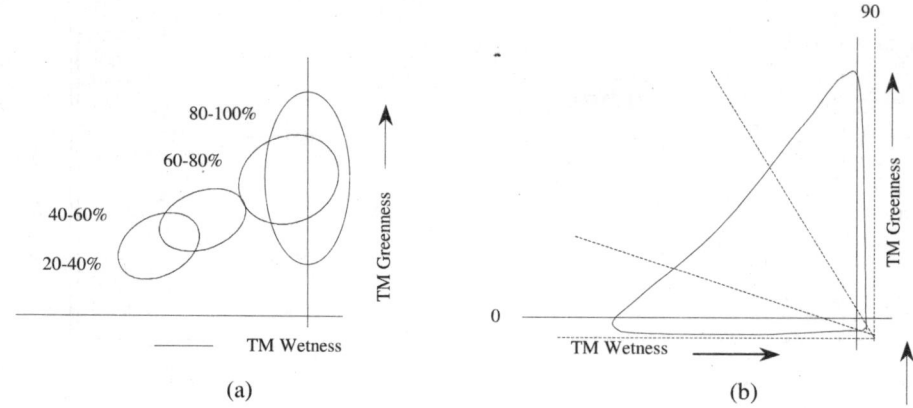

Fig. 6.9 Angular measure of percent cover of vegetation (*Source: Crist and Cicone, 1986*)

Principal Component Analysis

Different bands of multispectral data are often highly correlated and thus contain similar information. For example, Landsat MSS bands 4 and 5 (green and red, respectively) typically have similar visual appearances since reflectances for the same surface cover types are almost equal. Image transformation techniques based on complex processing of the statistical characteristics of multi-band data sets can be used to reduce this data redundancy and correlation between bands. One such transform is called *Principal Components Analysis*. The objective of this transformation is to reduce the dimensionality (*i.e.*, the number of bands) in the data, and compress as much of the information in the original bands into fewer bands. The new bands that result from this statistical procedure are called *components*.

This process attempts to maximize (statistically) the amount of information (or variance) from the original data into the least number of new components. As an example of the use of principal components analysis, a seven band Thematic Mapper (TM) data set may be transformed such that the first three principal components contain over 90 percent of the information in the original seven bands. Interpretation and analysis of these three bands of data, combining them either visually or digitally, is simpler and more efficient than trying to use all of the original seven bands. Principal components analysis, and other complex transforms, can be used either as an enhancement technique to improve visual interpretation or to reduce the number of bands to be used as input to digital classification procedures.

To perform Principal Component Analysis (also known as *Karhunen-Loeve Transform*) is applied to a correlated set of multispectral data. This transformation results into an uncorrelated multispectral data that has certain ordered variance properties. The original axes of the data set may not necessarily be having the best arrangement in a multi feature space to analyse the data in different bands. The primary objective of principal component analysis is to translate and/or rotate the original axes on to a new axes system such that there is no loss of information and that the data is compressed into less number of bands.

Fig. 6.10a shows a schematic representation of a 2 band data set. Fig. 6.10b shows the translation of the origin of axes to its new position as define by (μ_1, μ_2) while Fig. 6.10c shows the rotation of the axes by ϕ degrees to yield a new axes system. Here PC1 is the first axis known as the first principal component and is located along a direction having maximum amount of variance of information. The next axis PC2 is perpendicular to the first axis PC1.

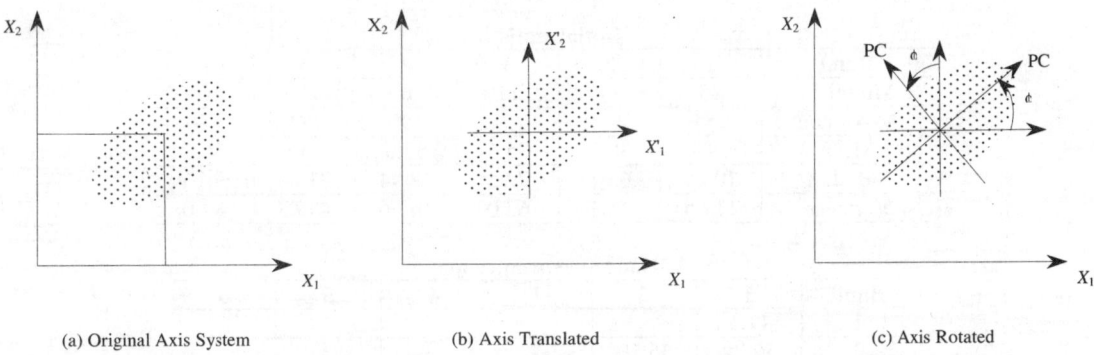

Fig. 6.10 Schematic representation of Principal Component Analysis (*Source: Jensen, 1986*)

In order to transform the original data into its new axis system, transformation coefficients are derived from the covariance matrix of the original data. Thus this transformation is a data dependent process and hence has to be recalculated for each data set. A linear function is used to obtain the new values in the new axis system.

The transformation is performed as follows:-

(*i*) Compute the covariance matrix (Cov) for a *n*-dimensional data set.
(*ii*) Calculate the eigen values. E = [λ_{11}, λ_{22} .... λ_n] and eigen vectors EV = [a_{np}] for all *n* band and *p* components of the covariance matrix.

Eigen value *E* contains information regarding the total variance explained by each principal component. Generally, it is found that the first 3 principal components account for nearly 95-98% of the information of the original of a 6 band TM data set (excluding the thermal band). Eigen vector EV matrix gives the correlation of each band with each component, and is used to compute the new brightness values in the component images. Table 6.4 shows an example of principal component analysis steps.

6.9 IMAGE CLASSIFICATION AND ANALYSIS

An analyst attempts to classify features in an image by using the elements of visual interpretation to identify homogeneous groups of pixels that represent various features or land cover classes of interest. In *digital image classification*, the analyst uses the spectral information represented by the digital numbers in one or more spectral bands, and attempts to classify each individual pixel based on this spectral information. This type of classification is termed *spectral pattern recognition*. In either of the case, the objective is to assign all pixels in the image to particular classes or themes. The resulting classified image is comprised of a mosaic of pixels, each of which belongs to a particular theme, and is essentially a thematic map of the original image.

Table 6.4 Principal component analysis (*Source: Ghosh, 1991*)

Initial statistics

Band	1	2	3	4	5	6
Minimum	45	14	11	6	1	0
Maximum	254	240	254	199	181	140
Mean	68.7	30.5	35.5	57.6	90.8	36.1
Std. dev.	10.56	6.01	7.93	12.94	21.89	9.16
Variance	111.5	36.1	63.0	167.6	479.3	84.0

Variance-covariance matrix

Band	1	2	3	4	5	6
1	111.48					
2	59.72	36.10				
3	72.85	45.24	62.96			
4	61.62	47.88	64.20	167.57		
5	77.43	65.77	114.53	193.67	479.27	
6	40.96	30.93	51.80	67.26	190.54	83.96

Computation of Eigen values from covariance matrix

	Principal component					
	1	2	3	4	5	6
Variance	716.29	144.10	71.41	4.68	3.96	0.91
Total variance	941.34					
Percentage	76.09	15.31	7.59	0.50	0.42	0.10
Cumulative	76.09	91.40	98.99	99.49	99.91	100.1

Eigenvectors computed from covariance matrix

	Component$_p$					
	1	2	3	4	5	6
Band$_k$	0.205	0.637	0.327	−0.054	0.249	−0.079
	0.127	0.342	0.169	−0.077	0.012	0.821
	0.204	0.428	0.159	−0.076	−0.075	−0.562
	0.443	−0.471	0.739	0.107	−0.153	−0.004
	0.742	−0.177	−0.437	−0.300	0.370	0.011
	0.106	0.033	−0.080	0.887	0.424	0.005

When talking about classes, it is important to distinguish between *information* classes and *spectral* classes. Information classes are those categories of interest that the analyst is actually trying to identify in the imagery, such as different kinds of crops, different forest types or tree species, different geologic units or rock types, etc. Spectral classes are groups of pixels that are uniform (or near-similar) with respect to their brightness values in the different spectral channels of the data. The objective is to match the spectral classes in the data to the information classes of interest. However, it is rare that there is a simple one-to-one match between these two types of classes. Many times it is found that 2 to 3 spectral classes merge to form one informational class, while some classes may not be of any particular interest. It is the analyst's job to decide on the utility of the different spectral classes and their correspondence to useful information classes.

Common classification procedures can be broken down into two broad subdivisions based on the method used: *supervised* classification and *unsupervised* classification. In a supervised classification, the analyst identifies in the imagery, homogeneous representative samples of the different surface cover types (information classes) of interest. These samples are referred to as *training areas*. The selection of appropriate training areas is based on the analyst's familiarity with the geographical area and knowledge of the actual surface cover types present in the image. Thus, the analyst is *supervising* the categorization of a set of specific classes. The numerical information in all spectral bands for the pixels comprising these areas, are used to *train* the computer to recognize spectrally similar areas for each class. The computer uses special programs or algorithms to determine the numerical *signatures* for each training class. Once the computer has determined the signatures for each class, each pixel in the image is compared to these signatures and labeled as the class it closely *resembles* digitally. Thus, in a supervised classification, the analyst is first identifies the information classes based on which it determines the spectral classes which represent them.

Unsupervised classification, in essence, reverses the supervised classification process. Spectral classes are grouped, first, based solely on the numerical information in the data, and are then matched by the analyst to information classes (if possible). Programs called *clustering algorithms* are used to determine the natural groupings or structures in the data. Usually, the analyst specifies how many groups or clusters are to be looked for in the data. In addition to specifying the desired number of classes, the analyst may also specify parameters related to the separation distance amongst the clusters and the variation within each cluster. The final result of this iterative clustering process may result in some clusters that the analyst would like to subsequently combine, or that some clusters have been broken down, each of these require a further iteration of the clustering algorithm. Thus, unsupervised classification is not completely without human intervention. However, it does not start with a pre-determined set of classes as in a supervised classification.

Supervised Classification
In order to carry out supervised classification the analyst may have to adopt a well defined procedure in so as to achieve a satisfactory classification of information. The important aspects of conducting a rigorous and systematic supervised classification of remote sensor data are as follows:

(*i*) Selection of an appropriate classification scheme.
(*ii*) Selection of representative areas as training sites.
(*iii*) Extraction of training data statistics
(*iv*) Testing of training data for separability in order to identify the best possible combination of bands for classification.
(*v*) Selection of an appropriate classification algorithm.
(*vi*) Classification of image into appropriate defined classes.
(*vii*) Evaluation of classification accuracy.

If the above considerations, as listed above, are carefully addressed and if real differences exist between the spectral reflectance characteristics of the various land cover classes, it would yield a useful thematic map. A detailed discussion of these considerations is discussed below for the benefit of the readers.

Classification Scheme

Classification schemes have been developed so that they can readily be incorporated as meanful land use and land cover data as obtained by interpreting remote sensing data. Some of the important are U.S. Geological Survey Land Use/Land Cover Classification System, the Michigan Classification System, and the Cowardin Wetland Classification system (Jensen, 1986).

Major points of difference between various classification schemes are their emphasis, and ability to incorporate information obtained using remote sensing data. The U.S. Geological Survey classification system is *resource* oriented in contrast with various *people* or *activity* oriented systems, such as the Standard Land Use Coding (SLUC) Manual. The system identifies nine Level I categories of land use and land cover classes (Table 6.5). The system has been developed such that it is structured for use in interpretation of remote sensor data obtained at various scales and resolutions (Table 6.6) and not for data collected in situ. It was developed to include land-use data derived through visual interpretation, yet it has been widely used for digital classification.

On the other hand, the Standard Land Use Coding Manual (*SLUC*) is land-use *activity* oriented and depends largely on in-situ observations to obtain specific land-use information. However, it is felt that there is a need to merge both the approaches to produce a hybrid classification system such that it incorporates both land use interpreted from remote sensing data and precise land-use information obtained through ground survey.

Michigan Land Use Classification (*MLUC*) scheme is based on this philosophy, and has served as a guideline for many other schemes. Both the schemes are similar that there are nine categories at level I, but differ at level II, III, and IV as categories of MLUC are consistent with the environmental and cultural conditions of Michigan State. Further, MLUC is only land cover based up to Level III, but at Level IV it is both land cover and activity oriented in order to meet the needs of user at local level.

Training Site Selection and Statistics Extraction

Once a classification scheme has been adopted, the analyst may identify and select sites within the image that are representative of the land cover classes of interest. Training data should be of value if the environment from which they obtained is relatively homogenous. The image coordinates of these sites are identified and used to extract statistics from the multispectral data for each of these areas. For each feature class c, the mean value (μ_{ci}) for each and variance-covariance matrix (V_c) are calculated in a similar manner as explained in Section 6.3.

Table 6.5 U.S. Geological Survey Land Use/Land Cover Classification

Level I	Level II
1. Urban or built-up land	11 Residential
	12 Commercial and services
	13 Industrial
	14 Transportation, communications, and services
	15 Industrial and commercial complexes
	16 Mixed urban or built-up land
	17 Other urban or built-up land
2. Agricultural land	21 Cropland and pasture
	22 Orchards, groves, vineyards, nurseries, and ornamental horticultural areas
	23 Confined feeding operations
	24 Other agricultural land
3. Rangeland	31 Herbaceous rangeland
	32 Shrub and brush rangeland
	33 Mixed rangeland
4. Forest land	41 Deciduous forest land
	42 Evergreen forest land
	43 Mixed forest land
5. Water	51 Streams and canals
	52 Lakes
	53 Reservoirs
	54 Bays and estuaries
6. Wetland	61 Forested wetland
	62 Non-forested wetland
7. Barren land	71 Dry salt flats
	72 Beaches
	73 Sandy areas other than beaches
	74 Bare exposed rocks
	75 Strip mines, quarries, and gravel pits
	76 Transitional areas
	77 Mixed barren land
8. Tundra	81 Shrub and brush tundra
	82 Herbaceous tundra
	83 Bare ground
	84 Mixed tundra
9. Perennial snow and ice	91 Perennial snowfields
	92 Glaciers

Table 6.6 The four levels of the U.S. Geological Survey Land Use/Land Cover Classification System and the type of remotely sensed data typically used to provide the information (*Source: Jensen, 1986*)

Classification level	Typical data characteristics
I	Landsat or IRS data
II	High-altitude data acquired at 12,400 m or above; resulting imagery having a scale of 1:80,000 or less.
III	Medium-altitude data acquired between 3100 and 12,400 m; resulting into images having scales between 1:20,000 and 1:80,000.
IV	Low-altitude data acquired below (3100 m); results in imagery that is larger than 1:20,000 scale

The success of a supervised classification depends upon the training data used to identify different classes. Hence selection of training data has to be done meticulously keeping in mind each training data set has same specific characteristics. These characteristics are discussed below.

Number of pixels: This is an important characteristic regarding the number of pixels to be selected for each information class. However, there is no guideline available, yet in general, the analyst must ensure that sufficient number of pixels is selected.

Size: The training sets identified on the image should be large enough to provide accurate and reliable information regarding the informational class. However, it should not be too big as large areas may include undesirable variation.

Shape: It is not an important characteristic. However, regular shape of training area selected provides ease in extracting the information from the satellite images.

Location: Generally informational classes have small spectral variability, thus it is necessary that training data are should be so located that it accounts for different types of conditions within the image. It is desirable that the analyst undertakes a field visit to the desired location to clearly mark out the selected information. In case of inaccessible or mountainous regions, aerial photographs or maps can provide the basis for accurate delineation of training areas.

Number of training areas: The number of training areas depends upon the number of categories to be mapped, their diversity, and the resources available for delineating training areas. In general, five to ten training samples per class are selected in order to account for the spatial and spectral variability of informational class. Selection of multiple training areas is also desirable as it may be possible that some training areas of a class may have to be discarded later. It is found that it is usually better to define many small training fields than to have a few in number but large training areas.

Placement: The training area should be placed in such a way that it does not lie close to the edge of the boundary of the information class.

Uniformity: This is one the most critical and important characteristics of any training data for an information class. The training data collected must exhibit uniformity or homogeneity in the information. If the histogram displays one peak, *i.e.*, unimodal frequency distribution for each spectral class, the training data is acceptable. If the display is multimodal distribution, then there is variability or mixing of information and hence must be discarded.

Idealised sequence for selecting training data: In order to selecting training data, no fixed or well defined procedures can be laid out. However, as a guideline, the key steps in selection and evaluation can be enumerated as follows:

(i) Collect information, including maps and aerial photographs of the area under study. If any previous study has been carried out, then acquire the necessary documents, maps, and reports.
(ii) Conduct field trips to acquire first hand knowledge to selective and representative sites in the study area. The field trips must coincide with the date and time of data acquisition. If not possible, then it should be at the same time of the year.
(iii) Conduct to preliminary examination of the digital data, in order to make assessment for the quality of the image.
(iv) Identify prospective training areas. These locations may be defined with respect to some easily identifiable objects on the image. Further, the same may be identified on the map and aerial photographs if readily available.
(v) Extract the training data areas from the digital image.
(vi) For each informational class, display and inspect the frequency histogram for all bands. In case of multimodal frequency distribution, identify the training areas which are responsible for the same and discard them.
(vii) Compute the training data statistics in the form of minimum and maximum, mean, standard deviations, variance-covariance matrices.
(viii) Now ascertain the separability of the informational classes using feature selection.

Feature selection: Once the training statistics are calculated for each class in each band, the next step is to identity those bands that are most effective in discriminating each class from all other classes. This process is commonly called as *feature selection*. The goal is to eliminate those bands from the analysis that provide only redundant information. In this way, the dimensionality of the data set may also be reduced. It minimizes the cost of the digital image classification. Feature selection may involve both statistical and/or graphical analysis to determine the degree of between-class separability in the remote sensing training data. Combinations of bands are normally ranked according to their potential ability to discriminate each class from all other using n bands at a time.

Statistical methods of feature selection are most commonly used to quantitatively select the subset of bands (or features) that provides the greatest degree of statistical separability between any two classes, c and d in order to minimize the classification error. If two classes overlap over each other as shown in Fig. 6.11, then this overlap can be interpreted as follows.

(i) A pixel may be assigned to a class to which it does not belong *i.e.* an error of commission.
(ii) A pixel is not assigned to its appropriate class *i.e.* an error of ommission.

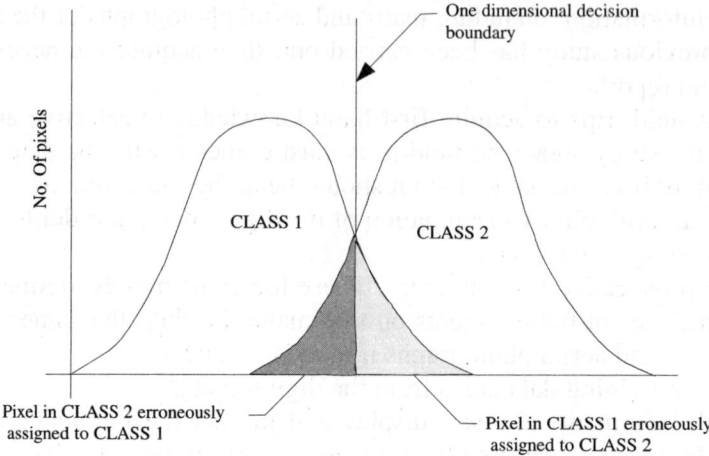

Fig. 6.11 Schematic representation of overlapping of information from training data set
(*Source: Jensen, 1986*)

If the training data for each class from each band are normally distributed as shown in Fig. 6.11, it is possible to use either a divergence or transformed divergence equation to identify the optimum subset of bands to use in the classification procedure.

Divergence is statistical separability measure used in the digital processing of remote sensing data, and is widely used for feature selection. This approach helps in deciding the best q band subset from n bands to be use in the supervised classification process. The number of combination, C, of n bands taken q at a time is given by Eq. 6.9.

$$C\left(\frac{n}{q}\right) = \frac{n!}{q!(n-q)!} \qquad \ldots(6.9)$$

For example, if there are six Thematic Mapper bands (*e.g.*, n = 6) and analyst is interested in using the best three band combination (*e.g.*, q = 3) for classification, then the number of combination

$$C\left(\frac{6}{3}\right) = \frac{6!}{3!(6-3)!}$$

= 20 combinations

If the best two-band combinations are to be used, then 15 possible combinations have to be evaluated.

Divergence is computed using the mean and covariance matrices of the class statistics collected in the training phase of the supervised classification. The *degree of separability* or *separability* is a measure of statistical separability between two classes c and d. The degree of divergence or *separability*, $Diverg_{cd}$, is computed for each pair of classes c and d, as:

$$\text{Diverg}_{cd} = 0.5\, Tr\,[(V_c - V_d)(V_d^{-1} - V_c^{-1})] + 0.5\, Tr\,[(V_c^{-1} + V_d^{-1})(M_c - M_d)(M_c - M_d)^T\,] \qquad \dots(6.10)$$

where Tr [.] is the trace of a matrix (i.e., the sum of the diagonal elements), V_c and V_d are the covariance matrices and M_c and M_d are the mean vectors for classes c and d, respectively.

However, when there are more than 2 classes, the common solution is to compute the average divergence, (Diverg$_{avg}$). This involves computing the average value of divergence over all possible pairs of two classes, c and d at a time for a given band combination. This is continued for all possible sets of band combination. Then another subset of bands, q is selected for the m classes and analyzed. The set of band combination (features) having the maximum value of average divergence is selected at the superior set of bands to be used in the classification algorithm. This can be expressed as:

$$\text{Diverg}_{avg} = \frac{\sum_{c=1}^{m-1} \sum_{d=c+1}^{m} \text{Diverg}_{cd}}{C} \qquad \dots(6.11)$$

The above value of Diverg$_{avg}$ does not saturate and hence no definite boundary condition for good separability between classes can be defined. Kumar and Silva (1977) suggested that it is possible to take the divergence logic one step further and compute transformed divergence, Diverg_{cd}^T, which can be expressed as

$$\text{Diverg}_{cd}^T = 2000\left(1 - \exp\frac{(-\text{Diverg}_{cd})}{8}\right) \qquad \dots(6.12)$$

This gives an exponentially decreasing weight to increasing distances between the classes. It also scales the divergence values to lie between 0 and 2000. A value of 2000 suggests excellent inter- class separability, while a value of 1900 provides good separability. If the value is less than 1700, then separability between classes is considered to be low.

There are other methods of feature selection also based on determining the separability between two classes at a time. For example, the Bhattacharyya distance assumes that the two classes, c and d, are Gaussian in nature and that the means, M_c, M_d and covariance matrices V_c and V_d are available. It is computed as:

$$\text{Bhat}_{cd} = \frac{1}{8}(M_c - M_d)\frac{V_c + V_d}{2}(M_c - M_d) + \frac{1}{2}\ln\left[\frac{det[\{(V_c)+(V_d)\}/2]}{\sqrt{det(V_c)}\sqrt{det(V_d)}}\right] \qquad \dots(6.13)$$

To select the best q features (i.e., combination of bands) from the original n bands in an m-class problem, the Bhattacharyya distance is calculated between each of the $m(m-1)/2$ pairs of classes for each of the possible ways of choosing q features from n dimensions. The best q features are those dimensions whose sum of the Bhattacharyya distance between the $m(m-1)/2$ classes is highest.

Selection of Appropriate Classification Algorithm

Various supervised classification methods have been used to assign an unknown pixel to any one of the classes. The choice of a particular classifier or decision rule depends on the nature of the input data and the desired output. Parametric classification algorithms assume that the observed measurement vectors X_c obtained for each class in each spectral band during the training phase of the supervised classification are Gaussian in nature (*i.e.*, they are normally distributed). Nonparametric classification algorithms make no such assumption. It is instructive to review the logic of several of the classifiers. Among the most frequently used classification algorithms are the Minimum Distance Parallelepiped, and Maximum Likelihood classifier.

Minimum-Distance to Means classification algorithm

It is one the simplest and most commonly used decision rule classifier. Here the analyst provides the mean vectors for each class in each band μ_{ck}, from the training data. To perform a minimum distance classification, the distance to each mean vector, μ_{ck}, from each unknown pixel (BV_{ijk}) is computed. Using Euclidian distance based on the Pythagorean theorem.

$$\text{Dist} = \sqrt{(BV_{ijk} - \mu_{ck})^2 + (BV_{ijl} - \mu_{cl})^2} \qquad \ldots(6.14)$$

To which ever class the unknown point has the smallest distance, to that class the unknown pixel is assigned to. It can result in classification accuracies comparable to other more computationally intensive algorithms, such as the maximum likelihood algorithm.

The Parallelepiped classification algorithm

This algorithm is based on simple 'and/or' Boolean logic. Training data statistics in n spectral bands are used in performing the classification. Brightness values from each pixel of the multispectral imagery are used to produce and n-dimensional, mean vector $M_c = (\mu_{c1}, \mu_{c2}, \mu_{c3}, \ldots, \mu_{cn})$ with μ_{ck} being the mean value of the training data obtained for class c in band k out of m possible classes as previously defined. σ_{ck} is the standard deviation of the training data class c of band k out of m possible classes.

Using a one-standard deviation threshold, a parallelepiped algorithm decides BV_{ijk} is in class c if, and only if,

$$\mu_{ck} - \sigma_{ck} \leq BV_{ijk} \leq \mu_{ck} + \sigma_{ck} \qquad \ldots(6.15)$$

where $c = 1,2,3, \ldots, m$, number of classes, and

$k = 1,2,3, \ldots, m$, number of bands.

Therefore, if the low and high decision boundaries are defined as

$$\text{Low}_{ck} = \mu_{ck} - \sigma_{ck} \qquad \ldots(6.16)$$

and

$$\text{High}_{ck} = \mu_{ck} + \sigma_{ck} \qquad \ldots(6.17)$$

The parallelepiped algorithm becomes

$$\text{Low}_{ck} \leq BV_{ijk} \leq \text{High}_{ck} \qquad \qquad ...(6.18)$$

These decision boundaries form an n-dimensional parallelepiped in feature space. If the pixel value lies between the low and the high threshold for a class in all n bands evaluated, it is assigned to that class, otherwise it is assigned to an unclassified category.

The Maximum Likelihood Classification

The classification strategies considered so far do not consider the variation that may be present in spectral categories and also do not address the problems arising when spectral classes overlap. Such a situation arises, frequently, as one is interested in classifying those pixels that tend to be spectrally similar, rather than those which are distinct enough to be easily and accurately classified by other classifiers.

The essence of the maximum likelihood classifier is to assign a pixel to that class which would maximize the likelihood of a correct classification, based on the information available from the training data. It uses the training data to estimate the mean measurement vector M_c, for each class and the variance-covariance matrix of each class c for band k, V_c.

It decides, if x is in class c if, and only if,

$$p_c > p_i, \text{ where } i = 1,2,3, ..., m \text{ possible classes}$$

where

$$p_c = [-0.5\log_e \{\det(V_c)\}] - 0.5[(X-M_c)^T (V_c)^{-1}(X-M_c)] \qquad ...(6.19)$$

and

p_i = probability of that class existing.

Theoretically, p_i for each class, is given equal weightage, if no knowledge regarding the existence of the features on the ground is available. If the chance of a particular class existing is more than the others, then the user can define a set of priori probabilities for the features and the equation can be slightly modified.

Decide x is in class c if, and only if,

$$p_c(a_c) > p_i(a_c), \text{ where } i = 1,2,3, ..., m \text{ possible classes}$$

and

$$p_c(a_c) = \log_e (a_c) - [-0.5\log_e \{\det(V_c)\}] - 0.5[(X-M_c)^T (V_c)^{-1}(X-M_c)] \qquad ...(6.20)$$

The use of priori probability helps in incorporating the effects of relief and other terrain characteristics. The disadvantage of this classifier is that it requires a large computer memory space and computing time, and yet sometimes may not produce the best results (Jensen, 1986).

Unsupervised Classification

In contrast to supervised classification, unsupervised classification requires only a minimal amount of initial input from the analyst. It is a process whereby numerical operations are performed that search for natural groupings of the spectral properties of pixels, as examined in multispectral feature space. The user allows the computer to select the class means and covariance matrices to be used in the classification. Once the data are classified, the analyst attempts, to assign these natural or spectral classes to the information classes of interest. This may not be easy. Some of the clusters may be meaningless as they represent mixed classes of earth surface materials. This ambiguity is resolved by the analyst who understands the spectral characteristics of the terrain in order classify clusters into information classes. Many methods of clustering have been developed for a wide variety of purposes apart from pattern recognition in remote sensing. Clustering algorithms used for the unsupervised classification of remotely sensed data, generally, vary according to the efficiency with which the clustering takes place. An example of a conceptually simple but not necessarily efficient clustering algorithm has been used below to demonstrate the fundamental logic of unsupervised classification known as CLUSTER.

This algorithm operates in a two-pass mode. In the first pass, the algorithm sequentially builds class clusters. In the second pass, a minimum-distance classifier is applied to the whole data set on a pixel-by-pixel basis, where each pixel is assigned to one of the mean vectors created in pass 1. The details of this algorithm are given below.

Pass 1: Cluster Building

During the first pass, the analyst may be required to supply four types of information:

(i) R, radius of the cluster,
(ii) C, a distance parameter for merging clusters,
(iii) N, the number of pixels to be evaluated between each merging of the clusters, and
(iv) C_{max}, the maximum number of clusters to be identified by the algorithm

To start the process of building of cluster centres, the first pixel of the image is considered to be the cluster centre of the first class. Then the second pixel is taken up and its membership for the first cluster is found by computing the distance between this point and the cluster centre of class 1. If the distance between the pixel and the cluster centre of class 1 is less than or equal to R, then this pixel belong to class 1. Now the class 1 has two points within its cluster and the cluster centre of class 1 is modified by taking the average value of both the pixels. Now the third pixel is taken up for examination. If the distance between this pixel and the cluster centre of class 1 less than or equal to R, then the pixel belongs to class 1 and the cluster centre of class 1 will be again modified by taking the average values of all the three pixels. If the distance of the third pixel exceeds the distance R, then this pixel does not belong to the class 1, hence this pixel now becomes the cluster centre of a new class *i.e.* class 2. This process of building cluster continues till N pixels have been examined for their membership to cluster of different classes. At this point, the cluster building process stops temporarily and the distance between class clusters are examined for their separability.

The class clusters that have now been identified have to be checked such that the cluster centres of all classes are separated by a minimum value C. Those clusters, which are lying at a distance less than C, have to be merged together as they belong to the same cluster. The new cluster centres of the merged cluster are found by taking weighted average value of the old cluster centres being merged. Once the cluster centres have been checked for proper separability, the building of clusters start from the point where it had stopped. It is found that the centres of the cluster, which have been identified, tend to move in its position in the initial phase, and as more points are examined, the positions of the clusters start to stabilize before converging into a fixed position. This process of cluster building continues till the maximum number of cluster centres (C_{max}) have been identified or the end of image is encountered. Finally, the separability of each cluster is checked before proceeding to Pass 2.

Pass 2: Classification of Image
Having identified the cluster centres of all the classes, the classification of the image starts. Each point is assigned a class membership on the basis of minimum distance to means classifier. When the whole image has been classified, the analyst now examines the classified image. Since the classes that have been identified are basically spectral class and not informational classes, hence the analyst now has to undertake the process of converting the spectral classes into informational classes. In this process of convergence it is found that two or more spectral classes may combine together to yield a single information class. This process is rather a tedious, cumbersome, and complex, hence requires a great amount of expertise on the part of the analyst in merging many spectral classes into one informational class. However, it may be possible that a cluster size becomes too large. Then, in that case, the cluster is split into two smaller clusters. In CLUSTER algorithm, it only allows for merging of classes, and has no option for splitting.

Another algorithm known as ISODATA (Iterative self Organizing Data Analysis Technique A) has both the options of splitting and merging of clusters. Further, this algorithm is a single pass *i.e.* both cluster centre identification of each class and the classification are performed simultaneously. However, the algorithm is computationally exhaustive, and requires an experienced analyst to successfully use it. Details of this algorithm are given in Haralick and Fu (1983). Apart from this, several other algorithms have been developed, namely AMEOBA, FORGY. These algorithms are similar in approach, and hence have not been discussed.

Classification Accuracy Assessment
No classification task using remote sensing data is complete till an assessment of accuracy is performed. The analyst and the user of a classified map would like to know as to how accurately the classes on the ground have been identified on the image. The term *accuracy* correlates to *correctness*. In digital image processing, accuracy is a measure of agreement between standard information at given location to the information at same location on the classified image. Generally, the accuracy assessment is based on the comparison of two maps; one based on the analysis of remote sensing data and second based on information derived from actual ground also known as the *reference map*. This reference map is often compiled from detailed information gathered from different sources and is thought to be more accurate

than the map to be evaluated. The reference map consists of a network of discrete parcels, each designated by a single label.

The simplest method of evaluation is to compare the two given maps with respect to areas assigned to each class or category. This yields a report of the areal extents of classes, which agree to each other. The accuracy assessment is presented as an overall classification of map or as site-specific accuracy.

Overall classification accuracy represents the overall accuracy between two maps in terms of total area for each category. It does not take into account the agreement or disagreement between two maps at specific locations. The second form of accuracy measure is *site-specific accuracy*, which is based upon detailed assessment of agreement between the two maps at specific locations.

In case of supervised classification, the simplest strategy is to compare the classified data to the training samples used to generate the training statistics. However, this may yield a higher value of accuracy, and strictly speaking the accuracy should be 100%, as one would expect correct classification of all the training data. The best approach would be to select a set of data of known identity, out of which, one portion is used as training data and the rest to assess the accuracy of classification.

Error Matrix

The standard form for reporting site-specific accuracy is the error matrix, also known as the *confusion matrix* or the *contingency table*. An *error matrix* not only identifies the overall error for each category, but also the misclassification for each category. An error matrix is essentially consists of an n by n array, where n is the number of class or categories on the map reference. Here the rows of the matrix represent the *true classes* or information on the reference map, while column of the matrix represent the classes as identified on the classified map. Fig. 6.12 shows a schematic representation of the error matrix, while Table 6.7 shows an actual error matrix. The values in the last column give the total number of true points per class used for assessing the accuracy. Similarly, the total at the bottom of each column gives

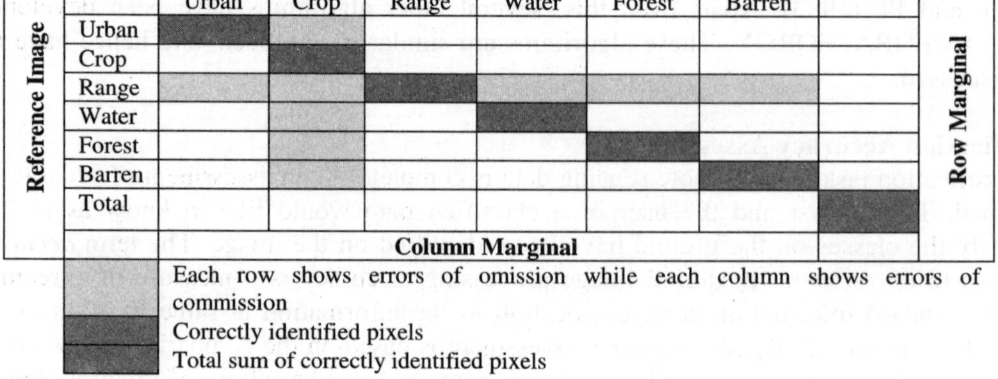

Reference Image	Class	Classified Image						Total	Row Marginal
		Urban	Crop	Range	Water	Forest	Barren		
	Urban								
	Crop								
	Range								
	Water								
	Forest								
	Barren								
	Total								
		Column Marginal							

Each row shows errors of omission while each column shows errors of commission
Correctly identified pixels
Total sum of correctly identified pixels

Fig. 6.12 Schematic representation of an error matrix (*Source: Campbell, 1986*)

information regarding the number of points/pixels per class in the classified map. The diagonal elements of the error matrix indicate the number of points/pixels correctly identified both the reference and classified maps. The sum total of all these diagonal elements is entered at the right hand side bottom most element, *i.e.*, total number of points/pixels correctly classified both in the reference and classified maps. The off-diagonal elements of the error matrix give information regarding the errors of ommission and commission.

Table 6.7 A sample Error Matrix (*Source: Ghosh, 1991*)

Actual class	Predicted class						
	Heather	Water	Forest 1	Forest 2	Bare soil	Pasture	Total
Heather	826	0	0	5	27	0	858
Water	0	878	0	0	0	0	878
Forest 1	0	0	720	183	0	7	910
Forest 2	33	0	21	878	2	0	934
Bare soil	61	0	0	0	560	0	621
Pasture	0	0	0	1	0	219	220
Total	920	878	741	1067	589	226	4081

In Table 6.7, a total of 4421 pixels have been selected from five informational classes for the assessment of classification accuracy. Of these, 4081 pixels have been identified on correctly on the classified image, hence an overall accuracy (*p'*) of 92.3% has been achieved in the classification of the image. This statistic is useful however it does not report the confidence of the analyst. For this, a 95% 1-tailed lower confidence limit test for a binomial distribution can be determined as given below:

$$p = p' - \left[1.645 \sqrt{\frac{(p')(q)}{n} + \frac{50}{n}} \right] \qquad \ldots(6.21)$$

where
  p = Overall accuracy at 95% confidence level
  p'= the overall accuracy,
  q = 100- p', and
  n = the sample size.

If this value of *p* exceeds the defined criterion at the lower limit, then it is possible to accept this classification with 95% confidence limit. Normally, the defined criterion for confidence limit is set at 85%. For the above given example, the accuracy at the lower is 91.6% and hence it is acceptable as the classified map has met or exceeded the defined accuracy standards.

The off-diagonal elements of the error matrix provide information on errors of *ommission* and *commission* respectively. Errors of ommission are found in the upper right half of the matrix and for each class it is computed by taking the sum of all the non-diagonal elements along each row, and dividing it by the row total of that class. In the example given above, 32 pixels of heather class have been omitted by the classified map, and out of these 27 pixels have been identified as bare soil and 5 pixels as forest 2 class. This has resulted in a 3.7%

error of ommission. Table 6.8 shows the percentage error of ommission for all classes. Similarly, the errors of commission are computed by taking the sum total of all the non-diagonal elements in the error matrix in the column direction of that class. It is found that the classified image has identified 94 extra pixels as heather class, out of which 61 pixels and 33 pixels actually belong to bare soil and forest 2 classes, respectively. This has resulted in a 10.2% error of commission. Table 6.8 shows the percentage error of commission for all classes.

Table 6.8 Percentage errors of Ommission and Commission (*Source: Ghosh, 1991*)

Class	Ommission			Commission		
	No. of pixels omitted	Total no. of pixel	% error	No. of pixels commission	Total no. of pixels	% error
Heather	32	858	3.7	94	920	10.2
Water	0	878	0	0	878	0
Forest 1	190	910	20.9	21	741	2.8
Forest 2	56	934	6.0	189	1067	17.7
Bare soil	61	621	9.8	29	589	4.9
Pasture	1	220	0.5	07	226	3.1

A single accuracy measure is adequate to describe the confidence limit of the entire map, however, it is necessary to determine the accuracy of individual classes. For this, a two-tailed 95% confidence limit test, for each category, can be determined as:

$$p = p' \pm \left[1.96 \sqrt{\frac{(p')(q)}{n}} + \frac{50}{n} \right] \qquad \ldots(6.22)$$

Table 6.9 shows the 95% confidence limits for errors of ommission and commission.

Table 6.9 Confidence limits of Ommission and Commission (*Source: Ghosh, 1991*)

Class	Ommission				Commission			
	Points correct	n	% correct	95% confidence limits	n	% correct	95% confidence limit
Heather	826	858	96.3	95.0 – 97.6	920	89.8	87.8 – 91.8
Water	878	878	100.0	99.4 – 100.0	878	100.0	99.4 – 100.0
Forest 1	720	910	79.1	76.4 – 81.8	741	97.2	95.9 – 98.5
Forest 2	878	934	94.0	92.4 – 95.6	1067	82.3	80.0 – 84.6
Bare soil	560	621	90.2	87.8 – 92.6	589	95.1	93.3 – 96.7
Pasture	219	220	99.5	98.3 – 100.0	226	96.9	94.4 – 99.1

Using 85% as the criteria, it is seen that Forest 1 has failed the test as both the upper and lower limits of accuracy at 95% confidence limit is less than 85%. Similarly, when errors of commission are evaluated, it is found that Forest 2 fails to meet the criteria. This is also evident from Table 6.7. It can be seen that 183 pixels have not been identified as Forest 1, and has been identified as Forest 2 class. It implies that even though the overall classification accuracy has exceeded the defined criterion of acceptability, the training samples of classes Forest 1 and Forest 2 have not been able to provide the correct information to the classification process, and hence the training samples have to be collected with caution.

In the above procedure for determining accuracy of classification, it is highly dependent upon the training samples used for classification and assessment of classification accuracy. In order to assess the agreement between two maps, Kappa (κ), which is a measure of the difference between, observed agreement between two maps (as reported by overall accuracy) and the agreement that might be contributed solely by chance matching of two maps. It attempts to provide a measure of agreement that is adjusted for chance and is expressed as follows:

$$\kappa = \frac{\text{Observed} - \text{Expected}}{1 - \text{Expected}} \qquad \ldots(6.23)$$

Observed is the overall accuracy, while *expected* is an estimate of chance agreement to the observed percentage correct. Expected is computed by first taking the products of row and column totals to estimate the number of pixels assigned to each element of the matrix, given that pixels are assigned by chance to each class.

Table 6.10 shows the sample computation of κ for the error matrix as given in Table 6.7.

Table 6.10 Sample computation of κ

789360	753324	635778	915486	505362	193908	858
807760	770884	650598	936826	517142	198428	878
837200	798980	674310	970970	535990	205660	910
859280	820052	692094	996578	550126	211084	934
571320	545238	460161	662607	365769	140346	621
202400	193160	163020	234740	129580	49720	220
920	878	741	1007	589	226	

Total of diagonal element = 3046621
Total of all elements = 28866023

Expected Agreement by chance = $\dfrac{\text{Sum of diagonal elements}}{\text{Total of all elements}}$

= 0.126

$$\hat{K} = \frac{0.916 - 0.126}{1 - 0.126}$$

= 0.904.

The value of $\hat{K} = 0.904$ means that the classification has achieved an accuracy that is 90% better than would be expected from random assignment of pixels to classes. Apart of this many other measures have been suggested. Table 6.11 gives the list of other accuracy measures for assessment of classification.

Table 6.11 List of accuracy measures

Measure	Abbreviation	Explanation	Formula	Base reference (s)
Overall accuracy	OA	Percent of samples correctly classified	$\frac{1}{N}\sum_{i=1}^{q} n_{ii}$	Story and Congalton (1986)
User's accuracy	UA	Index of individual class accuracy computed from row total	n_{ii}/N_i	Story and Congalton (1986)
Producer's accuracy	PA	Index of individual class accuracy computed from column total	n_{ii}/M_i	Story and Congalton (1986)
Average accuracy	AA_u	Average of all the individual user's accuracies.	$\frac{1}{q}\sum_{i=1}^{q}\frac{n_{ii}}{N_i}$	Fung and LeDrew (1988)
	AA_p	Average of all the individual producer's accuracies.	$\frac{1}{q}\sum_{i=1}^{q}\frac{n_{ii}}{M_i}$	
Combined accuracy	CA_u	Average of overall accuracy and average user's accuracy	$\frac{1}{2}[OA + AA_p]$	Fung and LeDrew (1988)
	CA_p	Average of overall accuracy and average user's accuracy	$\frac{1}{2}[OA + AA_p]$	
Kappa coefficient of agreement	K	Proportion of agreement after removing the proportion of agreement by chance	$\frac{P_o - P_e}{1 - P_e}$	Congalton et al. (1983)
Weighted Kappa	K_w	Proportion of weighted disagreement corrected for chance	$1 - \frac{\sum v_{ij} P_{oij}}{\sum v_{ij} P_{eij}}$	Rosenfield and Fitzpatrick-Lins (1986)
Conditional Kappa	K_{+I}	Conditional Kappa computed from the ith column in error matrix (Producer's)	$\frac{P_{o(+i)} - P_{e(i+1)}}{1 - P_{e(+i)}}$	
Tau coefficient	T_e	Tau for classifications based on equal probabilities of class membership	$\dfrac{P_o - \frac{1}{q}}{1 - \frac{1}{q}}$	Foody (1992), Ma and Raymond (1995)
	T_p	Tau for classifications based on unequal probabilities of class membership	$\frac{P_o - P_r}{1 - P_r}$	
Conditional Tau	T_{i+}	Conditional Tau compute from the ith row (User's)	$\frac{P_{o(i+)} - P_i}{1 - P_i}$	Naesset (1996)
	T_{+I}	Conditional Tau computed from the ith column (Producer's)	$\frac{P_{o(+i)} - P_i}{1 - P_i}$	

APPLICATION OF REMOTE SENSING 7

7.1 INTRODUCTION

With the launching of Landsat 1, in 1972, remote sensing technology has opened new vistas in the field of planning, surveying, monitoring and management of natural resources. It has provided easier techniques to undertake effective and efficient mapping of land, water, soil, forest, agriculture, urban area growth, flood plain mapping, crop acreage estimation etc. Detailed information can be extracted from satellite data on a temporal basis, and can be used as an input into Geographical Information system (GIS) for effective decision making. In order to carry out mapping activity using remote sensing data, the identification of features/objects is based either on the principles of visual interpretation techniques (Chapter 5) or digital image analysis (Chapter 6).

In this chapter, some important applications and case studies have been discussed in order to provide an insight regarding the methodology and approach to be adopted for different studies. The readers may note that visual interpretation techniques are equally important as in digital interpretation techniques. In reality, the success of digital interpretation is largely dependent on visual interpretation skill of the analyst.

Some of the applications discussed below as examples will help the readers in understanding the processes involved in application of remote sensing technique in a particular project.

   (*i*)   Land use and land cover mapping
   (*ii*) Urban growth studies
   (*iii*) Crop identification
   (*iv*) Ground water mapping
   (*v*)   Flood plain mapping
   (*vi*) Hydro morphological studies
   (*vii*) Wasteland mapping
   (*viii*) District level planning
   (*ix*) Disaster management.

7.2 LAND USE AND LAND COVER MAPPING

Land is one of the critical natural resource on which most developmental activities are based. For success of any planning activity, detailed and accurate information regarding the land cover and the associated land use is of paramount importance. In order to undertake a proper, systematic, and structured land cover/land use mapping, it is important to identify land cover/land use classes as per a classification scheme, such as USGS Land cover/Land use Classification scheme or Cowardin Wetland classification scheme, or develop a new one.

A study on land use and land cover for a part of Haridwar district was carried out for the area lying between 78°07'13" E and 78°16'14" E longitudes and 30° N and 30°08'53" N latitudes covering an area of nearly 260 km². The area is primarily covered with forest, vegetation, built up, water, sand etc. IRS-1C LISS-III image (Plate 7.1) of April 3, 2000, was used along with PAN image (Plate 7.2) of the same date. The methodology adopted is shown in Fig. 7.1. On the basis of field visits, 11 classes were identified. These classes are (*i*) thin forest, (*ii*) medium forest, (*iii*) dense forest, (*iv*) fallow land, (*v*) shrubs, (*vi*) open land, (*vii*) shallow water, (*viii*) wet sand, (*ix*) dry sand, (*x*) built-up area, and (*xi*) deep water.

Training data were identified and extracted using ERDAS Imagine software (Plate 7.3). Based on the training data statistics, it was found that Average Transformed Divergence was 1999, which is an indication that all the classes had good spectral separability. Two classifiers, namely, Minimum distance to Means and Maximum Likelihood were used for classifying the land use and land cover of the area and the respective results are presented in Plates 7.4 and 7.5. The classified image was tested for accuracy, and it was found that Minimum Distance to Mean classifier gives an overall accuracy of 90% while Maximum Likelihood classifier gave an overall accuracy of 94.44% with Kappa coefficient of 0.89 and 0.94 for Minimum Distance to Mean and Maximum Likelihood classifiers, respectively.

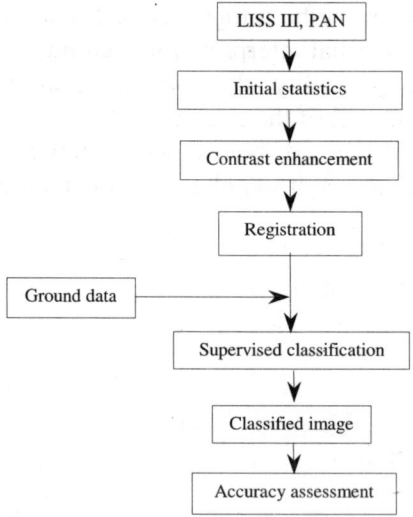

Fig. 7.1 Flow chart of methodology adopted (*Source: Reddy, 2003*)

7.3 CROP INVENTORY STUDIES

Remote sensing has proved to be an ideal data set for making crop inventory. Typical spectral reflectance of a crop shows absorption due to pigments in the visible region (0.4 to 0.7 μm), high reflectance in the near infrared region because of internal cellular structure of the leaves, and absorption at 1.45 μm, 1.95 μm and 2.6 μm spectral bands due to water content. Spectral response of a crop canopy is influenced by the leaf-area index (LAI) and per cent ground cover, growth stages, differences in cultural practices, stress conditions, and canopy architecture. Background soil/water is an important influencing factor. Each crop has its own architecture, growing period, etc., thus enabling discrimination through remote sensing data. If there are two crops with similar spectral signatures on a given data (confusion crops), multidata data may be used to discriminate them. Vigour of the crop is responsible for high absorption in the red and reflectance in the near infrared. It has been observed that the ratio of near infrared to red radiance is a good indicator of the vigour of the crop. Some of these properties are utilized in crop identification, yield forecasting and crop condition assessment.

Acreage Estimation
Identification and discrimination of various crops/land cover classes require quantitative use of subtle differences in their spectral data, and hence rely mostly on digital image processing techniques. The acreage estimation procedure broadly consists of identifying representative sites of various crops/land cover classes on the image based on the ground truth collected, generation of signatures for different training sites and classifying the image using training statistics.

District level acreage estimation has been generally carried out by analyzing the complete data. In this case, the administrative boundary of the district is digitized; a mask is generated and superimposed on the scene. All the data elements (pixels) within this mask are extracted for further analysis. Either all the pixels are classified or a systematic sample is used for supervised maximum likelihood procedure. Such a procedure has been successfully used for wheat, rice, sorghum, oilseeds, and cotton in many districts of various states. Mulberry acreage estimates have been attempted at taluk level.

Estimation of crop acreages for large areas, like states, requires handling of a very large volume of data, larger efforts in ground truth data collection, etc. In such a case, it will be difficult to complete the entire data analysis in the short time available and provide preharvest estimates. To overcome this problem, sampling technique based procedures have been developed and successfully used (Fig. 7.2). Based on the crop concentration statistics, agrophysical and/or agroclimatic conditions, the study area is divided into homogeneous strata. Each stratum is divided into segments each of 10 km x 10 km size. Segment sizes of 7.5 km x 7.5 km and 5 km x 5 km are also being tried consisting of heterogeneous areas. Further stratification is done on the criterion of a sample segment having more than fifty per cent agricultural area or less. For digital data analysis, generally ten per cent of the total population is used. Second level stratification has been attempted using in-season data either by visual techniques or with near infrared to red digital ratio images. Appropriate statistical methods are employed to aggregate results at stratum and study area/state levels.

Yield Forecasting
Yield is influenced by a large number of factors such as crop genotype, soil characteristics, cultural practices adopted (*e.g.*, irrigation, fertilizer), weather conditions, and biotic influences, such as weeds, diseases and pests, etc. Spectral data of a crop is an integrated manifestation of the effect of all these factors on its growth. The two approaches adopted for yield modeling using remote sensing data are (*i*) remote sensing data or derived parameters which are directly related to yield, and (*ii*) remote sensing data which are used to estimate some of the biometric parameters, which in turn are input parameters to a yield model.

Spectral index of crop canopy (NIR/Red, Greenness, and NDVI) at any given point of time reveals the crop growth, and its decay as affected by various factors in the time domain. A typical analytical form for a greenness profile is shown in Fig. 7.3. The nature of the curve remains the same when greenness is replaced by any spectral index involving near infrared and red radiance of a crop.

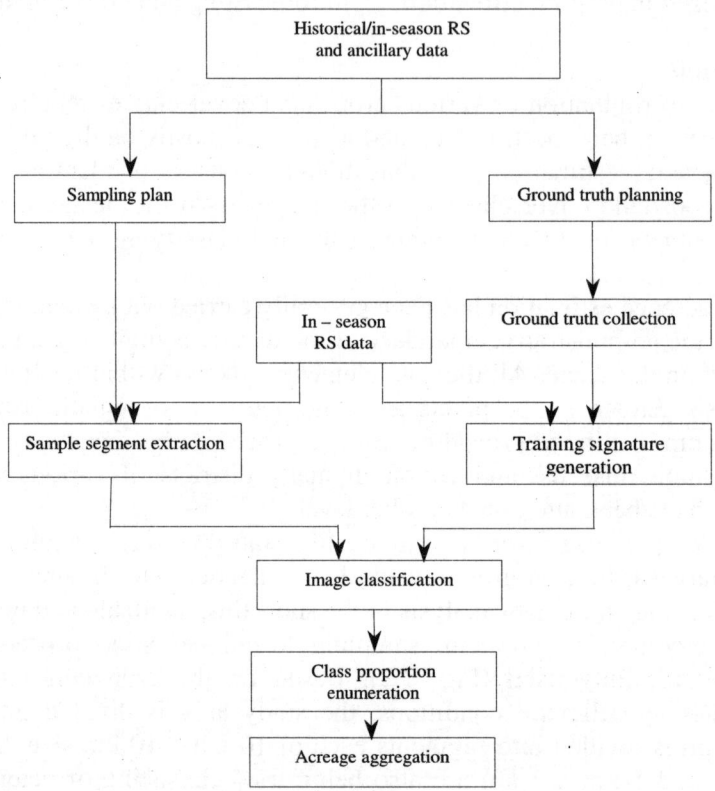

Fig. 7.2 Procedure for acreage estimation using sampling technique (*Source: Navalgund et al, 1991*)

$$G_{(t)} = G(t/t_o)^\alpha \exp \beta \, (t_0^2 - t^2) \quad \text{for } t > t_o$$
$$G_{(t)} = G_o \quad \text{for } t \leq t_o \quad \quad \ldots(7.1)$$

where $G_{(t)}$ is the spectral index at time t expressed in days after sowing, G_o refers to soil greenness, t_o is the date of spectral emergence. α and β refer to growth and decay of the plant. Profile parameters may be related to physiologically significant processes and hence yield.

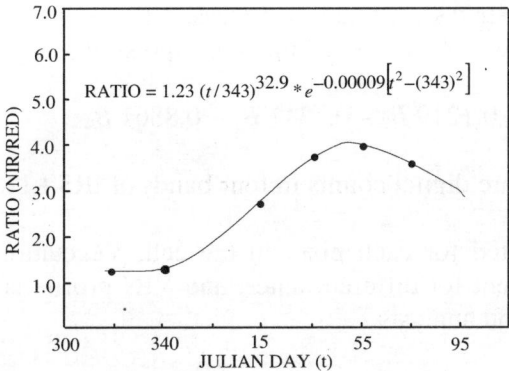

Fig. 7.3 A typical growth profile for wheat (*Source: Kauth and Thomas, 1976*)

Efforts have been made to develop a relationship between yield and spectral index using spaceborne data at maximum vegetative cover, and also using different parameters of the profile. The two feasible approaches making use of these parameters, are (*i*) single data spectral approach, and (*ii*) growth-profile approach.

Single date spectral index approach: Various steps adopted in this approach are given in Fig. 7.4. Two corrections found extremely important in deriving a yield model are: (*i*) correction for different dates of data acquisition using a growth profile (Fig. 7.3), and (*ii*) sensor-to-sensor transformations for radiances (Fig. 7.5). Historical yield values obtained using crop-cutting experiments are used in developing regression relation.

Growth profile approach: The spectral growth profile approach provides complete information on the canopy development. However, 16/22 day repetitively of Landsat/IRS does not provide sufficient temporal resolution to obtain growth profiles. NOAA-AVHRR provides an opportunity for estimation of growth profiles, although spatial resolution is coarse for crop identification (Dubey, *et al*, 1991)

Crop Condition Assessment

Condition of the crop is affected by factors, such as supply of water and nutrients, insect/pest attack, disease out-break, and weather conditions. These stresses cause physiological changes, which may alter the optical properties of leaves, and bring about changes in canopy geometry. The task of crop condition assessment requires (*i*) detection of stress, (*ii*) differentiation of stressed crop from the normal crop at a given time, (*iii*) quantification of extent and severity of stress, and (*iv*) assessment of the production loss. Condition assessment is normally done on a grid-cell basis using multiband satellite data. The area of interest is divided into geographically referenced grid cells of appropriate size, and each grid is monitored individually. Various steps involved in the crop condition assessment procedure used for cotton crop are given in the flow diagram (Fig. 7.6). Vegetation Index (VI) for the pixels of the crop of interest is computed for the selected sample segments and VI statistics are generated. The VI is either tasseled cap greenness, NIR to red radiance ratio or normalized difference vegetation index (NDVI). Greenness is expressed as:

Greenness = $- 0.172 B_1 - 0.1219 B_2 - 0.4341 B_3 + 0.8861 B_4$ ...(7.2)

where B_1, B_2, B_3, and B_4 are digital counts in four bands of IRS LISS-II data, respectively.

From the VI computed for each pixel in the cell, Vegetation Index Number (VIN) is computed for each segment for different dates, and VIN profile is generated and stored in a data base to carry out trend analysis.

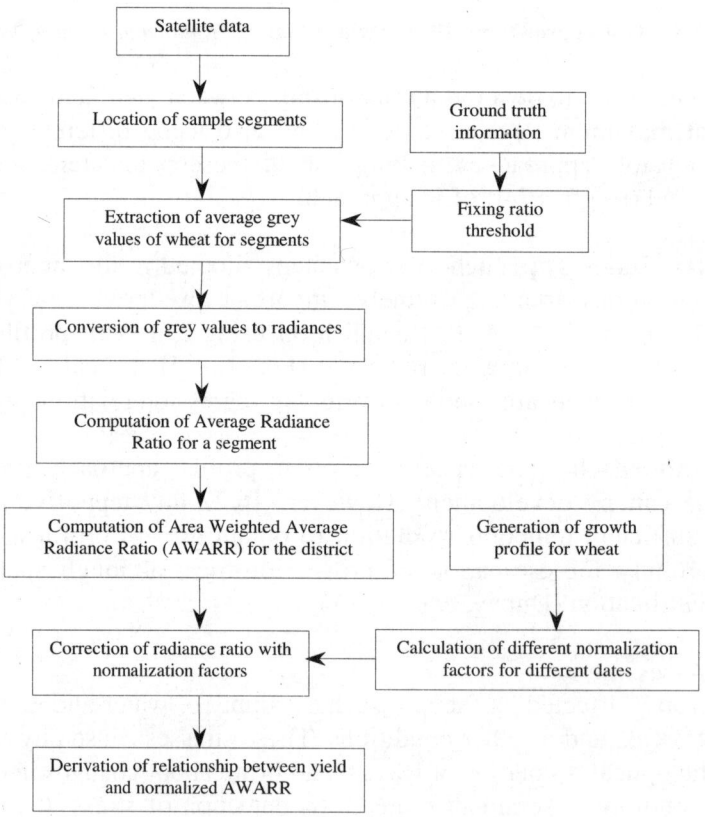

Fig. 7.4 Steps involved in Remote Sensing based yield forecasting (*Source: Navalgund et al., 1991*)

Application of Remote Sensing 121

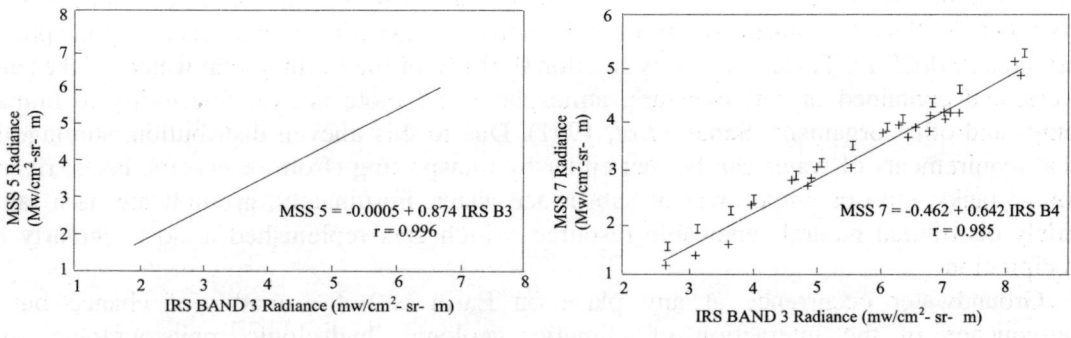

Fig. 7.5 Spectral radiance value of various surface features (*Source: Sharma, 1991*)

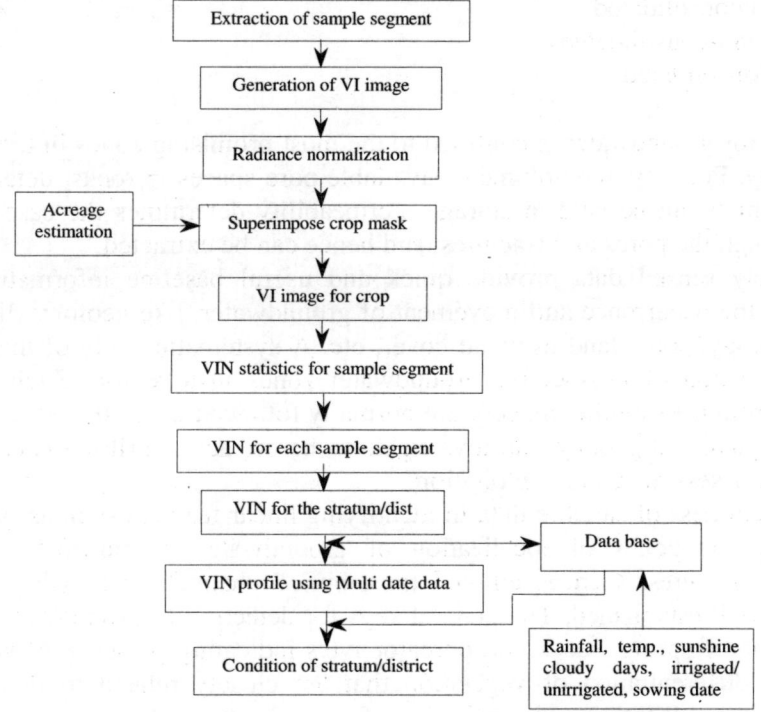

Fig. 7.6 Procedure for crop condition assessment using remote sensing data (*Source: Navalgund, et al., 1991*)

7.4 GROUND WATER MAPPING

Water is an extremely important resource. Life on the planet Earth, evolved in aquatic environment, depends on water for sustenance. However, distribution of this resource is highly uneven. About 97.41 % of the water on the Earth is contained in the oceans as saline water. The balance of 2.59 % as fresh water is contained in ice caps and glaciers (1.953%), groundwater (0.614%), lakes (0.008%), soil moisture (0.005%), and in rivers, atmosphere, and biota (0.0005%). Thus, only a tiny fraction 0.014 % of the Earth's total water in lakes and rivers, and contained in soil moisture, atmosphere and biota is available easily to human beings and other organism (Sahai, *et al*, 1991). Due to this uneven distribution, shortage in local requirements of water can be met either by transporting (from reservoirs, lakes, rivers, storage tanks, etc.) or withdrawal of subsurface water. Fortunately, groundwater is a fairly widely distributed natural renewable resource, which gets replenished almost regularly by precipitation.

Groundwater occurrence at any place on Earth is not a matter of chance but a consequence of the interaction of climatic, geologic, hydrologic, physiographic, and ecological factors. The groundwater in India occurs mainly in three types of geological formations:

(*i*) Unconsolidated
(*ii*) Semi-consolidated
(*iii*) Consolidated.

Search for groundwater is confined to the most promising zones in terms of porosity and permeability. Porosity, the volume of available pore spaces in rocks, determines the amount of water which can be held in storage. Permeability determines the ease with which water moves through the pores and fractures, and hence can be extracted.

Remotely sensed data provide quick and useful baseline information on the factors controlling the occurrence and movement of groundwater, like geology (lithology/structure), geomorphology, soils, land use/land cover, etc. A systematic study of these factors leads to better delineation of prospective groundwater zones in a region. Such prospective zones identified from the satellite imagery are normally followed up on the ground through detailed hydrogeological and geophysical investigations before actual drilling is carried out for exact quantitative assessment and exploitation.

The usefulness of satellite data in identifying linear features such as fractures/faults, that are usually the zones of localization of groundwater in hardrock areas, and certain geomorphic features, such as alluvial fans, buried channels, etc., which often form good aquifers is well established. The general keys for detection of groundwater aquifers relate to identification of spring, seeps and phreatophytes indicating presence of shallow water table conditions, differentiation of vegetation that are closely related to depth and salinity of groundwater, and location and monitoring of groundwater systems under stress. Groundwater potential maps using Landsat data in hardrock areas have been prepared in India using either the geomorphology as the base, or through an integrated approach of studying various themes, depending upon the complexity of control over the groundwater occurrence and movement.

A major thrust to the utilization of IRS-1A data on a national level was provided by the National Drinking Water Mission (NDWM), one of the technology missions launched by the Govt. of India in 1986. The primary objective of NDWM was to provide potable drinking water to all the villages @ 40 litres per capita per day (lpcd), and an additional 30 lpcd for cattle in desert areas. It was stipulated that the source should be located within a distance of 1.6 km from a village

The methodology adopted for the preparation of districtwise hydro-geomorphological maps on 1:250,000 scale as shown in Fig. 7.7, comprised the following six steps:

(i) Data procurement,
(ii) Base-map preparation,
(iii) Preliminary interpretation,
(iv) Ground Checks,
(v) Final interpretation, and
(vi) Final map preparation.

The data used initially was Landsat Thematic Mapper imagery in the form of FCC using bands 2, 3, and 4 with 30 m spatial resolution, and later IRS-1A LISS-II (FCC) imagery with 36 m spatial resolution. In some cases, LISS-I imagery with 73 m spatial resolution has also been used. The imagery used was either in the form of paper prints on 1:250,000 scale or transparencies which could be enlarged to 1:250,000 scale.

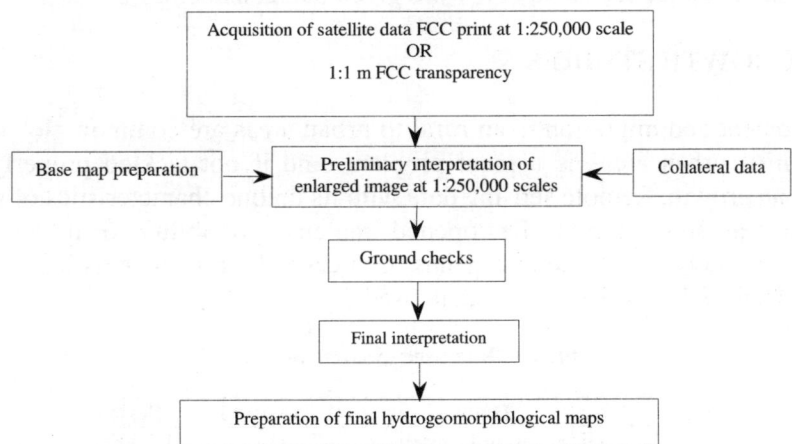

Fig. 7.7 Steps in hydrogeomorphological mapping using satellite data (*Source: Sahai, et al, 1991*)

To start with, districtwise base maps were prepared on 1:250,000 scale, showing major drainage courses and important localities, roads, rail lines and a few other cultural features like canals, etc. The next step involved a systematic visual interpretation of the satellite imagery to delineate various geomorphic features/landforms and their depiction on to the base map. These geomorphic features/landforms were then evaluated critically in terms of the broad lithology they are comprised of, the associated structural features, the development of drainage around and on them, the broad soil type and its effective depth, the thickness of weathered mantle in case of hardrock country, and the type of land use/land cover prevailing in the area to finally arrive at the groundwater prospects. Existing geological maps and other collateral data, supplemented by supportive field checks were also used. District wise hydrogeomorphologic maps were thus prepared for all the districts of the country.

The synoptic, near orthographic, multispectral view of the terrain under uniform illumination conditions provided by the IRS data has the unique capability of providing regional level information on all the above mentioned factors with a fair degree of accuracy and reliability. Though no detailed lithological interpretation was possible from the IRS imagery, differentiation of broad rock groups was made based on factors like spectral reflectance of rock types (when exposed), in conjunction with associated structural, drainage, and landform characteristics, etc. Fractures, faults and their intersections in hardrock terrain as well as certain synclinal structures which at places are the loci for groundwater occurrence in sedimentary rocks were interpreted carefully from the IRS imagery. Such features are often missed in ground surveys for want of a synoptic view. While certain landforms like buried channels, present day valley fills, alluvial fans, bazadas, etc., having known relationship with groundwater occurrence were readily mapped from the IRS imagery, the utility of the satellite data in studying the drainage pattern/density etc. including the paleo drainage (which have a bearing on the recharge conditions and localization of groundwater) were also fully harnessed. The distribution of soils with regard to their broad textural classes and land use/land cover pattern mappable from the IRS imagery, were given due consideration.

7.5 URBAN GROWTH STUDIES

Urban development and migration from rural to urban areas are common global phenomena. The expansion of urban areas is a great problem, and if not tackled properly can lead to haphazard urban growth. Remote sensing data with its unique characteristics of synoptic view, repetitive coverage, and reliability has opened immense possibility for urban area mapping and its growth. A study of Dehradun city has been carried out using existing guide maps and satellite data. Table 7.1 lists the various data used.

Table 7.1 Data used (*Source: Soni, 2003*)

S.No.	Data	Year	Scale
1	Guide map of Dehradun	1945	1:20000
2	Guide map of Dehradun	1965	1:20000
3	Topo sheet 53J/3	1984	1:50000
4	Topo sheet 53F/15	1965	1:50000
5	IRS LISS II	1988	Digital data
6	IRS 1D PAN	1997	Digital data
7	IKONOS	2001	Digital data

Plate 7.6 shows the different land use of Dehradun and its surrounding areas for the year 1945, 1965, 1988, 1997 and 2001. With the help of these land use maps, the urban expansion of Dehradun during the period of 1945-2001 has been studied (Plate 7.7). Table 7.2 shows the different types of land use during the period 1945-2001, while Table 7.3 shows the land use change matrix for the same period. It may be noted the in the case of Dehradun city, the major urban growth has taken place on barren land (22 km^2) followed by agricultural land

(11 km²). One of the encouraging aspects of urban growth in Dehradun is that very little forest land has been encroached. However, there has been a continuous encroachment along the length of rivers Rispana Rao and Bindal Rao passing through the city. The settlements in this region are under high risk of damage during floods. Plate 11.1 illustrates the change in land use from 1945-2001.

Table 7.2 Area for different land use in different years (*Source: Soni, 2003*)

Land use/ Land cover	Area (km²)				
	1945	1965	1988	1997	2001
Urban	12.57	27.83	34.96	46.16	50.31
Agriculture	39.08	34.77	33.60	30.18	29.04
Forest	15.04	14.10	11.79	11.26	10.35
Vacant/Scrub	41.33	31.89	28.81	24.13	22.28
River	6.21	5.64	5.07	2.5	2.25
Total	114.23	114.23	114.23	114.23	114.23

Table 7.3 Land use change matrix (*Source: Soni, 2003*)

Land use class	Change in area (1945 – 2001) (km²)				
	Urban	Agriculture	Forest	Vacant/scrub	River
Urban	-	-	-	-	-
Agriculture	11.38	-	-	0.57	-
Forest	1.76	0.56	-	2.37	-
Vacant/scrub	22.15	0.63	-	-	-
River	2.46	0.73	-	0.78	-

7.6 FLOOD PLAIN MAPPING

Floods have become an unfailing event almost every year in the Indian states of Assam, Bihar, Uttar Pradesh, North Bengal, etc. Millions of hectares of land are inundated, especially by Brahmaputra, Ganga, and their tributaries, resulting in damages to crops worth millions of rupees. It is recognized that floods cannot be totally controlled, but with suitable basin wise flood management measures, like construction of detention storages with or without flood storage, embankments and drainage channels, anti-erosion, river training and protection works, watershed afforestation and soil conservation, flood hazard zoning, flood forecasting, disaster preparedness, etc. the flood damages can be vastly minimized. For implementing these measures and providing flood relief, it is necessary to acquire, besides historic conventional data, latest and reliable information about extent of flood inundated areas, duration of flooding, river configuration during and after floods, areas of silt deposits and shoals, drainage congested areas and blocked outfalls, watershed characteristics, vulnerable areas of bank erosion, etc. It is here that remote sensing has more efficient and effective role over ground based survey methods that are arduous, time consuming and beset with limitations. In fact, flood mapping using remote sensing is considered as one of the important applications of remote sensing technology in real-time operational mode, since maps on 1:50,000 scale can be prepared within a week's time from the date of acquisition of the satellite image (Ramamoorthi *et al*, 1991).

Both visual interpretation and digital analysis techniques have been adopted in flood studies. In the digital analysis, the low radiance evaluation of water bodies in the near infrared region is done by scanning of pixels as a 'water' and 'no water' classes, which gives an accurate distribution of the water features. Subsequent analysis of 'water' pixels could provide further information about quality, depth, sediment loading, etc. Visual interpretation, being fast and less expensive has been most preferred approach.

Flood Plain mapping using satellite data

In order to carry out flood plain analysis, certain flood susceptibility indicators, as visible on single band satellite imagery can be identified (Sollers, 1974). These are:

(*i*) Upland physiographic
(*ii*) Water characteristics (such as shape, drainage density, etc).
(*iii*) Degree of abandonment of natural levees
(*iv*) Occurrence of stabilized sand dunes on river terraces
(*v*) Channel configuration
(*vi*) Backswamp areas
(*vii*) Soil moisture availability
(*viii*) Soil difference
(*ix*) Vegetation differences
(*x*) Land use boundaries
(*xi*) Agricultural development
(*xii*) Flood control measures on the flood plain

Subsequently, flood area can also be identified using temporal satellite data. In this approach, two infrared band images are taken, the first one belongs to pre-flood period and the second belongs to post-flood period. In infrared band image, water reflects little or no incident energy and hence appears dark in the image. When the pre-flood infrared image is projected in green and flood/post flood infrared image in red, then the flood area appears in red in the composite image. Further, standing water appears very dark, while dry soil appears bright.

When a non-flood image is projected as red, in combination with a flood image projected as green, the composite colour image is composed of the following elements (Deutsch and Ruggles, 1974):

(*i*) When surface water is present on both images, the composite image receives little or no light and is therefore, essentially black. This depicts the area normally covered by river and other surface water bodies.
(*ii*) When there is surface water in the flood image, and dry soil in the non-flood image the composite image receives only red light, and is therefore, a highly saturated red colour. This depicts the *flood inundation* area.

(*iii*) When the ground is not covered with water in both images, the composite image receives relatively equal amounts of red and green light, and is therefore yellow. This depicts the area unaffected by flood waters.

(*iv*) When there is water-saturated soil in the flood image, and dry soil in the scene projected as red, the composite image receives red light combined with a small amount of green light, and thus results in a colour continuum between yellow and red.

Table 7.4 Flood mapping using temporal composites (*Source: Deutsch and Ruggles, 1974*)

Feature in non-flood image	Feature in flood image	Appear as on colour composite	Result
Surface water	Surface water	Black	River & Surface Water bodies
Dry soil	Surface water	Red	Flood inundated areas
Soil	Soil	Yellow	Not affected by floods
Dry soil	Saturated soil	Shades of yellow and red	Flood affected areas

Interpretation of Flood Area by Using Colour Composites

The flooding becomes more explicit on the false colour composite of two visible and one infrared band images. The additive colour viewer is used to produce two and three-colour composite highlighting condition of the flood plains. From among the numerous additive-colour combinations the following are deemed to be best for detailed flood plain analysis or mapping (Deutsch and Ruggles, 1974).

Interpretation A

Here the first visible band is projected in blue while the second visible band is projected in green and infrared band, filtered to about 60% transmission, as red. This interpretation is best suited for interpreting the extent of flooded area in relation to rural land use. Morphologic and geologic features are enhanced, and water details are well preserved. Areas with standing water appear as blue, and wet or saturated soils as brown.

Interpretation B

Here the second visible band is projected as red, filtered to about 60% transmission and infrared band as green. This pictorial rendition is considered to be the best for differentiating areas of varying degrees of inundation and wetness. Standing water appears as red and the wet or saturated flood plain as green. This rendition, therefore, is used as the basic source of information from which the flood maps are prepared.

Interpretation C

The blue visible band is projected as blue, the green visible band as green while infrared band is filtered to about 40% transmission as red. This pictorial rendition is best suited for interpreting flooded areas in relation to urban pattern. While preserving water detail, it enhances cultural details. In this rendition, standing water is shown in shades of blue or green, and areas of wet or saturated soil appear as brown.

7.7 HYDRO MORPHOLOGICAL STUDIES

A river is a dynamic system. It adjusts its channel roughness, geometry, pattern, and profile with changing climatic, geological, hydrological, hydraulic, sediment, and man-made constraints. Thus fluvial hydraulics and river morphology are complicated subjects. The planform of a reach of an alluvial river reflects the hydrodynamics of flow within the channel and the associated process of sediment transport and energy dissipation. Braided streams occur in high energy environments with large and variable discharges carrying heavy sediment loads on steeper gradients. These streams are characterized by a random pattern of multi-thread channel network due to appearance of braided bars within the overall waterway of the river. The presence of braid bars increases the total flow resistance and the energy losses along the boundary, thereby promoting the development of a network of hydraulically inefficient channels.

River Kosi is one of the largest braided streams of the world. The river discharge varies 10 times of its mean annual discharge during the monsoon and carries a very high concentration of sediment load. The river has shifted 115 km westwards during 1776-1950. It has been confined by the construction of eastern and western and embankments in 1959 and a barrage in 1963 at Bhimnagar. It is still shifting its bed within the embankments and can be termed as unstable. Keeping the above phenomena in view, a 165 km reach of the river between Chatra and Koparia has been selected for hydromorphological study using satellite imageries and hydrographic data. The objective of the present study is to study the changes in river pattern and the lateral extent of channel migration to define the intensity of braiding over different time space.

Landsat 2, MSS Band 5 dated March 30, 1975, IRS-IA LISS band 4 of Dec 5, 1989 April 7, 1989, along with a topographical map at a scale of 1:200,000 for the year 1966 (Kosi Project Authority, Birpur) was used for the study. Further planform maps of river Kosi for the year 1956, 1966 and 1993 prepared by Kosi Project Authority, Birpur were used. Further, 39 cross section details of the river were also available. Using visual interpretation techniques, planform map of the river stretch was prepared at a scale of 1:200,000 (Mishra, 1997).

From the plan forms maps for the years 1956, 1966, 1975, 1989 and 1993, the following parameters were extracted:

(i) L_R = the overall length of the channel belt reach measured along a straight line,
(ii) $\sum L_i$ = the length of all land/bars in the reach,
(iii) L_r = the length of reach measured midway between banks of the channel belt,
(iv) L_{ctot} = the sum of the mid channel length of all segments of primary channels in a reach, and
(v) L_{cmax} = the mid channel length of the widest channel through the reach.

Fig. 7.8 illustrates the various parameters used for assessing the morphological nature of a river. Based on these parameters, the following morphological parameters were computed.

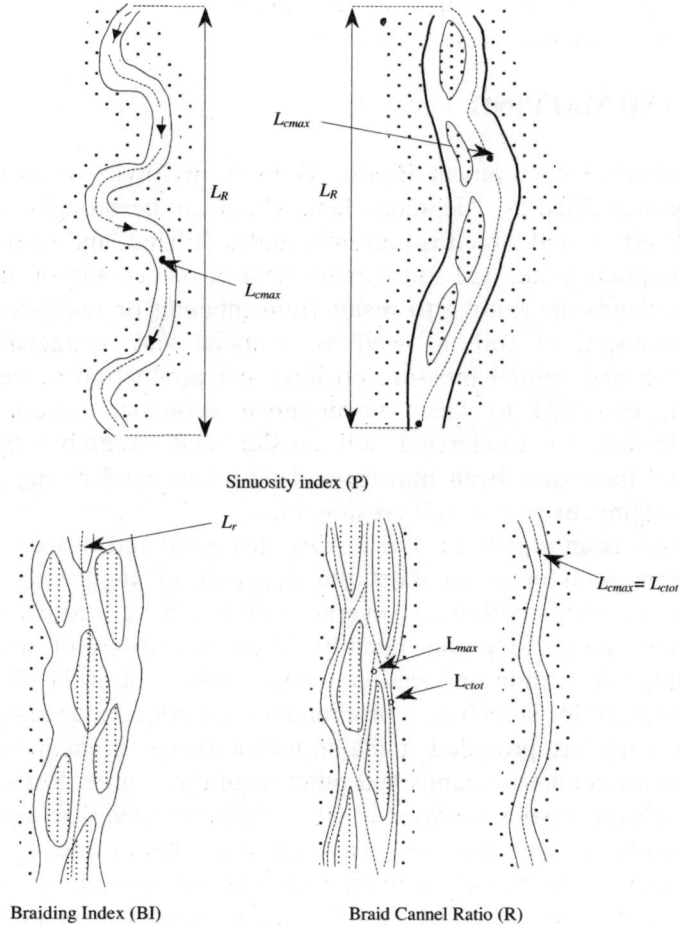

Fig. 7.8 Parameter definition of braiding (*Source: Friend and Sinha, 1993*)

(i) Sinuosity Index (P) = $\dfrac{L_{cmax}}{L_R}$

(ii) Braiding Index (BI) = $\dfrac{2 \times (\sum L_i)}{L_r}$

(iii) Braid channel Ratio (B) = $\dfrac{L_{ctot}}{L_{cmax}}$

Table 7.5 shows the computed sinuosity and braiding parameters for river Kosi. It is found that sinuosity index for all years of analysis is ranging between 1.0 and 1.3 which is less than 1.5. This is clear indication that River Kosi is not a sinuous river. A river is considered to be a braiding one if the value of BI or B is greater than 1. In this case it is found that for all the years the value of BI or *B* is more than 2, thus giving a clear indication that the river is braided in nature. It is found that maximum braiding occurs 100 km downstream of Birpur barrage and is continuously increasing.

7.8 WASTELAND MAPPING

National Wastelands Development Board, Ministry of Environment and Forests, Govt. of India, describes wasteland as, "degraded land which can be brought under vegetative cover, with reasonable effort, and which is currently under-utilized, and land which is deteriorating for lack of appropriate water and soil management or on account of natural causes" (Rao, *et al*, 1991). Wastelands are known to result from inherent or imposed disabilities related to location, environment, or soil, as well as financial and management constraints. With increasing human and animal pressure on land, the production of vegetation for food and other uses has extended to areas under great ecological stress and less favourable environment, leading to accelerated soil erosion and excessive land degradation. Vast stretches of land have thus been transformed into wasteland owing to desertification, soil salinity, waterlogging, excessive soil erosion, etc.

In India, there is an urgent need to reverse this trend and restore the wastelands to their production potential in order to meet the demands of increasing population and other developmental activities. Although several agencies have estimated the total extent of wastelands, their figures vary considerably. Their definitions of wastelands were equally variable. Reliable information on location, nature, and extent of the different wastelands on a large scale is essential for launching a programme on wasteland development.

Remote sensing has provided a major technological breakthrough in the method of acquiring information on wastelands and other natural resource. Remote sensing from space with its unique characteristics of synoptic view, repetitive coverage and reliability has opened immense possibilities for resource mapping and monitoring, resource targeting and management to achieve optimization in resource utilization and conservation.

A Task Force was identified by the Planning Commission in 1986, to evolve a suitable wasteland classification system. The classification system evolved and approved by the Planning Commission comprises of the following categories (NWDB, 1987):

   (*i*) Gullied and/or ravenous land,
   (*ii*) Upland with or without scrub,
   (*iii*) Water-logged and marshy land,
   (*iv*) Land affected by salinity/alkalinity (coastal or inland),
   (*v*) Shifting cultivation area,
   (*vi*) Sandy (desert or coastal),
   (*vii*) Mining/industrial wastelands,
   (*viii*) Under-utilized/degraded notified forest land,
   (*ix*) Degraded pastures/grazing land,
   (*x*) Degraded land under plantation crops,
   (*xi*) Barren rocky/stony wastes/sheet-rock areas,
   (*xii*) Steep-sloping areas, and
   (*xiii*) Snow covered and/or glacial areas.

Table 7.5 Computation of sinuosity and braiding parameters (Source: Mishra, 1997)

Distance along the River (km)	Sinuosity index (P)				Braiding index (BI)				Braid channel ratio			
	1956	1966	1975	1989	1993	1956	1966	1975	1989	1993	1956	1966	1975	1989	1993
10	1.08	1.15	1.07	1.01	1.19	3.18	3.53	3.51	2.20	1.73	2.48	2.85	2.93	2.38	2.55
20	1.05	1.03	1.07	1.01	1.01	4.06	5.66	7.30	6.61	3.61	3.83	4.59	4.95	4.67	3.68
30	1.03	1.07	1.16	1.34	1.08	4.81	12.84	10.81	6.33	5.48	4.04	7.78	6.50	3.57	3.78
40	1.08	1.17	1.20	1.30	1.17	5.24	9.01	10.56	7.18	4.00	3.76	5.88	5.63	4.28	3.54
50	1.07	1.37	1.27	1.13	1.10	3.15	9.28	6.65	7.42	2.56	3.02	4.94	4.37	4.92	2.65
60	1.09	1.09	1.09	1.17	1.02	1.00	7.30	8.11	6.82	2.90	1.36	5.09	4.82	4.29	2.55
70	1.08	1.14	1.11	1.23	1.14	9.59	9.59	7.89	9.75	4.42	6.28	6.27	5.60	5.87	3.09
80	1.09	1.13	1.06	1.05	1.04	13.24	7.78	11.65	12.63	6.66	8.34	5.30	7.33	7.93	4.40
90	1.14	1.06	1.28	1.00	1.03	20.23	13.63	7.40	8.40	4.89	11.88	8.67	4.63	5.51	3.53
100	1.11	1.16	1.02	1.23	1.13	18.99	19.46	14.05	7.70	4.88	12.77	10.21	8.94	4.63	3.62
110	1.21	1.13	1.03	1.19	1.18	16.45	10.55	8.17	7.75	3.54	10.51	6.84	6.34	4.83	3.20
120	1.31	1.10	1.15	1.25	1.02	15.90	10.74	9.03	5.58	2.52	8.58	6.38	5.65	3.51	2.54
130	1.04	1.27	1.03	1.22	1.05	18.08	6.55	8.87	4.46	2.46	12.10	4.08	5.42	3.01	2.33
140	1.05	1.20	1.11	1.20	1.13	12.03	1.00	7.81	4.37	4.02	7.67	1.00	4.80	2.71	3.03
150	1.07	1.11	1.06	1.15	1.06	6.65	1.45	5.12	2.24	2.00	4.77	1.819	3.70	2.20	2.04
160	1.05	1.06	1.13	1.05	1.02	3.50	1.25	1.31	1.57	1.67	1.95	1.682	1.18	1.43	1.48
165	1.08	1.22	1.07	1.04	1.00	2.43	1.77	1.00	1.00	1.00	2.31	1.51	1.00	1.00	1.00

The methodology of wasteland mapping using IRS-1A data is shown in Fig. 7.9. IRS-1A geocoded FCC products on 1:50,000 scale were used for visual interpretation. In order to get the optimum spatial information about wasteland, Rabi season data (November to March, when there is maximum vegetation cover) are used. The image characteristics, such as colour, tone, texture, pattern, shape, size, location, and association enable one to identify and delineate different types of wastelands. These delineations, however, are tentative and subject to confirmation in the field. Therefore, ground truth forms a vital input to mapping with remote sensing data. It may be mentioned here that the key for interpretation is subject to changes depending upon the season, scale, and resolution of the imagery. In the present study, certain categories of wastelands like salt-affected land, water-logged/marshy land, and sandy areas were easily delineated by virtue of their spectral separability, pattern, and location, whereas gullied or ravenous land, shifting cultivation, etc., were delineated with moderate success. However, the category of undulating upland with or without scrub, which is widely prevalent throughout the country, could not be easily delineated due to its merging with fallow lands and others having similar spectral reflectance patterns. However, this category was identified and delineated, as far as possible, by using multidate imagery.

The ancillary data that are used consist of Survey of India topographic maps at 1:50,000 scale. These maps provide not only a base for plotting thematic details of wastelands interpreted from IRS imagery, but also provide an accurate reference for the land cover, *e.g.*, boundaries of reserved forest, types of forests, gullied area, barren rocky area, and agriculture area. However, the thematic information on these maps is outdated since the surveys were conducted about 10-15 years back. Revenue and Census maps obtained from various sources provide information on village boundaries and are transferred to the maps. The final wasteland map has generated on 1:50,000 scale with its base derived from Survey of India topographical map, village boundaries drawn from Revenue and Census maps and spatial limits of different categories of wastelands derived from IRS-1A imagery.

7.9 DISTRICT LEVEL PLANNING

District level planning is an integrated area planning aimed at all-round development of the area. Each district plan should specifically deal with its problems, typical to its natural resources, terrain and environmental condition, agro-climatic set-up, and socio-economic profile.

The developmental plan should meet the basic needs and aspirations of the local people within the overall frame-work of the multi-level planning. Therefore, the district level plan should supplement the overall goals of state and national plans which are necessary for economic growth through higher production, generation of employment, alleviation of poverty, fulfillment of the basic needs of the people, and maintenance of ecological balance. Development of remote sensing techniques in the last three decades and their proven application, in the efficient survey and mapping of natural resources, have provided quick assessment of natural resources in the context of district level planning.

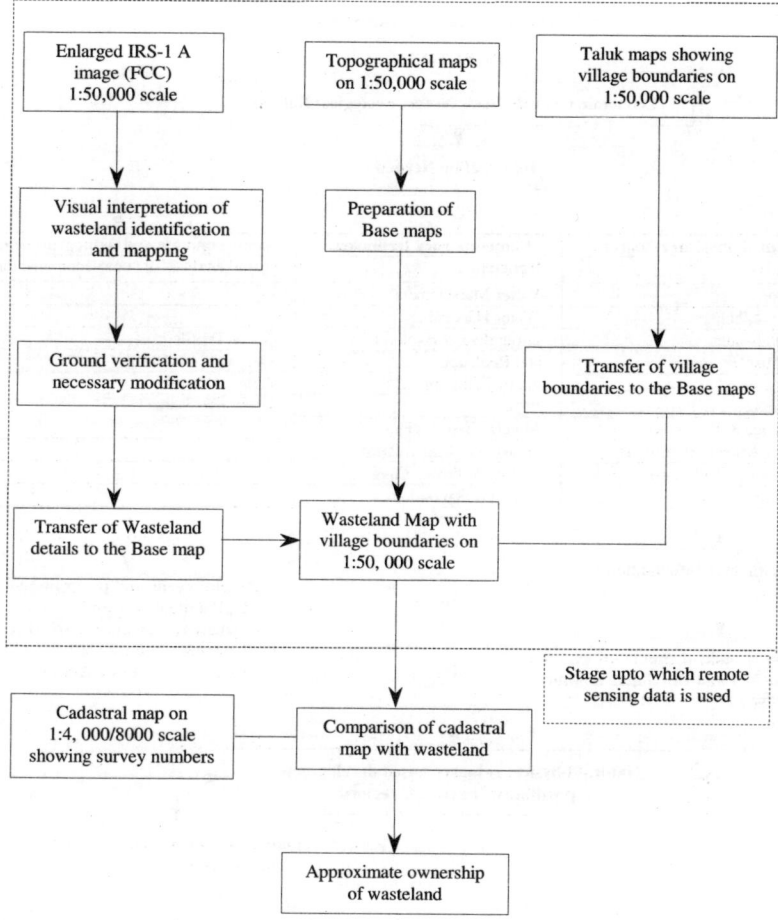

Fig. 7.9 Methodology for wasteland mapping (*Source: Rao et al, 1991*)

Based on applications of IRS-1A and GIS, a methodology has been developed by the scientists of the Department of Space for district level planning. This methodology adopts the guidelines of the Planning Commission Working-Groups on District and Block Level Planning, and provides for resource data integration from available sources, and socio-economic and demographic database devised by the National Informatics Centre (NIC) under its DISNIC programme. The scheme is presented in Fig. 7.10.

The most essential pre-requisite for a planning process is an accurate database on natural/physical resource, socio-economic and demographic set-up, and contemporary technology, as shown in Table 7.6.

IRS-1A data interpreted by specialists in respective disciplines facilitates quick assessment and mapping of soils, geology, geomorphology, groundwater and land use/land cover.

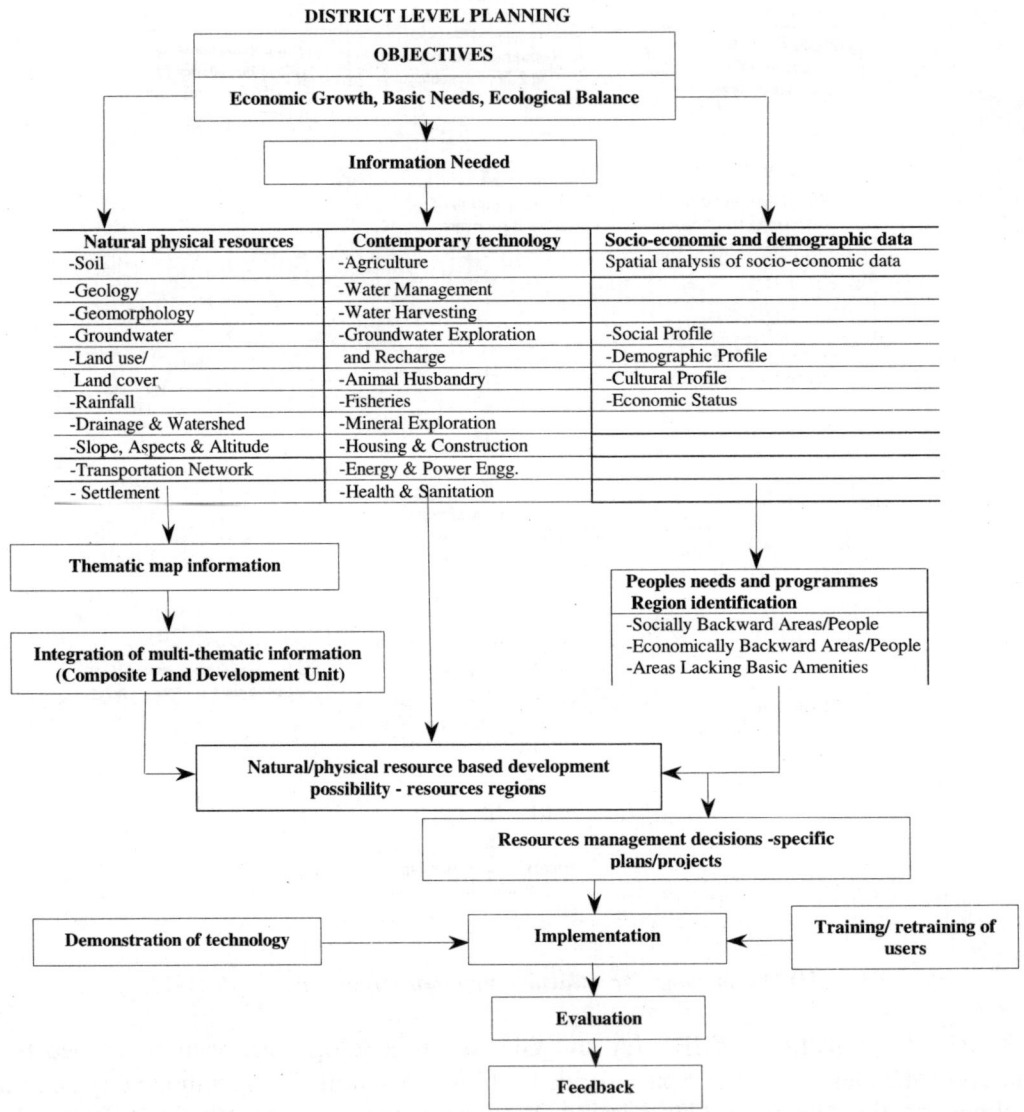

Fig. 7.10 Methodology for integrated district level mapping (*Source: Rao, et. al, 1991*)

Information on terrain conditions, such as slope, altitude, drainage and watershed is derived from the topographical maps, whereas rainfall and other climatic details are collected from meteorological data recording stations. Population census, agricultural census, economic surveys, and the recent efforts of NIC are the main sources for socio-economic and demographic database.

Data analysis and integration is designed at two levels, namely *resource region analysis*, mainly to ascertain resource potential and limitations for the suitability of various resource management activities adopting appropriate on temporary technology, and *programme region analysis*, aimed mainly at the identification of socially and economically backward areas lacking in basic amenities of life.

Table 7.6 Database for district level planning (*Source: Rao et al, 1991*)

Natural/Physical data	Socio-economic and demographic database	Contemporary technology
<ul><li>Soil and Land</li><li>Geomorphology</li><li>Geology</li><li>Hydrogeomorphology</li><li>Land use/Land cover</li><li>Rainfall</li><li>Settlement</li></ul>	<ul><li>Economic services</li><li>Road and Transport</li><li>Communication</li><li>Power and Energy</li><li>Extension services</li><li>Supply of input (seeds fertilizers etc.)</li><li>Storage</li><li>Market</li><li>Cooperative societies</li><li>Health services</li><li>Educational facilities</li><li>Drinking water supply</li><li>Sanitation</li><li>Demography</li><li>Level of income</li><li>Land holding</li></ul>	In the area of agriculture, horticulture, forestry, soil and water management, ground and surface water development, irrigation practices fisheries, animal husbandry, poultry, road construction housing, power and energy conservations, health and sanitation

The resource region analysis is attempted by recording the land characteristics in respect of various resources for each land parcel, refered as '*Composite Land Development Unit*' (CLDU) using GIS approach. Then each of the CLDUs is evaluated in terms of its suitability for a particular use in the context of contemporary technology, irrespective of its present use. A particular CLDU may be suitable for more than one project/scheme implying alternate land use possibilities. These alternatives can be further analyzed in the context of programme regions in order to assess the possible benefits and beneficiaries so that appropriate priorities can be attached to such schemes/projects in accordance with pre-determined goals/objectives of planning.

7.10 DISASTER MANAGEMENT

Disasters are events that cause misfortune, damages, and adverse effects on property and environment of human beings. Disaster is defined as an abnormal condition of the environment which can exert a serious and damaging effect on human, animal and plant life beyond a certain critical level of tolerance (Barrett and Curtis, 1982). However, it is difficult to define the critical level of tolerance since it involves considerable knowledge on various aspects including economic conditions and human psychology. Some of the disasters that occur worldwide on a regular basis are earthquake, floods, cyclones, avalanche, landslide, tsunami, drought, forest fire, etc. The time scale indicating their general nature of occurrence and duration of the disaster is given in Table 7.7. Severe disasters like earthquake, landslide, and tsunami are geological based events.

Table 7.7 Duration of disasters (*Source: Rao et al, 1983*)

Disaster	Time scale	Duration
Earthquake	Very sudden	Seconds to minutes
Avalanche	Sudden	Hours
Landslide	Sudden	Hours to days
Tsunami	Sudden	Minutes
Cyclones	Not sudden	Days
Floods	Not sudden	Days to weeks
Forest fire	Not sudden	Hours to days
Drought	Very slow	Months

Satellite technology can help in the disaster preparedness by providing repetitive and synoptic up to date information on the locally available resources and by facilitating the forecast of the event in time so that alternative arrangement could be provided. Disaster prevention measures can be improved through satellite technology in three ways:

(*i*) mapping the disaster prone areas,
(*ii*) prediction/forecasting of impending disasters, and
(*iii*) monitoring the phenomena to predict their onset and progress.

Satellite data can help in disaster relief operations by providing the information on the extent of areas affected, magnitude of the damages, and the needs of the local population. For an effective disaster management, time is a crucial factor, and hence information on a near real time should be available. Many a times low spatial but high temporal resolution data are valuable in certain phenomenon like drought and floods. Geostationary satellite data is capable of providing information every half an hour and are useful in monitoring short term disasters like cyclones and tornadoes. It is felt that a combination of high spatial, temporal, and spectral resolution data would certainly be beneficial in disaster management. Here one example of disaster management using satellite data related to tsunami has been discussed below.

A tsunami (pronounced soo-nahm'ee) is a series of waves generated by the sudden movement or disturbance of the seafloor. Tsunamis are fast-moving, low, and long waves that radiate out from a triggering event, such as an undersea earthquake, volcanic eruption, landslide, or asteroid impact. When these waves reach shallow water, they change becoming towering, powerful walls of water that slam ashore. Although tsunamis travel as low, very fast waves, often over 500 mph, but when they reach shallow water, they slow down and bunch up. It is here that the true danger lies. Sometimes the water along the beach recedes right before a tsunami hits. Seconds later, one or more towering walls of water crashes on the shore, and not only are the direct hit of the waves extremely dangerous, but also the currents generated by the retreating water as the water flows swiftly back out to the sea.

A severe earthquake measuring 8.9 Ritchter scale, was felt in Andaman and Nicobar Islands having epicentre at 3.7°N and 95°E off the island of Sumatra between 0630 to 0635 hrs (IST) in the morning on 26th December, 2004 (Fig 7.11). The severe earthquake was followed by high tidal waves which caused extensive damages. This tsunami caused more casualties than any other in recorded history. In total, more than 220,272 people were killed, 22,352 have been listed as missing and 1,076,350 were displaced in South Asia and East Africa. At least 173,981 people were killed by the earthquake and tsunami in Indonesia. Tsunamis killed at least 29,854 people in Sri Lanka, 10,749 in India, 5,313 in Thailand, 150 in Somalia, 82 in Maldives, 68 in Malaysia, 59 in Myanmar, 10 in Tanzania, 3 in Seychelles, 2 in Bangladesh and 1 in Kenya. Tsunamis caused damage in Madagascar and Mauritius, and also occurred in Mozambique, South Africa, Australia and Antarctica. The tsunami crossed into the Pacific and Atlantic Oceans and was recorded in New Zealand and along the west and east coasts of South and North America.

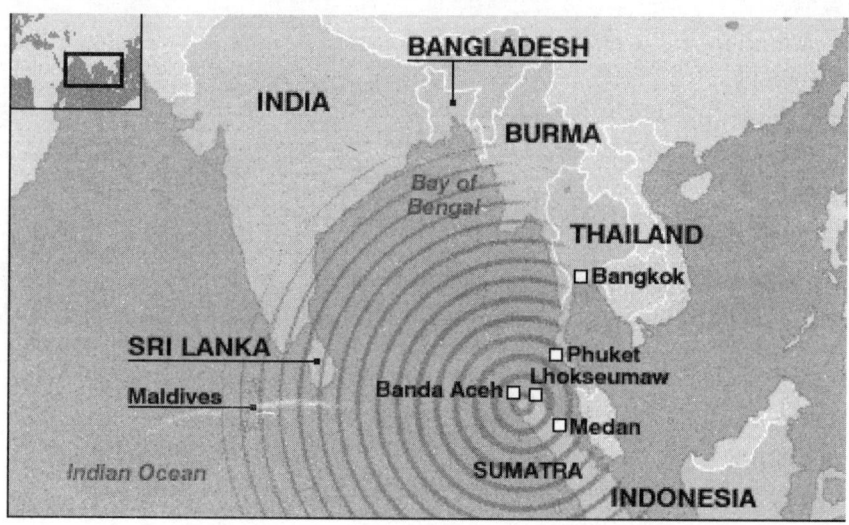

Fig 7.11 Locational map of Asian Tsunami of Dec 26, 2004

Satellite images have been found to provide excellent information regarding the extent of damage caused by tsunami waves. This information tends to vary with resolution and sensor type. Plates 7.8 shows the extent of damage, as marked out by yellow circles, around the city of Karaikal as seen on LISS IV AWiFS image. Similarly, the damage to Chennai city can be seen on IRS-P6 Mx image (Plate 7.9). It can be seen that the mouth of River Adayar has undergone drastic changes. When the same area is seen on OrbView-3 image having a resolution of 1m, the extent of damage can be seen more clearly (Plate 7.10). In this image, the recent changes to river channel configuration and also the depositional pattern of bed material can be seen

7.11 CONCLUDING REMARKS

The above studies highlight the use of satellite data for determining different types of landcover/landuse, natural resources assessment, planning of infrastructure, disaster management etc. This information can now be utilized more effectively if interrelationship between other aspects could be linked so that a better understanding could be reflected during planning and decision making. This is possible through Geographical Information System (GIS) discussed in the following chapters.

Application of Remote Sensing 139

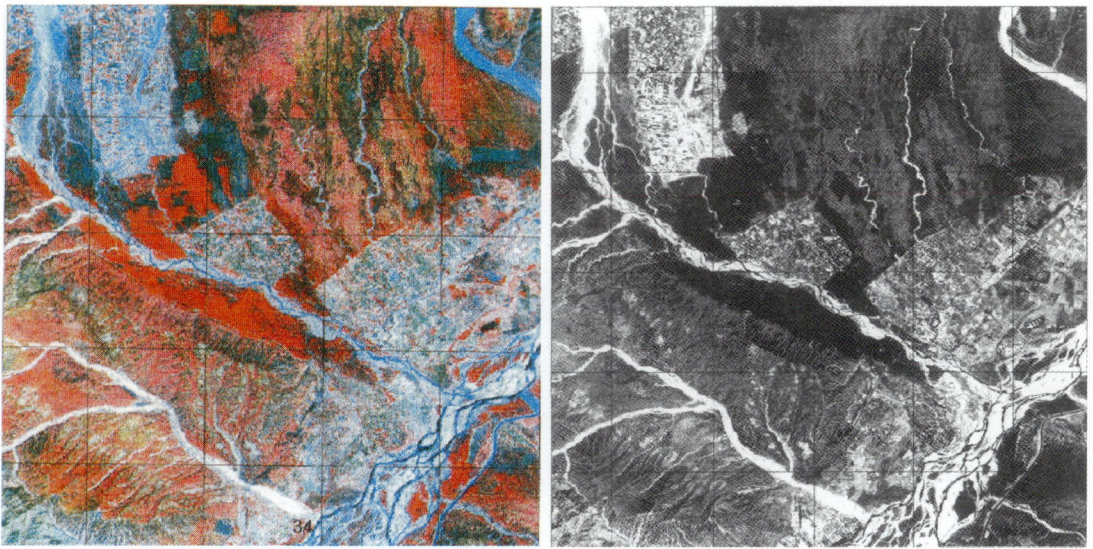

Plate 7.1 IRS-1C LISS III FCC   **Plate 7.2 IRS-1C PAN image**

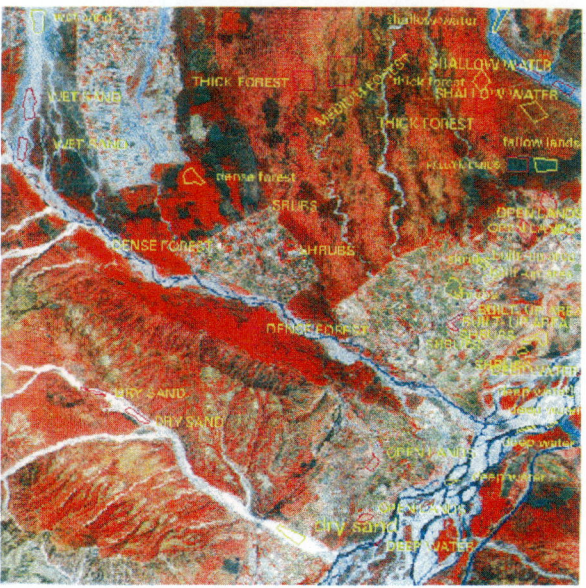

Plate 7.3 FCC image with training plots (*Source: Reddy, 2003*)

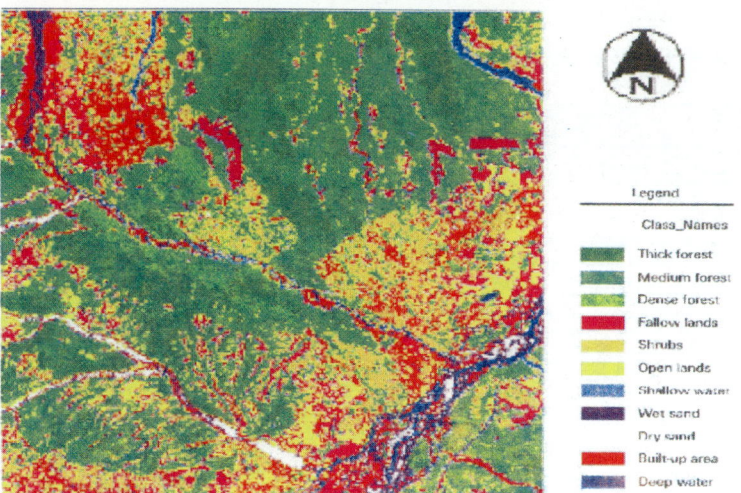

Plate 7.4 Classified image using minimum distance to mean classifier (*Source: Reddy, 2003*)

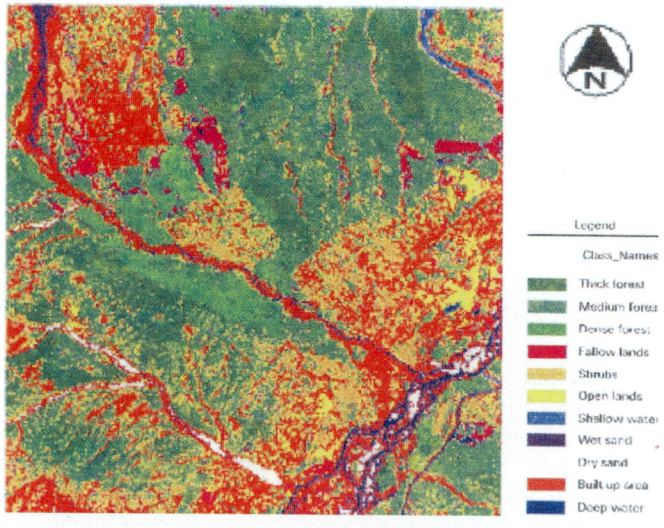

Plate 7.5 Classified image using maximum likelihood classifier (*Source: Reddy, 2003*)

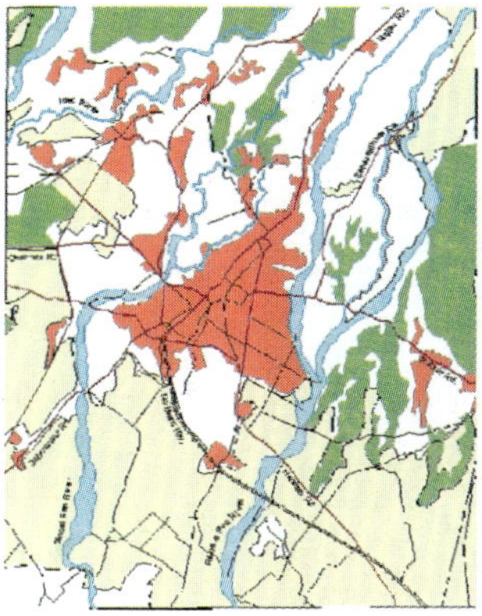

Application of Remote Sensing 141

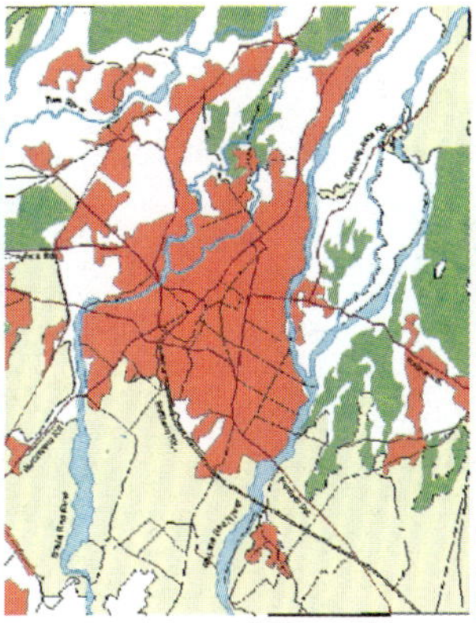

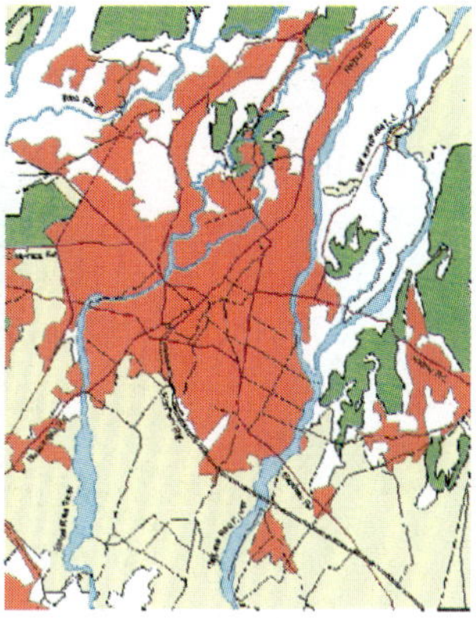

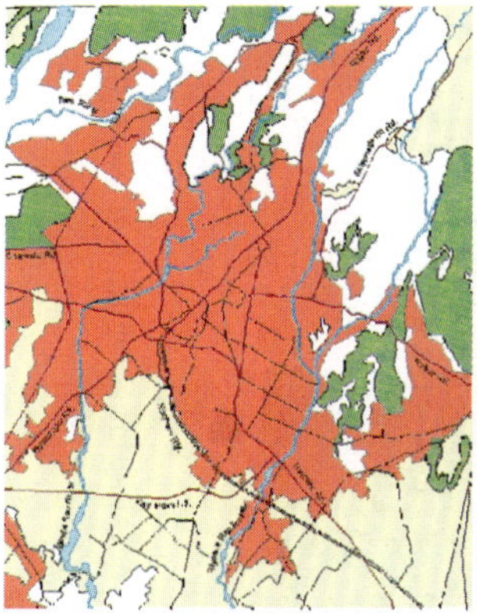

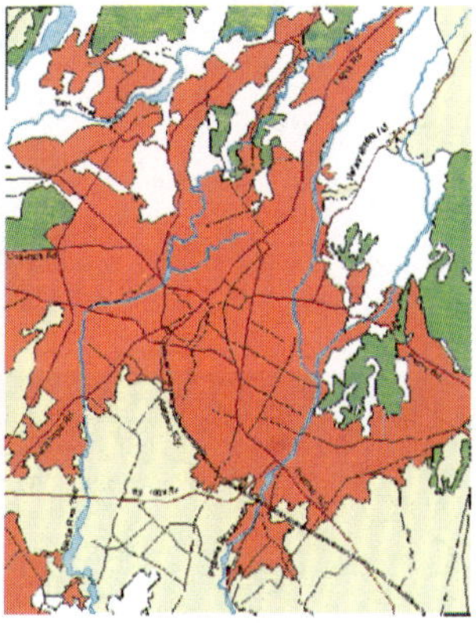

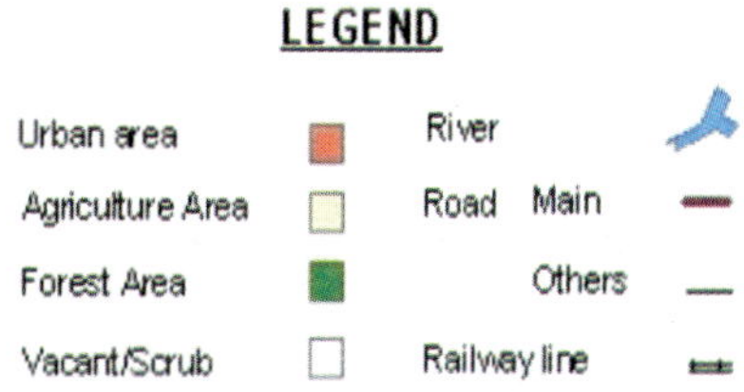

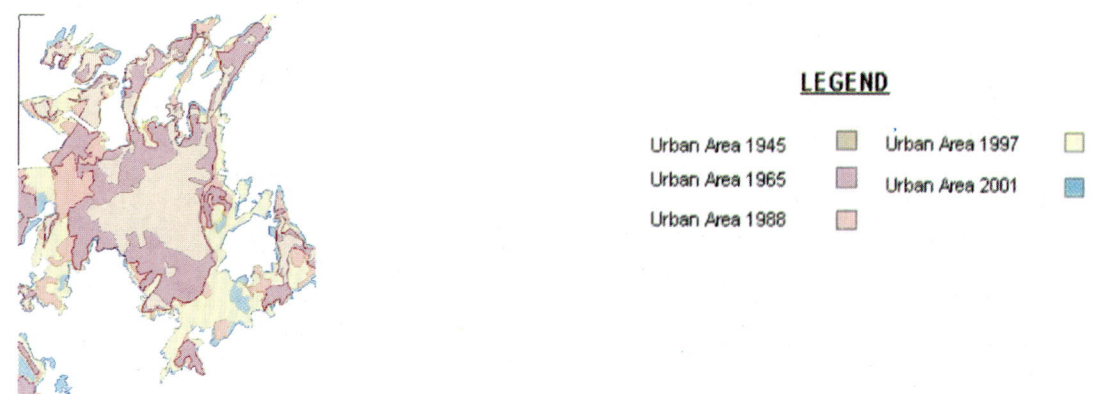

Plate 7.6 Land use of Dehradun (*Source: Soni, 2003*)

Plate 7.7 Urban expansion of Dehradun (1945-2001) (*Source: Soni, 2003*)

142 Remote Sensing and Geographical Information System

Plate 7.8 Tsunami damage along Karaikal coast, India

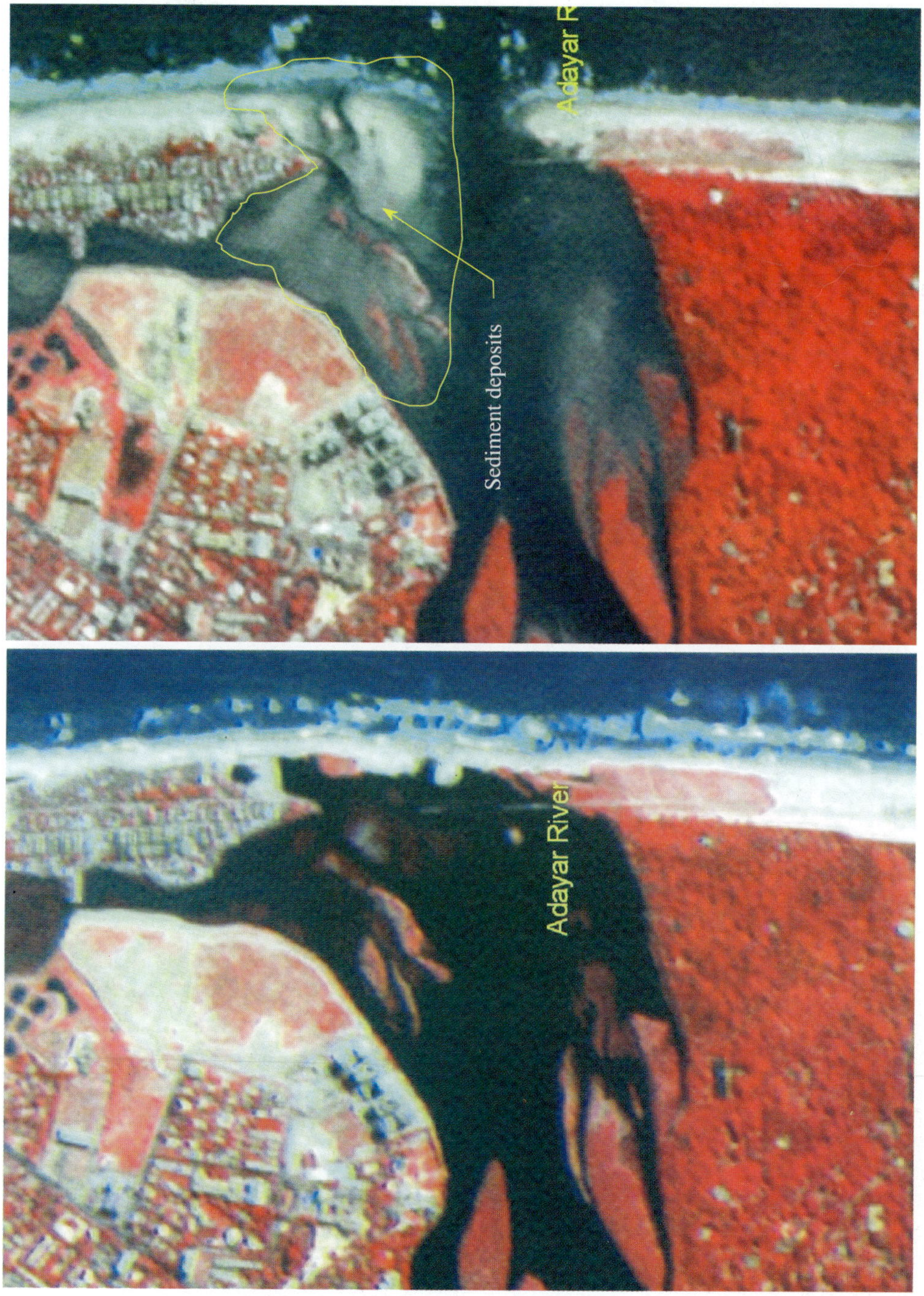

Application of Remote Sensing 143

Plate 7.9 Tsunami damage near Marina Beach, Chennai, India

Plate 7.10 Tsunami damage near Marina Beach, Chennai, India as observed by OrbView-3 satellite (*Courtesy: ORBIMAGE*)

GEOGRAPHIC INFORMATION SYSTEM 8

8.1 INTRODUCTION

The collection of data about the spatial distribution of significant properties of the Earth's surface in the form of maps by navigators, geographers, and surveyors has long been an important part of activities of organized society. Whereas the topographical maps can be regarded as general purpose maps, the thematic maps for assessment and understanding of natural resources are for specific purposes. The use of aerial photography and remote sensing has made it possible to map large areas with greater accuracy for producing thematic maps of large areas, for resource exploitation and management. Handling of large volume of data for quantitative spatial variation of data requires appropriate tool to process the spatial data using statistical methods and time series analysis.

With the introduction of computer assisted cartography, many new tools were developed to perform spatial analysis of the data and to produce maps in desired formats. These operations required a powerful set of tools for collecting, storing, retrieving, transforming, and displaying spatial data from the real world for a particular set of purposes. This set of tools constitutes a *Geographic Information System* (GIS). A Geographic Information System should be thought of as being much more than means of coding, storing, and retrieving the data about the aspects of earth's surface, because these data can be accessed, transformed, and manipulated interactively for studying environmental process, analyzing the results for trends, or anticipating the possible results of planning decisions.

8.2 DEFINITION OF GIS

Geographical Information System is associated with basic terms, Geography and Information system. The literal interpretation of geography is 'writing about the Earth'. In writing about the Earth, geographers deal with the spatial relationship of land with man. A key tool in studying the spatial relationships is the map which is a graphical portrayal of spatial relationships and phenomena over a small segment of the Earth or the entire Earth. On the other hand, an information system is a chain of operations that consists of from planning the observation to using the observation-derived information in some decision making process.

A GIS is an information system that is designed to work with data referenced by spatial or geographical coordinates. In other words, a GIS is both a database system with specific capabilities for spatially-referenced data as well as a set of operations for working with the data as shown in Fig. 8.1.

Some of the definitions of GIS given in different publications are

(*i*) "A system which uses a spatial database to provide answers to queries of a geographical nature" (Goodchild, 1991).

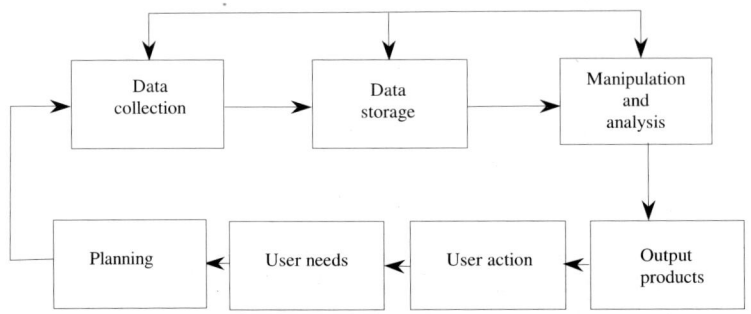

Fig. 8. 1 Simplified information system

  (*ii*) "A computer assisted system for the capture, storage, retrieval, analysis, and display of spatial data within a particular organization" (Clark, 2001).
  (*iii*) "A powerful set of tools for collecting storing, retrieving at will, and displaying spatial data from the real world" (Burrough, *et al*, 2000).
  (*iv*) "An organized collection of computer hardware, software, geographical data, and personnel designed to efficiently capture, store, update, manipulate, analyze, and display all forms of geographically referenced information" (ESRI).

The last definition given above is one of the most rigorous definitions of GIS. This definition includes requirement of personnel trained in the technology who can capture, store and update the data, and provide answers to the complex queries of the management by integrating information contained in various layers, through maps, tables, and charts.

A GIS is also result of linking parallel developments in many separate spatial data processing as shown in Fig. 8.2.

8.3 COMPONENTS OF GIS

The components of GIS can be defined in various ways, but very comprehensively, it can have the following components:

  (*i*) Computer system (hardware and operating system),
  (*ii*) Software,
  (*iii*) Spatial data,
  (*iv*) Data management and analysis procedures, and
  (*v*) Personnel to operate the GIS.

Hardware Components and Operating System
The hardware components of a GIS comprise of a Central Processing Unit (C.P.U.), disk drive, tape drive, digitizer, plotter, and visual display unit (V.D.U.) (Fig. 8.3). The disk drive and tape drive are basically data storage devices. The tape can be used for communicating with other systems. A digitizer is an input device to convert graphics into digital data, where as the plotter, an output device, converts the digital data into the graphical form. The scanner

is a graphic input device. The Visual Display Unit along with a keyboard or mouse is required to interact with the computer. The printer is required to get hardcopy of the reports, tables, charts, etc. The Central Processing Unit of a computer interacts with various hardware components, and performs computations and analysis.

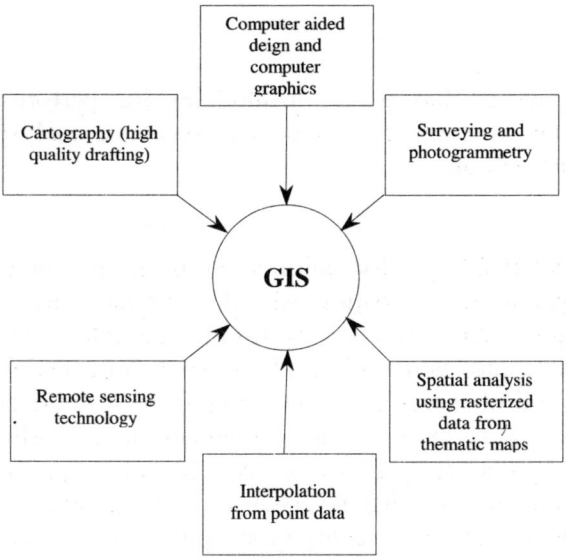

Fig. 8.2 Linking of efforts in several initially separate but closely related fields though GIS

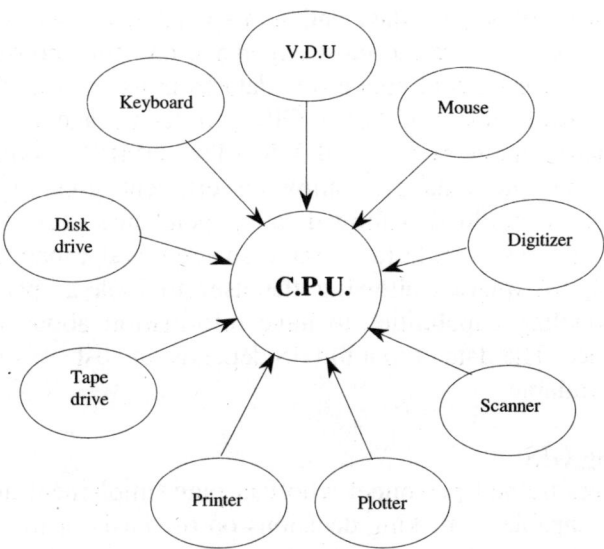

Fig. 8.3 Major hardware components of a GIS

An operating system of a computer is a kind of master control program to manage files, to check the various peripheral devices, and to help have a control over the computer. The operating system programs translate the instructions or queries to the machine language, and then translate the computer response from machine language to any language in which the operator works.

Software
The GIS software package has a set of modules for performing digitization, editing, overlaying, networking, vectorising, and data conversion, analysis and for answering the queries, and generating output.

Spatial Data
Spatial data are characterized by information about position, connections with other features and details of non-spatial characteristics. All GIS softwares are designed to handle spatial data. Spatial data require spatial referencing using a suitable geographic referencing system which should be flexible and lasting, since a GIS may be intended to last many years.

A traditional method of representing the geographic space occupied by spatial data in a GIS environment is in the form of a series of thematic layers. Alternatively, a space can be viewed as populated by discrete points or objects having empty spaces in between objects. The spatial data represented as either layers or objects are simplified by breaking down all geographic features with three basic entity types, points, lines and areas, before they can be stored in the computer.

Data Management and Analysis Procedures
Input data in the forms of spatial data and non-spatial data, and information about their linkages, and updating of data are the most expensive and time-consuming part of any GIS project. Data input is the process of converting data from its existing form to one that can be used by the GIS. The management of data in GIS includes storage, organization, and retrieval using a database management system (DBMS). The DMBS should provide support for multiple users and multiple databases allowing efficient updating and minimizing the redundant information. It should also allow data independence, security, and integrity.

GIS analysis procedures include (*a*) storage and retrieval capabilities for presenting the required information, (*b*) queries allowing the user to look at patterns in the data, and (*c*) prediction or modeling capabilities to have information about what data might be at different time and place. The data output in GIS depends on cost constraints, the type of users, and output devices available.

Personnel Operating GIS
A GIS project requires trained personnel who can plan, implement and operate the system. They should also be capable of making decisions on the basis of the output. The success of any GIS project depends upon the skill and training of the personnel handling the project.

8.4 GEOGRAPHICAL CONCEPTS

In the previous section, it has been said that the geographic features can be represented by three basic entity types, points, lines, and areas. In this section these entities have been defined in a graphical context and also some other terms frequently used. A spatial object represents a geographical area having a number of different kinds of associated attributes or characteristics. A spatial object with no area is a *point* that can be associated with a range of data, such as wells, rain gauge stations. One of the key attributes of a point is its geographical location represented in terms of coordinates, such as latitude and longitude.

When a spatial object is made up of a connected sequence of points, it is referred to as a *line*. Lines have only linear dimension, *i.e.*, they do no have width, and a specified location is given on one side of the line and not on the line itself. Attributes to a line could be the number of the wells that the line separates in an area having wells. *Nodes* are defined as the special kinds of points that usually indicate the junction between lines or the ends of line segments. A closed area is represented by a *polygon*. A polygon can be simple when it consists of undivided areas or complex when it is divided into areas of different characteristics. *Chains* are special kind of line segments which correspond to a portion of the bounding edge of a polygon.

In the context of spatial objects, the concepts of scale and resolution must also be clearly understood. *Scale* is the ratio of distances represented on a map or photograph to their true distances on the Earth's surface. A scale of 1:50,000 indicate that one unit of distance on a map is equal to 50,000 of the same unit, on the ground. A map may be a small-scale map or large-scale map. When one unit of distance on a map represents small distance compared to a larger distance on another map, the former map is a *large-scale map* compared to the later which is a *small-scale map*.

Resolution is an important concept when dealing with spatial data. Dictionary meaning of resolution is 'distinguishing the individual parts of an object', or 'the degree to which detail is visible in a photograph or on a television'. For the purpose of its application in spatial data, a more specific definition given by Tobler (1987) is "the content of the geometric domain divided by the number of observations, normalized by the spatial dimension". For a two-dimensional data set, such as maps and photographs, the area covered by the observations is referred to as *domain*. To determine the mean resolution element of a map the following formula may be used.

$$\text{Mean resolution element} = \sqrt{\left(\frac{\text{Area}}{\text{Number of observations}}\right)} \qquad \ldots (8.1)$$

For example, if an area of 4.8×10^6 km^2 consists of 30 states in a country, then the mean resolution element = $\sqrt{\frac{(4.8 \times 10^6)}{30}}$ = 400 km.

Smaller is the mean resolution element; higher is the resolution of dataset.

8.5 INPUT DATA FOR GIS

Input data for GIS cover all aspects of capturing spatial data and the attribute data. The sources of spatial data are existing maps, aerial photographs, satellite imageries, field observations, and other sources (Fig. 8.4). The spatial data not in digital form are converted into standard digital form using digitizer or scanner for use in GIS.

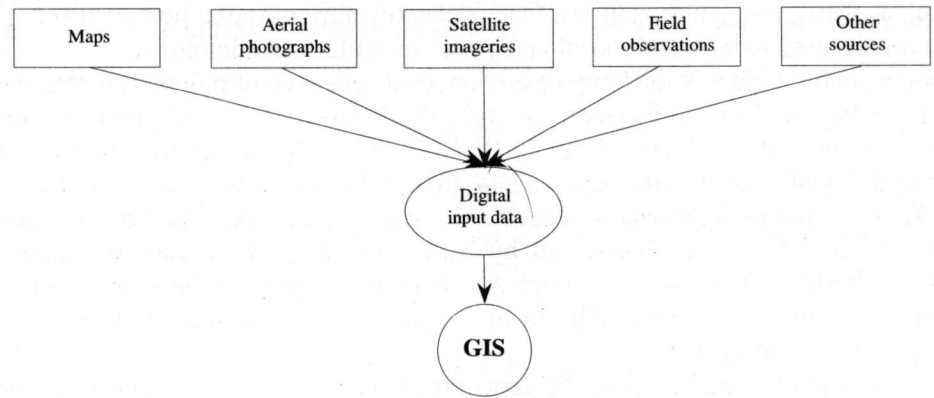

Fig. 8.4 Input data for GIS

The digital spatial data in an acceptable format and the attribute data are stored in the computer memory and managed by DBMS which is a part of GIS, for analysis and producing the results in users-desired formats.

8.6 TYPES OF OUTPUT PRODUCTS

GIS input data are collected from different sources in different formats and, their organization in digital form in desired structure for use in GIS requires a thorough understanding and the same has been discussed in Chapter 9. The output products can be of various kinds, and since these products are computer generated, the user/analyst should be aware of the desired forms of output options available in GIS software. A brief discussion on various kinds of output products of GIS is given below.

The most common graphics products produced by GIS are maps. A map is a two-dimensional model of a part of the Earth's surface, and it can be of various kinds. The common types of thematic maps are as below.

Thematic maps concentrate on spatial variations of a single phenomenon (*e.g.*, population) or the relationship between phenomena (*e.g.*, different classes of land cover).
Choropleth maps are typically used to communicate the relative magnitudes of continuous variables as they occur within the boundaries of unit areas (*e.g.*, average annual per capita income as it varies by country). In these maps, different tones, colours, and shading patterns are used to convey the variations in different areas.

Proximal or dasymetric maps focus on the location and magnitude of areas exhibiting relative uniformity (*e.g.*, land cover classes). Different colours and shading patterns are used to describe differences in the thematic values.

Contour maps represent quantities by lines of equal value to emphasize gradients among the values. Contour lines may be used to indicate variation in topography of a region, high and low pressure regions.

The other kinds of map required for a specific purpose are as under.

Dot maps depict spatial distribution of features by varying numbers of uniform dots (*e.g.*, population)

Line maps show the direction and magnitude of potential or actual flow (*e.g.*, to show sources and destinations as well as the volume of product transported from one state to others).

Land form maps depict the earth's surface as it were viewed from an oblique aerial point view.

Animated maps are generally used to display sequences through time (*e.g.*, growth of a city as its population and area increase through time).

Some users/analysts prefer to get the results of analysis displayed by means of non-map graphics. Some of the simple and common graphic presentation techniques are

Bar charts used to illustrate difference in an attribute between categories (*e.g.*, time-varying distribution of land use in an area such as urban, suburban, and rural).

Pie charts for displaying information by dividing a circle into sectors representing proportions of the whole (*e.g.*, in a state percentage of rural, suburban, and urban population).

Scatter plots for displaying behaviour of one attribute verses another attribute (*e.g.*, yield and applied fertilizer).

Histograms to show the distribution of a single attribute to examine the way the attribute is apportioned among the different possible values (*e.g.*, percentage of education at primary, secondary, higher, and other levels).

Sometimes users may require numerical products, such as mean, median, variance, standard deviation, maximum and minimum values of attributes. This statistical information may be presented in the form of tables and reports.

8.7 APPLICATIONS OF GIS

GIS finds its application in all those areas where professionals are involved in management and planning utilizing analysis of large amount of geographical data that relates to space, typically involving *positional data*. Positional data determine where things are or perhaps, where they were or will be. In other words, it is dealing with questions related to *geographic space*. Some of the typical applications may be as under.

(*i*) A forest manager may want to optimize production of timber using data on soil and current tree stand distribution under a number of operational constraints.

(ii) A geologist may be interested in identifying the best dam site by analyzing the earthquake data of the area, and rock formation characteristics.

(iii) A geoinformatics engineer may want to determine the best sites for his company's relay stations taking into account the land prices, topography, etc.

The areas of GIS applications are unlimited as it can be used for management and planning that may be required in any field, *e.g.*, civil engineering, urban planning, forestry, environmental management, flood control, natural disaster management, natural resources management, military, biology, geology, mining, hydrology, etc.

GIS is used to study the phenomena which are man-made or natural. Urban planning essentially requires a thorough understanding of the interrelationships between various entities such as road connectivity, schools, hospitals, police stations, etc., with respect to human settlements. On the other hand, in geomorphology, ecology, and soil studies, natural phenomena, such as rock formation, plate tectonics, distribution of soils, are analyzed. GIS applications also consist of study of both natural as well as man-made entities together. The study of effect of human activity on the environment, referred to as *environmental impact assessment* involves analysis of data about both natural and man-made features. A study on the growing industrial units in an area is a problem of environmental impact assessment.

GIS has distinct application in feasibility studies such as site suitability and simulation studies in erosion modeling.

GIS DATA 9

9.1 INTRODUCTION

Geographic data in digital form are numerical representations of the real world. It describes real-world features and phenomena coded in specific ways in support of GIS and mapping applications using the computer. The digital geographic data must be organized as a geographic database. Roughly two-thirds of the total cost of implementing a GIS involves building the GIS database which should be accurate and has a significant impact on the usefulness of the GIS.

9.2 GIS DATA TYPES

Geographic data consists of *spatial data* and *non-spatial* data. The spatial data give information about the geometrical orientation, shape and size of a feature, and its relative position with respect to the position of other features. Spatial data is described by its x and y coordinates. The non-spatial data, also known as *attribute data*, are information about various attributes like length, area, population, acreage, etc. Normally the spatial and non-spatial data are stored separately in a GIS, and links are established between the two at the time of processing and analysis.

The spatial data is normally available in analog form as maps but now the maps are also available directly in digital format. In GIS, both types of the spatial data are handled differently.

The non-spatial data describe the attributes of a point, along a line, or in a polygon. In other words they describe what is at a point (*e.g.*, a hospital), along a line (*e.g.*, a canal), or in a polygon (*e.g.*, a forest). The attributes of a soil category may be depth of soil, texture, type of erosion, or permeability. The non-spatial data, mostly available in tabular form, are also converted into digital format for use in GIS.

9.3 DATA REPRESENTATION

The data representation is in different kinds of variables, also known as scales that can be stored in a GIS. These variables are

  (*i*) Nominal,
 (*ii*) Ordinal,
(*iii*) Interval, and
(*iv*) Ratio.

Nominal Variables
Nominal Variables are used when the data are principally classified into mutually exclusive sets or levels based on relevant characteristics. The nominal variable is the commonly used measure for spatial data. It can be two types as below.

Dichotomous (Presence or absence) data are mainly logical definition of a data characteristic, and are also referred to as *Yes/No* data. It mainly applies where a particular data is to be classified into one of the two categories. For example, a village may or may not have hospital; a city may or may not have an airport.

Categorical data are used when it is required to classify the data into one of several categories by name with no specific order. Categories of land use such as residential area, recreational area, business areas, or trees such as Quercus agrifolia, Pinus Coulteri, Eucalyptas calophylla, are different kinds of categorical variables.

Ordinal Variables
Ordinal Variables are lists of discrete classes but with an inherent order or sequence. This representation of data is more sophisticated and orderly as the classes are placed into some form of rank order based on a logical property of magnitude. The ranking of data may be natural such as grades of agricultural land, or according to some criteria, such as population density. In general, class of streams may be first order, second order, and so forth, levels of education may be primary, secondary, college, post-graduate, are ordinal variables since the discrete classes have a natural sequence.

Internal Variables
Internal variables also have a natural sequence, but in addition, the differences between the values are quantified. For example, the elevation of points is an internal variable since the difference in elevation between two points having elevations 55 m and 65 m is the same as for other points having elevations 80 m and 90 m. The representation of population in same order is an example of internal variables.

Ratio Variables
Ratio variables have the same characteristic as internal variables, but in addition, they have natural zero or real origin (*i.e.*, starting point). Per capita income, the fraction of the weight of a soil sample that passes through sieve, are common ratio variables.

9.4 DATA SOURCES

The data for GIS collected from different sources as shown in Fig. 8.4, are discussed below.

Satellite Imagery
Remote sensing data in the form of satellite imagery is an important element of the organization of any GIS database as it makes possible repetitive coverage of large areas. Satellite imagery can be used as a raster backdrop on vector GIS data. Satellite images can

support numerous GIS applications including environmental impact analysis, site evaluation for large facilities, highway planning, development and monitoring of environmental baselines, emergency and disaster response, agriculture, and forestry. Satellite images are also useful for urban planning and management.

In addition to image analysis, satellite images are used to generate thematic information resulting into thematic maps.

Existing Maps

Paper maps are the most important source of data for GIS. Maps of various scales, sizes, formats, and time periods showing different features are available for large portion of the Earth, and these are major sources of data for the GIS database. The information available on a paper map is converted into digital form by the process of *digitization* for use in GIS. The advanced countries like U.S.A. also have the digital maps, which can directly be used in GIS without going into the process of digitization.

Aerial Photographs and Digital Orthophotographs

Another major source of data for a GIS application is the aerial photographs. Aerial photographs rectified for relief displacement or radial distortions are known as *orthophotos*. An orthophoto is geometrically equivalent to a conventional line map, and represents planimetric features on the ground in their true orthographic positions. Due to this, orthophotos possess the advantage of line maps, such as, ability to make measurements of distances, angles, and areas. However, orthophotos unlike line maps also contain the images of an infinite number of ground objects, and therefore, most of the time they need conversion into theme maps.

At present with the given computer power, its storage capacity, and speed, it has become possible to have digital orthophotos commercially. The digital orthophotos provide all information of a photograph, but at the same time allow the registration of vector maps used in GIS.

Attribute Data

Attribute data for a GIS are mainly tabular data collected by sampling. The tabular data which are tables consisting of rows representing samples and columns representing parameter values can be incorporated into GIS as rational tables.

Survey Data and Records

Some survey data and records about rock types, soil types, elevation, population, and other features are collected by the related national agencies of a country and maintained in the form of maps and tables. These data can be incorporated into a GIS.

Other Sources

Conventionally, terrain data can be obtained by field surveying using grid levelling, stadia tachometry or other field surveying methods. These methods have been replaced by the new generation surveying instruments, such as electronic tachometer or total station, and the

Global Positioning System (GPS) for collecting locational as well as attribute data. Another source of GIS data could be the internet/WWW.

Almost all analog or digital data available for use in a GIS may have limitations, and pose problems while organizing the GIS database.

9.5 TYPICAL GIS DATA SETS

By and large, the most common application of GIS is the effective management of natural resources and planning of regions at different levels. For such an activity a variety of data sets are required. These data sets can be broadly grouped as below.

Natural resource data consists of land use, crop type, cropping area, water bodies and drainage, soil types, forest types, groundwater potentials, mineral resources.
Demographic data consists of data related to population, age structure, sex ratio, urban and rural population, reserved caste population, occupational structure, and migration patterns.
Agro-economic data consists of information about cropped and irrigated area, agricultural production, land holdings, livestock population, livestock produce, market and pricing information.
Socio-economic data pertains to activities related to industrial, fishing, tourism development as well as beneficiaries of various schemes and programmes of development.
Infrastructure data consists of information related to various facilities, utilities and services, such as education, health, power, transport network, water supply, communication, general amenities, and drainage.

9.6 DATA ACQUISITION

Data acquisition in GIS refers to all aspects of collecting spatial data from all available sources as discussed in Sec. 9.5, and converting them to a standard digital form. This requires tools such as interactive computer screen and mouse, digitizer, word processors and spreadsheet programs, scanners (in satellites or aircrafts for direct recording of data or for converting maps and photographic images), and devices necessary for reading data already written on magnetic media as tapes or CD-ROMs.

Digital terrain data may be acquired by a variety of methods, depending on factors such as the location and the size of the area of interest, the purpose of terrain modeling, and the technical resources available. The ground survey methods are generally employed for large-scale terrain modeling for site planning and design. At smaller scales covering larger geographic areas, aerial photographs or satellite images are most suited.

Terrain Data from Satellite Remote Sensing
The terrain data acquired through sensors onboard satellite platforms being in digital format can be directly used after preprocessing for preparing a GIS database. These data are coded in picture elements called *pixels*, and stored in the form of a two-dimensional matrix that contains merely a number representing the amount of the reflected electromagnetic radiation

received in a given band. The digital images must be located properly with respect to a geodetic grid, otherwise the data they contain, cannot be related to their true ground position.

Terrain Data from Existing Maps

The acquisition of digital data by digitizing existing maps is comparatively cheaper, and requires less time compared to other methods. The topographic maps covering large part of a country are mostly available, and the required information for a particular job can be acquired from these maps by digitization. The elevation data extracted using contours have poor accuracy as compared to spot heights.

The digitized counters cannot be used for any useful GIS application other than regenerating the original contours themselves. For the digitized contour data to be used for digital terrain modeling it is necessary to carry out a series of post-digitized processes where by sampled points are turned into a *triangulated irregular network* (TIN) or a digital elevation model (DEM).

The digitization of paper maps is done using a spatial data capturing device called a *digitizer*. In recent years, the importance of the digitizer has diminished as a result of the growing use of scanners and the increasing availability of digital data from government agencies and commercial data suppliers.

Digitizers are available in different sizes with different resolutions (generally 0.025 mm). In addition to the resolution, digitizers are required to have a number of technical requirements such as stability, repeatability, skew, and accuracy.

Map data conversion by scanning and vectorization is often referred to as *screen digitizing* or *heads-up* digitizing to distinguish it from conventional map data conversion using digitizer. This approach of digital data conversion is capable of converting a large number of maps in a relatively short period of time and at a cost comparable to or lower than the conventional method of map digitizing. Scanning of map requires a considerable amount work to prepare the map for scanning, such as separating the captured map data into different layers, touching up thin line work, as well as closing gaps in line objects. After scanning, the raster image is converted into vector graphics to build layer topology (which refers to the spatial relationships of adjacency, connectivity, and containment among topographic features), and to link the resulting vector graphical elements to their associated attributes. For use in GIS data conversion, scanner has to meet a number of technical specifications such as resolution, accuracy, scan width and length, output file format, and quality of the accompanying software.

Terrain Data Collection by Photogrammetry

When the area of interest is too extensive or too rugged, the photogrammetric method is employed to collect the digital terrain data using appropriate photogrammetric instruments such as analog stereoplotter equipped with encoders, analytical plotter, or using digital photogrammetry methods.

Analog stereoplotter provide three-dimensional data from aerial photographs. The digitization of 3D models of the terrain formed by analog stereoplotter is done by equipping the stereoplotter with linear and rotary encoders. A variety of digitizing devices may be attached to the stereoplotter to convert the electronic signals from encoders into numerical coordinate values. The digitizing unit may be connected either to a dedicated microcomputer or to a mainframe. This method of photogrammetric digitization is largely manual where the operator has complete control over the selection of elevation points.

In analytical plotter, x, y, and z coordinates of points measured by operator are used to compute the corresponding photographic coordinates of the same points on each stereopairs of photographs in real time by the main processor of the stereoplotter, and then to move the positions of the photographs accordingly, so that the measured points can be viewed stereoscopically. The movement of the measuring point (or floating mark) of the analytical plotter may be programmed for easy implementation of predetermined sampling strategy for digital terrain data. Instead of sampling at regular grid intervals or along regularly spaced profiles, the density of the sampling points may be programmed to vary in different parts of the stereomodel according to the nature of the terrain.

Digital photogrammetry uses digital images instead of photographic diapositives. The digital images may be obtained either by scanning aerial photographs at extremely high resolution or by the use of remote sensing imaging systems. A digital photogrammetric workstation (DPWS), which produces digital images, includes the hardware and software, and peripherals, such as digital cameras, film scanners, plotters. A DPWS has capability of achieving accurate interactive stereo measurement, and it is equipped with a suite of computer programs that enables the DPWS to perform substantial part of photogrammetric operations. A DPWS also allows the operator to visualize, verify, and edit the data collected interactively. The digital photogrammetric technology has made it possible to use high-resolution stereoscopic satellite imagery for generating digital terrain data.

Terrain Data from Field Surveying Methods

Terrain data in digital form can be obtained directly by field surveying methods by employing instruments such as electronic tachometer or total station, and GPS.

Electronic tachometer or total station is capable of electronically measuring both angles and distances, and performing computation to obtain horizontal distance, slope distance, difference in elevation, coordinates, and elevation of points. These instruments are equipped with internal memory or external data recorder for temporary storage of data, which subsequently are transferred to dedicated microcomputer or mainframe computer.

Digital Terrain Data by GPS

GPS is a satellite-based surveying system to obtain highly accurate digital terrain data electronically in the form of x, y, and z coordinates. There are two basic field methods of GPS measurements: *static* and *differential*. Static GPS surveying is used to determine positions of survey control points in areas where geodetic control is lacking or unreliable. It is also used for accurate measurement of distance between two points using two or more dual-frequency GPS receivers to record the observations made on GPS satellites simultaneously for about six hours. Static GPS surveying is mainly employed for establishing geodetic control and measuring national and international networks, and not intended for ordinary detail terrain data acquisition. Differential GPS surveying is used to determine the positions and heights of ground points by making use of existing or newly established control points. The differential GPS surveying may be performed as a *kinematic GPS surveying* or *real-time GPS surveying*.

GPS is a very productive tool for acquiring digital terrain data in field where the necessary control is available and sky is visible.

Digital Terrain Data from Internet/World Wide Web
The Internet is a vast network of digital computers. They are linked by an array of different data transmission media such a satellite and radio links, and fiber optic, unshielded-twisted pair, co-axial, and telephone communication lines. Connecting these media involves an array of different devices ranging from complicated routers and data switches, through simple signal amplifying hubs to modems. The transfer of data is carried out by using a standard coupling protocol known as TCP/IP (Transmission Control Protocol/Internet Protocol).

The WWW (World Wide Web) is a term used synonymously with the Internet. It is in fact the main information tool of the Internet through which data written on 'web pages', are accessed. These are available to the user world wide through a 'web server', and viewed by users at remote locations using 'web browser' software.

For GIS users the WWW provides data, and is source of information. Whole libraries of vector, raster, and object data are being offered on the Internet as well as directory information on different data sets. At the moment speed of data transmission and Internet access are the main limitations of this data resource.

Attribute Data Tagging
The attribute data that are feature identifiers, feature codes, and contour labels are interactively added on screen to the graphical data after raster-vector conversion.

9.7 DATA VERIFICATION AND EDITING

It is important to check the acquired data for errors due to possible inaccuracies, omissions, and other factors. The errors in the spatial data are, generally checked by printing the data or by taking its computer plot, preferably on translucent or thin paper, at the same scale as the original. The print out or computer plot is placed over the original map. The two maps are compared visually and the discrepancies in the form of missing data, locational errors, and other errors, are clearly marked on the print out. If the map is a unique drawing, locational errors are only considered within the boundary of the map. If the map is one of the series of the maps covering large extent of area, or the digitized data is to be linked up with the map data already in the computer, then the spatial data must also be examined for spatial continuity across the map boundaries. Certain operations, such as polygon formation, may also indicate errors in the spatial data.

Checking of the attribute data is also done by visual inspection of the print out. A better method of checking the attribute data is to scan the data files with a computer program that can locate the gross errors such as text instead of numbers, numbers exceeding a given range, and so on.

Errors may arise during the capturing of spatial and attribute data in the following cases.

Spatial data are incomplete or double: When the data are entered manually, incompleteness in the spatial data may be due to omissions in the input of points, lines, or cells. In case of scanned data, this error of omission is usually in the form of gaps between lines where the raster-vector conversion process fails to join up all parts of a line. The raster-vector conversion of scanned data can lead to the generation of unwanted spikes. Sometimes one line may be digitized twice, and lines and nodes may be disjointed at the intersections.

Spatial data are in the wrong place: Spatial data may have minor placement errors to gross spatial errors due to mislocation of spatial data. Minor placement errors are usually the result of careless digitizing whereas the gross spatial errors are due to change of origin or scale that occurs during digitizing, or as a result of hardware or software faults.

Spatial data are defined using too many coordinate pairs: As a result of both digitizing and scanning process, lines in the database may be defined using too many points resulting into use of large storage space in a computer.

Spatial data are at the wrong scale: The digitization at the wrong scale results in erroneous representation of spatial data. In case of scanned data, the problem usually arises during the georeferencing process using incorrect values.

Spatial data are distorted: The spatial data may be distorted if the base maps used for digitizing are not at the correct scale.

Usually aerial photographs do not have uniform scale over the whole of image because of relief and tilt distortions, and sometimes due to aberration in the lens properties. Paper maps may suffer from *paper stretch*, which is usually greater in one direction than other. In addition, paper maps and field documents may contain random distortions as a result of having been exposed to rain, sunshine, frequent folding, etc. Transformation from one coordinate system to another may also cause error in the spatial data.

These errors are addressed through various editing and updating functions supported by most GIS software. Data editing is done visually by viewing the portion of the map containing the error on the computer monitor, and correcting them through the software by using a keyboard, mouse, or digitizer.

Data scaling problems may be overcome by applying simple numerical factors to the data. More complex rotating and translating operations are needed when fitting various data sets together such as a distorted thematic map to an accurate base map. The faulty map should be corrected with the base map, and a number of points on the original map linked by vectors to their correct position as shown in Fig. 9.1 where T is the transformation vector. Mathematical transformations stretch and compress the original map until the linking vectors have shrunk to zero length and the tie points are registered with each other. It is then assumed that all the other points on the original map have been relocated correctly. This process is known as *rubber sheeting* or *wrapping*. This method cannot be applied directly on rasterized data because of the rigidity of the fixed grid and the structure of the data. Attribute values and spatial errors in raster data are corrected by changing the value of the faulty cells.

GIS spatial data errors are discussed in detail in Sec. 9.9.

9.8 GEOREFERENCING OF GIS DATA

A spatial referencing system is required to handle spatial information. The primary aim of a reference system is to locate a feature on the Earth's surface or a 2D representation of this surface such as a map. A map portrays accurately real-world features that occur on the curved surface of Earth.

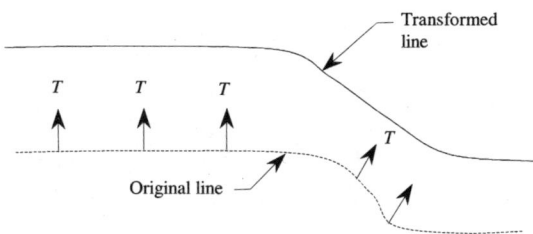

Fig. 9.1 Rubber sheeting

The objective of georeferencing is to provide a rigid spatial framework by which the positions of the real-world features are measured, computed, recorded, and analyzed in terms of length of a line, size of an area, and shape of a feature. Several methods of georeferencing exist, all of which can be grouped into three categories as under:

(*i*) Geographic Coordinate System,
(*ii*) Rectangular Coordinate System, and
(*iii*) Non-Coordinate System.

Geographic Coordinate System
The geographic coordinate system is the only system that defines the true geographical coordinates in terms of *latitude* and *longitude*. In this system of coordinates, the Earth is defined by a reference surface using latitude and longitude. As shown in Fig. 9.2, lines of longitude (also known as meridians) start at one pole and radiate outwards until they converge at the opposite pole, while lines of latitude lie at right angles to lines of longitude and run parallel to one another. The longitude λ of a place P is the angular distance of the meridian passing through P from the standard meridian passing through Greenwich, called *Greenwich meridian* or *Prime meridian*, measured positive in clockwise or anticlockwise direction from Greenwich meridian to 180°, in the equatorial plane. The latitude ϕ of the place P is the angle $P'OP$, O being the centre of the Earth, measured positive towards the North Pole and negative towards the South Pole from the equatorial plane. The value of latitude lies between 0° to ± 90°.

The geographical coordinate system of latitude and longitude assumes that the Earth is perfect sphere, which is far away from the reality. The Earth is actually an oblate spheroid and is also not a smooth and regular surface. At small scales, these factors causing blemishes and imperfections in shape can be ignored. However, when dealing with large-scale maps of a small portion of the Earth's surface it is essential to make local corrections for these factors.

The Quaternary Triangular Mesh referencing system tries to deal with irregularities in the Earth's surface by considering a mesh of regular-shaped triangles of same size and shape in place of latitude and longitude lines. In this system, each triangle occupies the same area of the Earth's surface. The flexible triangular mesh can be moulded to fit the slight bumps and blemishes that form the true surface of the Earth.

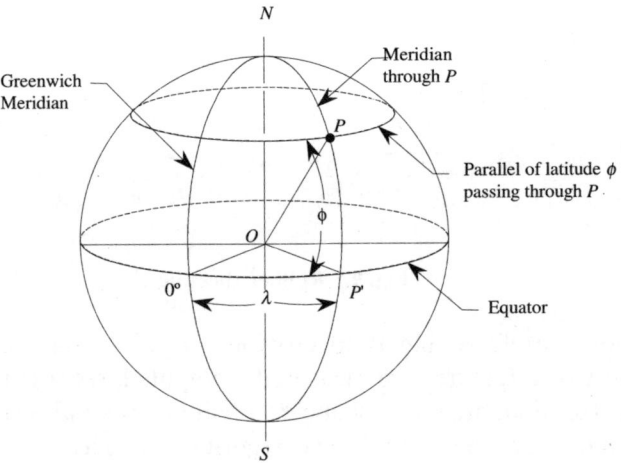

Fig. 9.2 Geographical coordinate system

Rectangular Coordinate System

Since most of spatial data available for use in GIS exist in 2D form, a referencing system that uses rectangular coordinates is most suited. This requires a map graticule or grid, placed on top of the map. The graticule is obtained by projecting the lines of latitude or longitude from our representation of the world as a globe onto a flat surface using a map projection. The function of map projection is to define positions on to the Earth's curved surface when transformed on to a flat map surface. There are several map projections, and a variety of these are in common use since no single projection can meet the requirements of all users.

The simplest regular square grid is the most widely used coordinate system for small areas. For larger areas, certain established cartographic projections such as the *Universal Transverse Mercator Projection* (UTM) are commonly used. This projection uses multiple cylinders that touch the globe at 6° intervals of longitude, dividing the globe into 60 projection zones, avoiding the pole (Fig. 9.3). To avoid extreme distortions that occur in the polar areas, the projection zones are limited between 84° N to 80° S. To improve the overall accuracy of measurements within a projection zone, the cylinder is made to intersect the globe at two standard meridians that are 180 km east and west of the central meridian resulting into true scale along the two standard meridians of longitude instead of one along the central meridian. Compensation of scale distortion introduced along the central meridian is obtained by applying a scale factor that is slightly less than unity of the order of about 0.9996 to all distance measurements. For the distance measurements near the zone boundaries, a scale factor slightly greater than unity of the order of 1.0004 is applied.

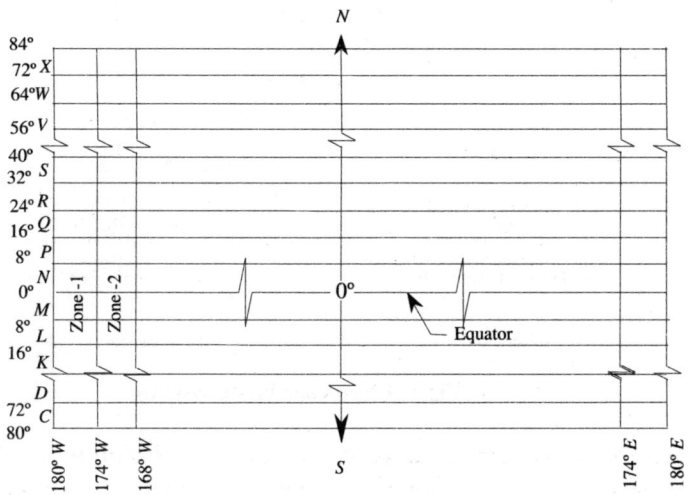

Fig. 9.3 Universal Transverse Mercator zones

The UTM projection has been adopted by many organizations for remote sensing, topographic mapping, and natural resource inventory. Realizing that the UTM is the most popular coordinate system amongst map users, most of the digital products in United States are being produced on UTM projection. Presently, many GPS receivers are adopting this coordinate system as an option, in fact, making it a de-facto standard coordinate system in the spatial data collection industry.

Non-coordinate System

In non-coordinate system, spatial referencing is done using descriptive codes rather than coordinates. Most widely used postal code which is appended to a postal address, is one of the examples of georeferencing using codes. These codes may be completely numeric, such as 267667 (PIN Code in India) or alpha numeric such as DL3 6KT (Postcode in UK) (Fig. 9.4). The basic purpose of such codes is to increase the efficiency of mail sorting and delivery rather than to be an effective spatial referencing system for GIS. This system has the following advantages:

(*i*) Provides coverage of all areas where people reside and work.
(*ii*) Individual codes do not refer to a single address.
(*iii*) Provide a degree of confidentiality for data released using this as referencing system.

Although postal codes are usually assigned to locate geographical areas, sometimes this is not the case; spatial codes may be assigned to institutions with large volumes of post, such as government agencies, and large commercial companies.

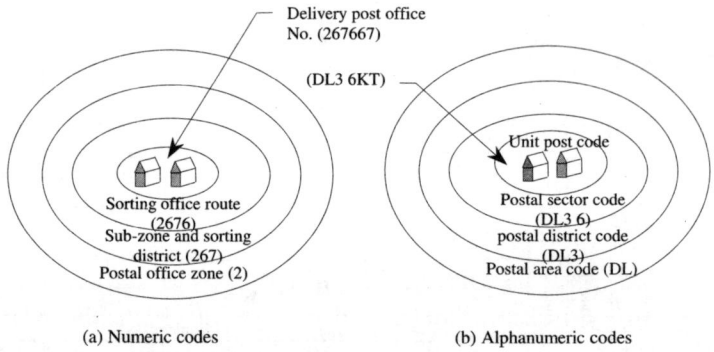

Fig. 9.4 Non-coordinate systems

All spatial referencing systems have problems associated with them. Some are specific to the referencing system, such as updating problems with postcodes or the difficulties caused by geographical coordinates with respect to map projections. However, some of the problems stemming from the nature of the spatial entities and requiring referencing are as below.

(i) When the spatial entities are mobile such as animals, cars, or people, any spatial reference to which they are tagged will only represent their known location at a particular time.

(ii) When the spatial entities such as river meander, roads, etc., change their location can be relocated and policy areas redefined.

(iii) The same object, such as a house, may be referenced in different ways as a point or an area, on maps of different scales.

Problem also arises due to use of different spatial referencing systems in different organizations supplying the data for GIS use, and therefore, choosing an appropriate referencing system may be difficult. This will require frequent integration of collected data having different referencing systems. GIS provides links to allow integration and conversion of data for some referencing systems.

9.9 SPATIAL DATA ERRORS

It is essential that the GIS products are high quality products. This can be achieved by making use of quality data with minimum errors. The data to be used in a GIS may have some inherent errors and some errors are produced by the system while working with the data. It is important for GIS users to document the limitations of their source data and the subsequent output generated.

Data Quality and Errors
The term quality is used for describing problems, and errors are associated with the sources, propagation and management of errors. Examining issues such as errors, accuracy, precision, and bias help in assessing the quality of data sets. Portrayal of features of interest is affected

by the resolution and generalization of source data, and data model used. Data sets used in GIS for analysis should be complete, compatible, and consistent.

Errors, Accuracy and Precision
The errors in a data set refer to the faults including the statistical concept of error meaning *variation*.

Accuracy is defined as the degree of conformity or closeness a data value to its true value (Chandra, 2000). An accurate GIS database represents real world. On the other hand, *precision* is the degree of conformity or closeness of repeated observed values of the same quantity in the data set to each other. It is perfectly feasible to have a GIS data set that is highly accurate but not very precise, and *vice versa*. Computers store data with a high level of precision which does not imply that the data set is of a high level accuracy.

Bias
Bias in GIS data indicates the systematic variation of data from reality, and can be referred to as a consistent error through out a data set. For example, a badly calibrated digitizer introduces a consistent overshoot in digitized data.

Resolution
Resolution being an important concept when dealing with spatial data, has been defined in Chapter 8. It is dependent on the scale of the original map, the point size, and line width of the features represented thereon, and the precision of digitizing.

Generalization
All spatial data are a generalization or simplification of real-world features. It is a process to simplify complexities of the real-world in producing maps and models at some scale. Thus, amount of generalization depends on the scale and limitations of the technical procedures used to produce data. Generalization introduces positional inaccuracies in representation of area features as points, exaggerated line thickness, and lateral displacement of adjacent features.

Completeness, Compatibility, Consistency and Applicability
A GIS data set should be complete spatially and temporally, including with respect to attribute information. Different data sets in the GIS database should be compatible to produce a sensible result. For example, overlaying of two maps, mapped originally at different scales, will produce worthless result because of incompatibility between the scales of the source maps. Compatibility can be ensured using similar methods of data capture, storage manipulation, and editing in producing ideal data sets. Consistency should also be observed within individual data sets. Inconsistency may arise within a data set when different sections of a data set come from different sources or different people digitize them. Applicability or suitability of a data set is viewed in terms of a set of commands, operations or analysis. For example, there is no matching between height data used for interpolation using Thiessen polygon method, as height data vary continuously while Thiessen polygon method assumes abrupt variations.

Source of Errors

Errors in GIS data may be categorized as:

(*i*) Conceptual errors,
(*ii*) Source data errors,
(*iii*) Data encoding errors,
(*iv*) Data editing and conversion errors,
(*v*) Data processing and analysis errors, and
(*vi*) Data output errors.

Conceptual errors: Conceptual errors arise from our understanding of the real world and how it is modeled. The perception of reality varies from person to person, and this affects the data. Whatever GIS model is adopted it is a simplification of the reality, and thus any simplification will introduce errors of generalization, completeness, and consistency.

Source data errors: GIS spatial and attribute data collected from various sources discussed in Sec. 9.6 are likely to include errors. The errors in survey data can be due to observational errors, instrumental errors, and personal errors, whereas the data collected by remote sensing and aerial photography can have the errors due to wrong spatial referencing, and mistakes in classification and interpretation. The temporal changes in the features also introduce errors due to time and date of data acquisition.

Maps, probably the most frequently used sources of data, contain errors in both spatial as well as attribute data caused by human or equipment failings. The cartographic process used in map-making introduces subtle errors in maps.

Data encoding errors: The process of transferring the data collected through maps, remote sensing, or ground surveys, into a GIS format is referred to as *data encoding*. Data encoding is probably the greatest source of error. Digitizing a map is one of the processes of data encoding in GIS. It can be done manually or automatically using suitable computer hardware. Despite of availability of hardware for automatic digitization, most of the digitizing work is done manually involving human judgment and limitations, which is one of the main sources of error in GIS. Translating a continuously curved line on a map into a digital image involves a sampling process. Out of infinite number of points on the line, a very small number of points along a curve are sampled due to human limitations. Representation of curved shape by nodes is shown in Fig. 9.5.

The errors are also introduced due to incorrect registration of map document before the digitizing commences both in manual and automatic digitizing. Raster scanners used for automatic digitizing suffer from resolution problems.

Editing and conversion errors: Since the data input by either manual or automatic digitizing, is never without errors it will almost always require editing and cleaning. It is difficult to locate the errors precisely and remove them, but many of them can be removed by careful scrutiny of data. Fig. 9.6 shows some of the common errors in digitizing and effect of editing and cleaning using automated procedures available in some of the vector GIS. The tolerance limit set for cleaning the errors has its own effect on cleaning depending on its value.

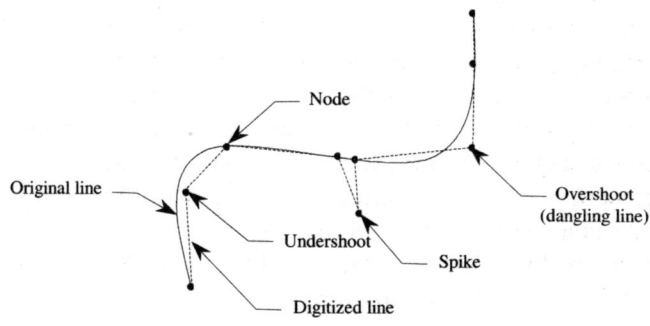

Fig. 9.5 Digitizing a line

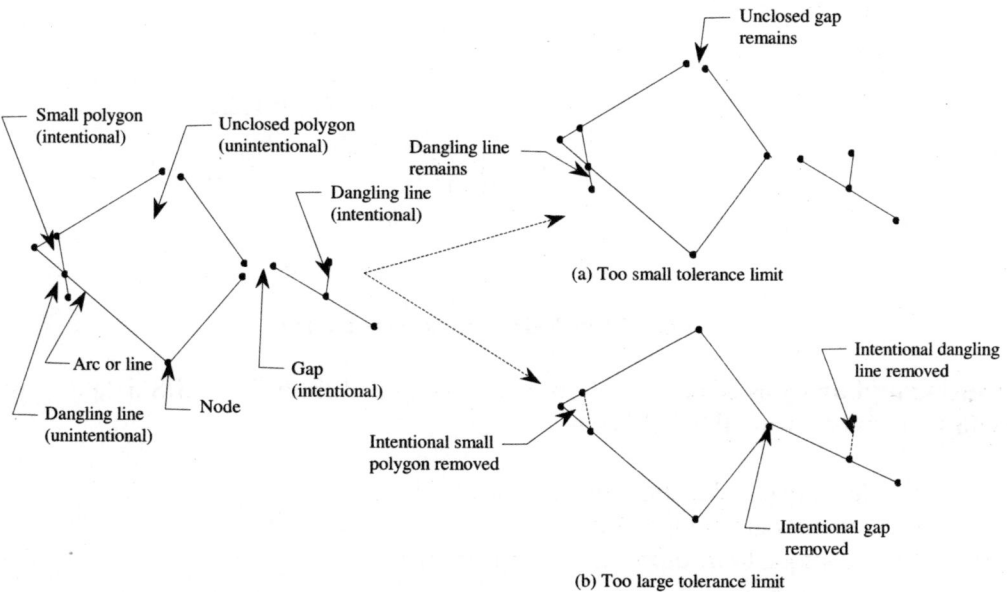

Fig. 9.6 Effects of tolerances on topological cleaning in vector GIS

While dealing with a raster GIS, a different problem arises when using the automated techniques for cleaning. In raster GIS, the noise, also referred to as misclassification of cells, can be either regular form which is easier to identify or scattered randomly in which case, it is difficult to locate. The noise errors are rectified by employing filters, which reclassify single cell or a group of cells by matching them with general trends in the data. It is important to choose an appropriate filtering method as the wrong method may remove genuine variations in the data or retain too much of the noise.

After cleaning and editing data it is required to convert vector data to raster data or *vice versa*. When raster data is converted into vector, topological ambiguities as shown in Fig. 9.7 are introduced. When converting vector data into raster data, both size of the raster cell and the method of rasterization used have important implications for positional error and, in some cases, attribute uncertainty. Positional and attribute errors resulting from generalization are found as classification error in cells along the vector polygon boundary (stepped appearance of the raster version when compared to the original vector form). Besides topological errors, conversion from vector to raster may lead to loss of small polygons and different raster map, due to incorrect placement of grid for rasterization in respect of orientation and origin.

Conversion of data is also required sometimes when the data generated by a system is to be read by another system. In such cases, the transfer of a GIS database from one software package to another can lead to errors usually termed as *mechanical errors*.

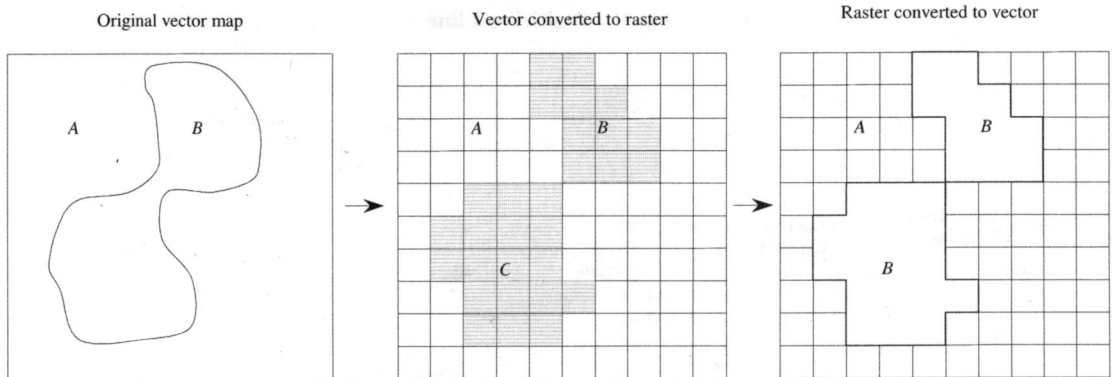

Fig. 9.7 Topological errors in conversion

Processing and analysis of errors: Before processing and analysis of GIS data are taken up, the GIS users must ensure the following:

- (*i*) The data are suitable and relevant for analysis.
- (*ii*) The data sets are compatible.
- (*iii*) The technique to be employed is appropriate.

GIS data processes that can introduce errors are mainly classification of data, aggregation or disaggregation of area data, and data integration by overlay techniques. A common error arising due to overlaying two polygon maps is *slivers*. These are very small polygons along correlated or shared boundaries of the two input maps (Fig. 9.8). The existence of slivers often results from digitizing errors. When the two maps of same area digitized or scanned separately at different time periods are overlaid, the digitized boundaries intersect to form slivers. Other causes of slivers include errors in the source map or errors in interpretation.

Also sometimes, the positional errors may cause attribute errors in overlay operations.

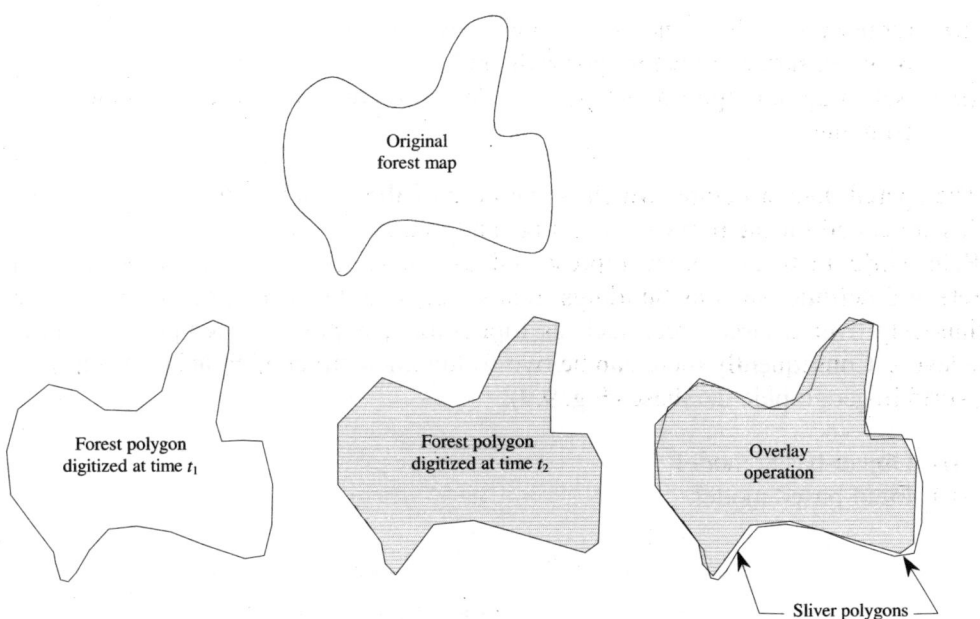

Fig. 9.8 Sliver polygons created in vector overlay

Data output errors: Due to inaccuracies in the GIS database and errors resulting in manipulation and analysis of data, it is inevitable that all GIS output, whether in the form of paper maps or digital database, will have inaccuracies, the extent depending on attention and care taken at all stages starting from construction, manipulation, and analysis of GIS database.

9.10 SPATIAL DATA MODELS

Digital geographic data represents real-world features and phenomena in numeric form coded in a specific way to support GIS and mapping applications using computer. To make the geographical data useful, it should be encoded in digital form, and organized as a digital geographic database that creates a perception of the real world similar to the perception created by the paper maps.

The conventional paper map represents a general-purpose snapshot or static view of the real world at a given time whereas the digital geographic database allows a range of operations such as sorting, processing, analyzing, and visualizing the spatial data thereby allowing the data to be a dynamic, together with the necessary tools for interacting with the data to perform certain specific objectives.

The ways of representing data are known as *data models*. The data model represents the linkages between the real-world domain of geographical data and the computer (or GIS representation of the features). The process of linkages involves

(*i*) identifying the spatial features from the real world that are of interest in the context to an application and choosing how to represent them in a conceptual model,

170 *Remote Sensing and Geographical Information System*

(*ii*) representing the conceptual model by an appropriate data model by choosing between raster or vector approach, and

(*iii*) selecting an appropriate spatial data structure to store the model within the computer.

The spatial data structure, which is the core of the model, is the physical way in which entities are coded for the purpose of storage and manipulation.

Real-world features can be represented as *object* and *phenomena*. While objects are discrete and definite, such as buildings, roads, cities, and forests, phenomena are distributed continuously over a large area, such as topography, population, temperature, rainfall, and noise levels. Consequently there can be two following distinct approaches of representing the real world in geographic database (Fig. 9.9).

(*i*) Object-based model
(*ii*) Field-based model.

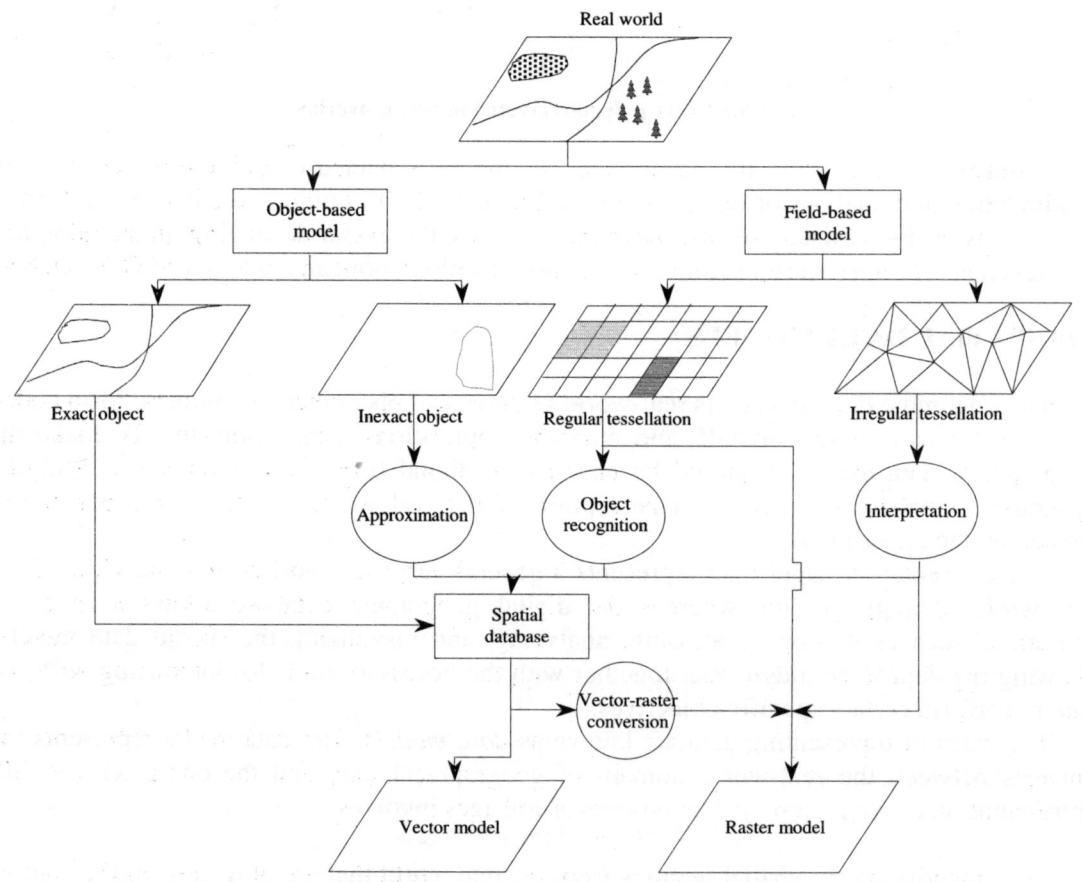

Fig. 9.9 Modeling approaches for the real world

Object-based Model
In object-based model, the geographic space is treated to be filled by discrete and identifiable objects. An object which is a spatial feature, has identifiable boundaries, relevance to some intended application, and can be described by one or more characteristics known as attributes. The spatial objects may be classified as

   (*i*) Exact objects
   (*ii*) Inexact objects or fuzzy entities.

Spatial objects are said to be *exact objects* if they represent discrete features such as buildings, roads, land parcels. The spatial objects, which have identifiable boundaries, but not well defined, are called *inexact objects* or *fuzzy entities*. The characteristics of the inexact objects change gradually across the assumed boundaries between neighbouring spatial objects. Landform features and natural resources belong mostly to this class of objects. Soil types, forest stands, wildlife habitats are some of the examples of inexact objects.

Data in object-based model are obtained by field surveying methods or photogrammetric method, map, aerial photointerpretation, remote sensing, or map digitization. Depending on the nature of the objects and the geographical scale of recording, spatial objects are represented as graphical elements of points, lines, and areas (polygons) (See Fig. 9.10).

Field-based Model
The field-based model treats geographic space as populated by one or more spatial phenomena of real-world features varying continuously over space with no obvious or specific extent. Data for spatial phenomena structured as fields, can be acquired either directly or indirectly by aerial photography, remote sensing, map scanning, and field measurements made at selected or sampled locations, such as topographic data for *triangulated irregular networks* (TIN). The data can also be generated using indirect data acquisition methods by applying mathematical functions, such as interpolation, reclassification, or resampling, to the measurements made at selected or sampled locations. Topographic data such as contours and Digital Elevation Model (DEM) are examples of data usually obtained by indirect methods of measurements. Spatial phenomena are represented as *surfaces*, which can be thought of being made up of spatial data units in the form of either regular tessellations or irregular tessellations.

At the database level, data in object-based spatial databases are mostly represented in the form of coordinate lists (*i.e.*, vector lines), and the spatial database is generally called as the *vector data model*. A spatial database when structured on the field-based model, the basic spatial units are the different forms of tessellation by which phenomena are depicted. The most commonly used type of tessellation is a finite grid of square and rectangular cells, and thus, field-based databases are generally known as the *raster data model*.

172 *Remote Sensing and Geographical Information System*

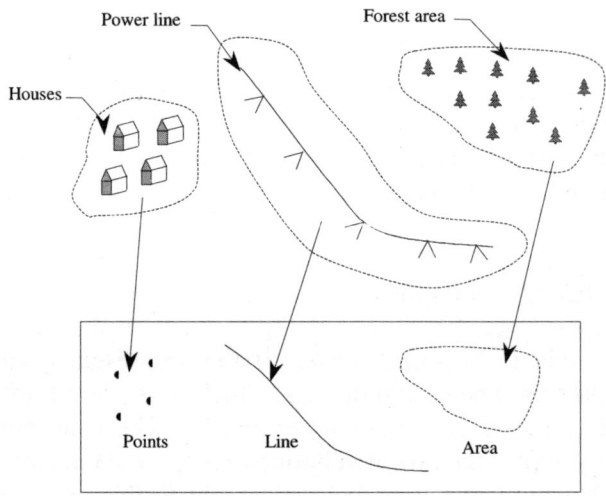

Fig. 9.10 Representation of exact objects

9.11 SPATIAL DATA STRUCTURES

Data structures provide the information that the computer requires to reconstruct the spatial data model in digital form. There are many different data structures in use in GIS. In general, data structures can be classified according to the type of data they use, *i.e.*, raster or vector data.

Raster Data Structures

In the raster world, a range of different methods is available to encode a spatial entity for storage and representation in the computer. Coding of raster data of the entity model, *i.e.*, an image (Fig. 9.11a) is done by having mirror image of the equivalent row of numbers in the file structure (Fig. 9.11c), in the cells in each line of the image (Fig. 9.11b) where the entity is present. The first line in the raster file is basically a header that gives information regarding the size of the file and the maximum value that a cell may have. In Fig. 9.11c, first two values of 10 indicate that there are 10 rows and each row has 10 pixels/columns, while the value 1 indicates the maximum value that a cell may have. Thus in this example, the value of 0 represent cells having no entity value while 1 indicates the presence of an entity.

In raster data structure, different spatial features, such as building, water bodies, contours, roads, are stored as separate data layers, *i.e.*, these features are stored in four separate data files, each representing a different layer of spatial data. However, if the entities do not occupy the same geographic location (or cell in the raster model) then it is possible to store them all in a single layer with an entity code given to each cell. This code indicates that which entity is present in a cell. An example of this case is shown in Fig. 9.12.

GIS Data 173

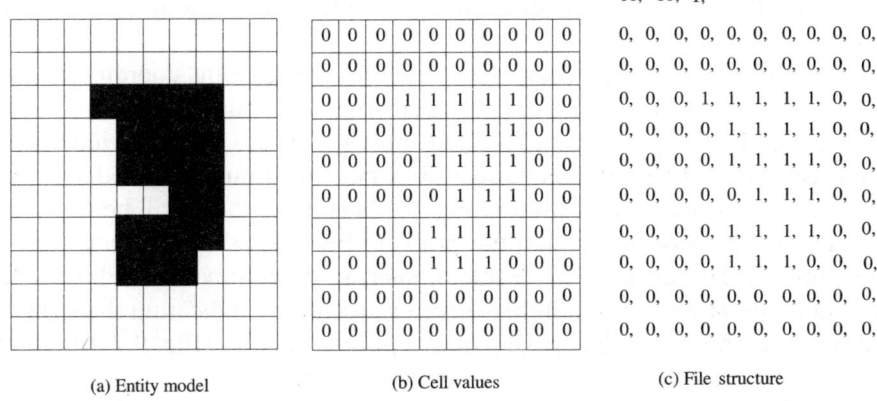

(a) Entity model (b) Cell values (c) File structure

Fig. 9.11 Raster data structure

In this figure; there are four different land use categories, cultivation, forest, water, and residential area (Fig. 9.12a) and they are designated as 1, 2, 3 and 4 respectively (Fig. 9.12b). The corresponding file structure is shown in Fig. 9.12c. The file line in the first structure has values 10, 10, 4 indicating that the image size is 10 rows by 10 pixels/columns and there are 4 different entities. Since a value must be recorded and stored for each cell in an image, raster data sets become large in size. Thus, a complex image made up of a mosaic of different features, such as a soil map with 10 distinct classes requires the same amount of storage space as a similar raster map of a single forest. To reduce the storage space, a range of data compaction methods is available.

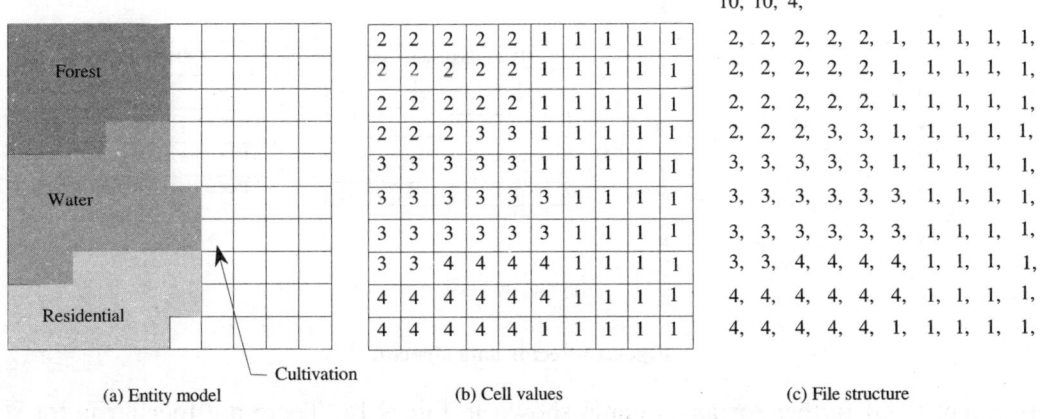

(a) Entity model (b) Cell values (c) File structure

Fig. 9.12 Feature coding of cells in the raster world

Vector Data Structure

The entities point, lines, and areas can be defined by a coordinate system. One of the simplest vector data structure to represent a geographical image in the computer is a file containing

(x, y) coordinates that represent the location of individual point features or the points used to construct lines and areas.

Fig. 9.13a shows a simple vector data structure using point coordinates to describe a parcel of land. The limitation of this approach emerges in dealing with more complex spatial entities. For example, when the parcel of land shown in Fig. 9.13a is divided into plots 1, 2, and 3 as shown in Fig. 9.13b, the individual plots become a number of adjacent polygons. If the simple vector data structure as used in Fig. 9.13a is used, there will be duplication of data to define each entity individually, and the problem becomes serious when there are a large number of entities. To address this problem, sharing of common coordinates for adjacent polygons is done point reference numbers given sequentially indicating that which points are associated with which points. This is known as a *point directory* (Fig. 9.13b). The problem

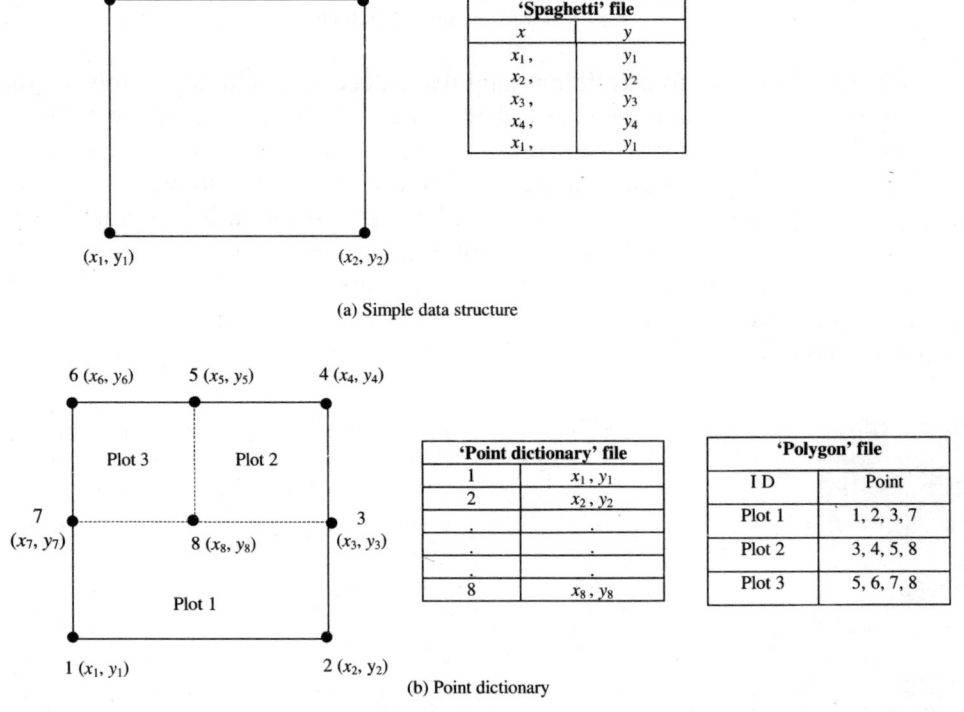

Fig. 9.13 Vector data structure

gets complicated further for an example shown in Fig. 9.14. There are four areas for stalls from different regions and in the centre there is an information centre. While the simple vector file structure can recreate the image of the whole area, it cannot inform the computer that the information centre lies within the larger area containing the stalls.

For the representation of points, lines, and polygons shown in Fig. 9.14, and the entities adjacent to them, a topological data structure that contains this information is required. There are numerous ways of providing topological structure, the example given in Fig. 9.14, illustrates the basic principles of topological data structure. Basically a topological data structure represents connectivity between entities and not their physical shape. Table 9.1 presents the basic requirements of the entities of point, line, and area to have topology.

The basic requirements of a topological data structure are that

(i) no node or line segment should be duplicated,
(ii) line segments and node can be referenced to more than one polygon,
(iii) all polygons should have unique identifiers, and
(iv) island and hole polygons can be adequately represented.

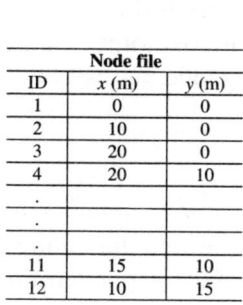
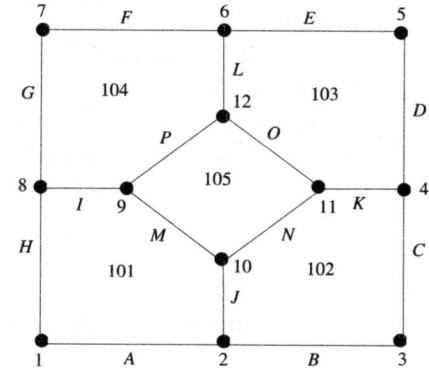

Node file

ID	x (m)	y (m)
1	0	0
2	10	0
3	20	0
4	20	10
.		
.		
.		
11	15	10
12	10	15

Polygon file

ID	Segment list
101	A, J, M, I, H
102	B, C, K, N, J
103	D, E, L, O, K
104	F, G, I, P, L
105	M, N, O, P

Segment file

ID	Start node	End node	Left polygon	Right polygon	Length (m)
A	1	2	101	Outside	10
C	2	3	102	Outside	10
C	3	4	102	Outside	10
D	4	5	103	Outside	10
.	.	.	.	.	.
.	.	.	.	.	.
O	11	12	105	103	7.07
P	12	9	105	104	7.07

Polygon attribute file

ID	VAR 1 (Name)	VAR 2 (Area in m²)
101	Punjab stall	87.5
102	Rajasthan stall	87.5
103	Maharashtra stall	87.5
104	Kerala stall	87.5
105	W. Bengal stall	50.0

Fig. 9.14 Topological structuring

Table 9.1 Topology requirements of basic spatial entities

Entity	Requirement
Point	Geographical reference to locate it with respect to other spatial entities.
Line	Ordered set of points (known as an arc, segment or chain) with defined start and end points (nodes) which also give the line direction.
Area	Data about the points and lines used in construction of the area, and how these are connected to define the boundary.

9.12 MODELING SURFACES

Surface entities such as elevation, population, pollution, and rainfall are also required to be represented in GIS, along with the entities such as buildings, road, water bodies, and forests. These surface entities vary in space, hence in order to show their variation, a convenient form is to represent them through trend or surface model, such as digital terrain model (DTM).

DTM is a set of digital data of (x, y, z) coordinates used to model topographic surface. More precisely when DTM represents height data as z coordinate, it is called the *digital elevation model* (DEM). Since it is not possible to have an infinite number of height values to model a surface accurately, the actual surface is approximated considering only a finite number of observations. Thus, to increase the accuracy of the DEM, an appropriate number of observations at appropriate locations, are required. DTM may be derived from a number of data sources, such as contour and height information from topographic maps, stereo photogrammetry, satellite images, and field survey. Digital terrain modeling can be achieved either by raster approach or vector approach as discussed below.

Raster Approach

In raster GIS, a DTM is represented by a grid in which each cell contains a single value of the height or the terrain covered by that cell. The accuracy of the terrain modeling by this approach depends on the spacing (resolution) of the grid and complexity of the terrain surface (Fig. 9.15a).

Vector Approach

A vector DTM is like a raster DTM with the difference that the vector DTM has regularly spaced set of spot heights (Fig. 9.15b). In vector GIS, to represent the terrain surface with irregular height data, the common practice is to use TIN (Fig. 9.15c), which creates a mosaic of irregular triangles by joining the points where spot heights are known by straight lines. The vertices of the triangles in a TIN represent the features such as peaks, depressions, and passes, and the sides represent ridges and valleys whereas the surface of the individual triangle provides area, gradient (slope), and orientation (aspect). Storing these values as TIN attributes helps in further analysis.

TIN model has the following advantages:

(*i*) Efficiency of data storage
(*ii*) Fewer points required to represent flatter areas.

To achieve efficiency in storage, TIN models use only *surface significant* points to create a terrain surface. These points are selected by the TIN model on the basis of their spatial relationship with their neighbours. Those points that cannot be closely interpolated from their neighbouring points are considered to be 'surface significant' points, and are used as TIN vertices.

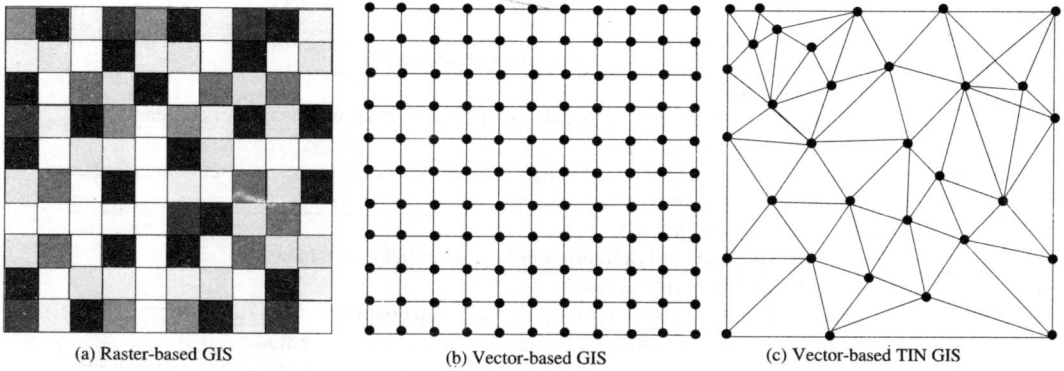

(a) Raster-based GIS   (b) Vector-based GIS   (c) Vector-based TIN GIS

Fig. 9.15 Modeling surfaces

9.13 MODELLING NETWORKS

A set of interconnected linear features such as roads, railways, telephone lines, through which goods and people are transported or along which communication of information is processed, is said to be a *network*. Since network models in GIS are abstract representations, the use of vector data model and raster data model is not suited for network analysis. Fig. 9.16 shows a vector network model made up of the same arc (line segments) and node elements as any other vector data model but with the addition of special attributes. Table 9.2, gives details of various attributes, such as arcs, nodes, centres, junctions, used in networks to represent different features.

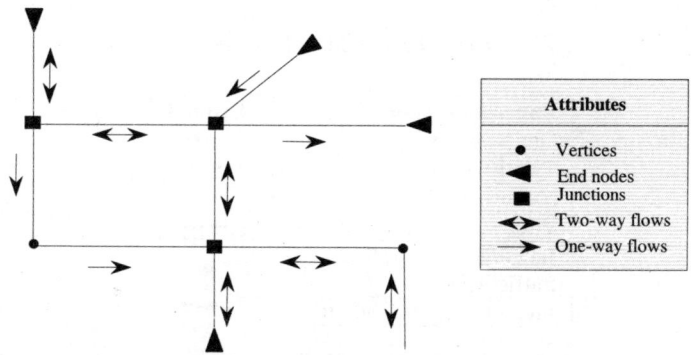

Fig. 9.16 Vector network model

Network has the following two key characteristics:

(*i*) Impedance
(*ii*) Supply and demand.

Table 9.2 Definition of attributes used in networks

Attributes	Features represented
Arc	Network links: Roads, railways, power lines, cables and pipelines of utilities, rivers, streams.
Node	End points of network links: Stops, Centres, junctions in transport networks, confluences in stream networks, switches and values in utility network.
Stop	Locations may be visited: Bus stop, pick-up and delivery points in a delivery system.
Centre	Discrete locations in a network: Shopping centres, airports, hospitals, schools, hospitals, cities (at smaller scale)
Turn	Transition from network link to network link at a network node: Turns across oncoming traffic on a road network take longer than turns down shipways, whereas turns that go against the flow of traffic on one-way roads are prohibited altogether.

Impedance is the cost associated in traversing a network link, stopping, turning, or visiting a centre. It may be the time to travel from one node to another, or in the case of a delivery van, time, fuel used, and salary of the crew. Different links have different impedance value as it depends on the factors such as traffic volume, traffic control system, and topography. The various forms of impedance values that may be linked to a network are given in given in Table 9.3.

Supply and demand is also important aspect of a network analysis. Supply is the quantity of a material available at a supply centre to meet the demand associated with the links of a network. For example, the number of LPG cylinders required in a certain area is the demand, whereas the number of LPG cylinders available at its store is the supply. Correct topology and connectivity are extremely important for network analysis.

Table 9.3 Impedence values

Impedence value	Symbol
One way flow	→
Two-way flow	↔
Narrow street (slow traffic)	─
Wide street (Faster traffic)	━
Junction controlled by traffic light	■
Junction with no right turn	►■
Van depot	D
Storage warehouse	W

9.14 GIS DATABASE AND DATABASE MANAGEMENT SYSTEM

GIS data are collected from various sources. A large portion of these data is managed in databases. The database approach provides a robust method to data organization and processing in the computer. However, it also requires more effort and resources to implement, and therefore, the design and building of GIS databases is an elaborate process by necessity.

Before knowing in detail the ways in which spatial entities may be stored efficiently in computer, it is essential to have some idea of the general issues of organizing data for optical storage and access. At the same time a little knowledge of data structuring is also required to understand the working of the system. All entities (point, line, or area) in GIS are spatial, and therefore, the conventional database is adapted to GIS in various ways, where the spatial data and attribute data can be handled efficiently.

Databases offer more than just a method of handling the attributes of spatial entities; they can help to convert data into information with a value. Information results from the analysis or organization of data, and in a database, data can be ordered, re-ordered, summarized, and combined to provide information. A database can perform sorting, ordering, conversion, calculation (maximum and minimum value, average), summary. With the additional mapping capabilities of GIS, the location of the points with particular information can be mapped. For making decisions, the GIS needs information derived from the stored data in database.

There are many definitions of a database, but perhaps the simplest definition may be that it is a set of structured data, which provides speedy data retrieval to perform certain decision-making operations.

GIS Database

A collection of multiple files, made up of three basic types of file structure for storage retrieval, and organization of data, is called a *database*. There can be following three ways of storing the data in a database:

(*i*) Simple list
(*ii*) Ordered sequential files
(*iii*) Indexed files.

Simple lists: The simplest form of database is a simple list of all the items in which any new item when added to the database, it is simply placed at the end of the file. In such files, is easier to add data but retrieving is not efficient, and for this reason, in a large data set searching a particular item takes long time. Supposing a list contains n data items, then for searching a single item, computer requires on an average of $(n + 1)/2$ operations. So if a computer takes 1 second to read a card having certain information, to search for a particular information from a pack of 200,000 cards, computer has to perform as many as $(200,000 + 1)/2$ search operations which requires more than 27 hours. It, obviously, shows the need of an organized database.

Ordered sequential files: In ordered sequential files, the data items can be arranged in alphabetical order as in a telephone directory. The normal strategy for search is based on *divide-and-conquer* approach. Instead of beginning the search at the beginning of the file, the record in the middle is examined first. If the desired values do not match, the search proceeds to find whether the required item occurs before or after the middle element. The appropriate half of the file is retained and the search repeated, until the item has been located.

Very often ordered sequential files are accessed by binary search approach. A binary search requires $\log_2 (n + 1)$ steps. So a file containing 200,000 data items will require nearly

2 hours as compared to 27 hours in simple lists, assuming each operation takes a second to perform.

In ordered sequential files, whenever a new item is added, an extra space for it is created before inserting it, but the advantage is that stored items can be searched faster.

Indexed files: In geographical information, the individual items (points, lines, or areas) carry information about the associated attributes in addition to an identification number or name of the key attributes. Typically, a search will consist of finding, the entities that match a selected set of attribute criteria. For example, it may be required to examine or analyse all study areas in poor condition that have slopes less than 20%. In most cases, a large numbers of attributes are linked to an entity; a more efficient method of search is required to find out specific entities with associated, cross-referenced attributes. This requires another database strategy of search procedure.

There are two ways of speeding up the access to the original data file as indexed file. If the data items in the files themselves provide the main order of the file, then the files are known as *direct files* as shown in Table 9.4a. Another approach of locating items in the main file may be specifying the location according to topic given in a second file, known as *inverted file* as shown in Table 9.4b.

Another approach of search could be to perform an initial sequential search on the data for each property (topic) by arranging the data in file called *index inverted file* (Table 9.4c). Though the indexed files permit rapid access to the database, they have inherent problems when used with files in which records are continually being updated by adding or deleting the records, such as in interactive mapping systems. Updating of records in direct files requires that both the file and its index must be modified. When new records are written to a file accessed by an inverted file index, the new record does not have to be placed at a special position. The new record may be simply added to the end of the file, but the index must be updated. File modification may be an expensive affair when dealing with large data files, particularly in an interactive environment. Another disadvantage of indexed files is that very often data can only be accessed through the key contained in the index files; other information may only be retrievable using sequential search methods.

Database Management System

A computer program that is designed to store and manage large amounts of data is called a *database management system* (DBMS). The overall objective of a DMBS is to allow users to deal with data without needing to know how the data are physically stored and structured in the computer. A DBMS must allow the definition of data and their attributes, and relationships, as well as providing security, and an interface between the end users and their applications and the data themselves. The functions of a DBMS can be summarized as:

(*i*) File handling and file management
(*ii*) Adding, updating, and deleting records
(*iii*) Information extraction from data.
(*iv*) Maintenance of data security and integrity
(*v*) Application building.

Fig. 9.17 shows the database approach to data handling.

Table 9.4 Indexed files

(a) Direct files

Index		File
Item key	Record No.	item
A	1	A_1
B	$n + 1$	A_2
C	$(n + 1) + 1$	.....
.....	.....	B_1
.....	.....	.....
.....	.....	C_1

(b) Inverted files

Soil profile No.	Attributes				
	SG	DR	DP	pH	ER
1	A	Good	Deep	5	Y
2	B	Good	Deep	6	N
3	C	Poor	Shallow	7	N
4	D	Poor	Deep	4	Y
5	E	Good	Shallow	5	Y

SG = Soil Group; DR = Drainage; DP = Depth; ER = Erosion; Y = Yes; N = No

(c) Index inverted file

Properties	Profiles (sequential No. in original file)				
Good drainage	1	2			5
Poor drainage			3	4	
Deep	1	2			
Shallow			3	4	5
Erosion (Yes)	1				5
pH (≤5)	1				5

Fig. 9.17 Database approach to data handling

A DMBS manages information that is organized using a database structure or model. This is analogous to the way in which spatial data are organized in a GIS according to a spatial data model (*e.g.*, raster or vector). Although new forms of database structure are being developed all the time, there are three fundamental ways of organizing information that also reflect the logical models used to model real world structures as below:

(i) Hierarchical database structure
(ii) Network database structure
(iii) Relational database structure.

The *object-oriented database structure* is an emerging trend in GIS and a topic of current research. Of these four database structures, the relational database structure is most widely used.

Hierarchical Database Structure
A data may have a multi-layered data with a direct relationship between each layer, similar to a tree like structure. The relationship between two successive layers is known as *parent-child* or *one to many* relationship. For such type of data, hierarchical database provides a quick and convenient means of data access. Here each part of the hierarchy can be reached using a *key* or criterion, and that there is a good correlation between key and associated attributes.

Fig. 9.18 illustrates the concept of hierarchical database. Here a map M consist of two polygons I and II. Each polygon consists of lines, and each line has a pair of points.

The main advantage of hierarchical database is that it is simple and provides easy access through keys defining the hierarchy. Further, it is easy to expand by adding more attribute and formulating new decision rules. The success of data retrieval in a hierarchical database depends upon the prior knowledge of structure of all possible queries.

One of the biggest disadvantages of hierarchical database is the repetitive data. Referring to Fig. 9.18, it can be seen that each pair of points is repeated twice and that for line l_3, the coordinates c and d are repeated four times. This is a simple wastage and causes large redundancy of data in case of large databases. In hierarchical structures, the access within the database is restricted to paths up and down the hierarchy levels.

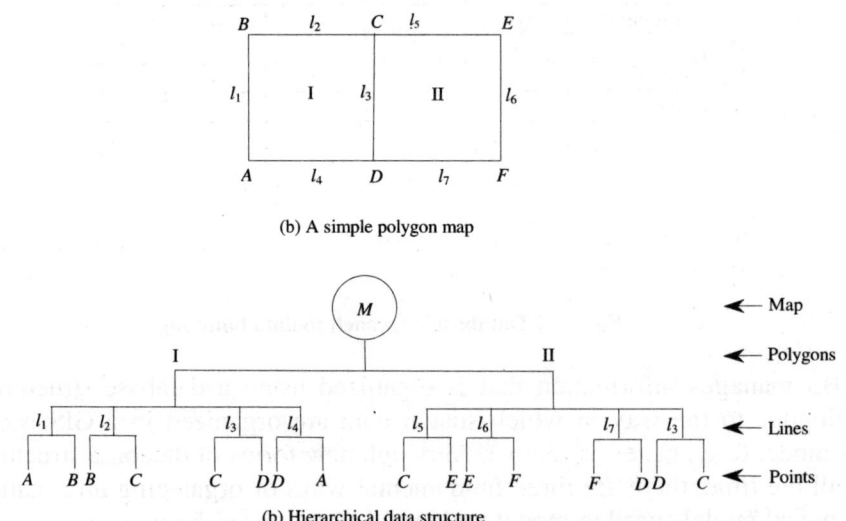

Fig. 9.18 Hierarchical organization of vector data

Network Data Structures

Many a times, a single entity may have many attributes, and each attribute is linked to many entities. To accommodate these relationships, each piece of data can be associated with an explicit computer structure called *pointer* which directs it to all the other pieces of data to which it relates (Fig. 9.19). Here rather than being restricted to a branching tree structure, each individual data is linked directly without the existence of a parent-child relationship. Such a data structure is called Network data structures.

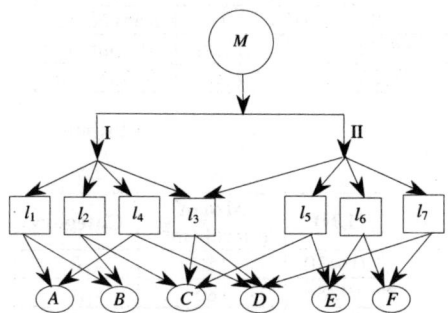

Fig. 9.19 Network data structures

Sometimes in order to reduce both redundancy and linkage, a compact network structure known as *ring pointer structures* are used, so that each entity appears once. In Fig. 9.19, each point is having many linkages or point. Fig. 9.20 shows the ring pointer structure, where the flow is simplified.

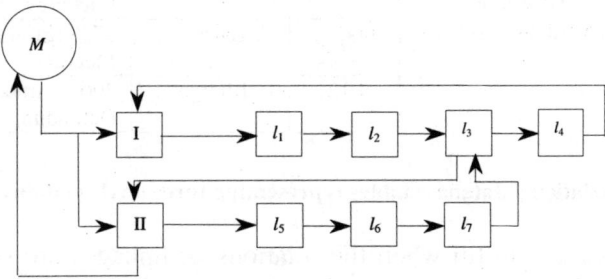

Fig. 9.20 Ring pointer structure

Relational Database Structure

In relational database structure, data are organized in a series of two-dimensional tables, each of which contains records for one entity (Fig. 9.21). These tables are linked by common data known as *key*. It is possible to make query on individual or a group of tables. Each table contains data for one entity. In the first table, the entity is 'landtitle'. In the second and third tables, the entities are 'parcel' and 'owner'. The data are organized into rows and columns, with columns containing the attributes of the entity. Each column has a distinctive name, and here every entry in a single column must be drawn from the same domain. A domain may be

184 *Remote Sensing and Geographical Information System*

all integer values, dates, phone, number, or a text). Within a table, the order of columns has no significance. These can be only one entry per cell, and each row must be distinctive to make possible use of entries in a row. The attribute values in the rows are called *tuples*. The cells may have null values if data values are not known.

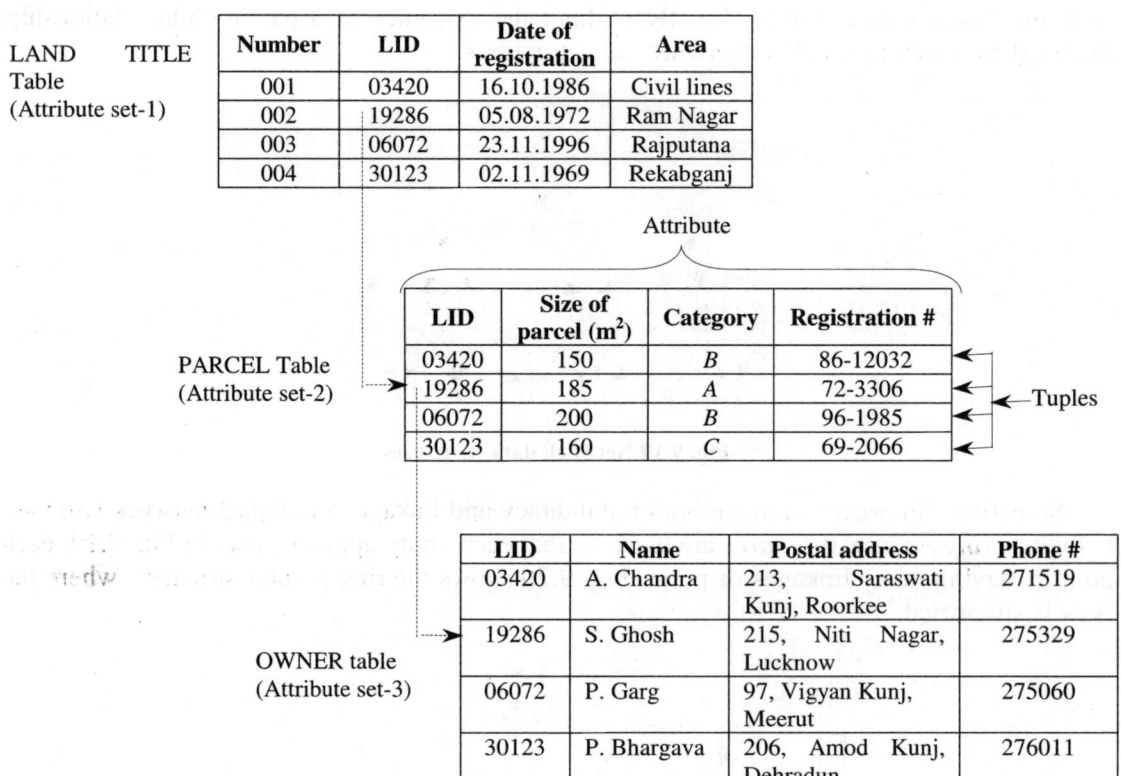

Fig. 9.21 Relational database tables representing three levels of data abstraction

Network structures are useful when the relations or linkages are specified before hand. They avoid data redundancy and make good use of available data. Hence, they allow greater flexibility of search than hierarchical structures. The disadvantage of network structure is that for large databases, the number of pointers can become large, and can become a substantial part of the database. Further, these pointers have to be maintained every time a change to the database is made. Building and maintenance of pointer structures may be a considerable overhead for the database system.

To avoid confusion in the terminology of relational databases, as different software vendors use different terms for the same function, relationship between relational database terminologies and the traditional table, or simple computer file, is illustrated in Table 9.5. The table structure in a relational database being extremely flexible, offers a wide variety of queries on the data. Relational databases are predominantly used for the handling of attribute data in GIS and not for the complex multidimensional spatial data. Many relational databases provide the facility of querying using menu systems or icons. Standard Query Language (SQL), which has the characteristics such as completeness, simplicity, and pseudo English-language style, has been developed to facilitate the querying of relational databases.

Table 9.5 Relational database terminology (*Source: Date, 1986*)

Paper version	File version	RDBMS
Table	File	Relation
Row	Record/case	Type
Column	Field	Attribute
Number of Columns	Number of fields	Degree
Number of rows	Number of cases	Cardinality
Unique ID	Primary key	Index
	Possible values	Domain

Some of the disadvantages of relational database structure are that

(*i*) it can not handle geographical concepts such as 'connected to', 'near to', or 'far from',
(*ii*) it always produces some data redundancy, and can be slow and difficult to implement,
(*iii*) it is difficult to handle complex objects such as found in GIS as there is a limited range of data types and difficulties with the handling of time.

Object-oriented Database Structure
Object-oriented concepts have its origin from computer programming languages, and have found application in solving redundancy and sequential search in relational structures. In GIS, object-oriented concepts have been found useful in handling complex spatial entities and for resolving problems of database modifications when operations such as polygon overlay are carried out.

Object-oriented database structures, developed by Object-Oriented Programming (OOP) language, combine the speed of hierarchical and network approaches with the flexibility of relational database structure by organizing the data around actual entities in comparison to functions being processed. In relational structure, each entity is defined in terms of its data records and logical relations between the attributes and their values.

In object-oriented databases, data are defined in terms of a series of unique and discrete objects having similar phenomena known as *object class* having some natural structuring criteria. The characteristics of an object may be described in the database in terms of its attributes (called its state) as well a set of procedures, which describe its behaviour (called operations or methods). These data are encapsulated within an object, which is defined by a unique identifier within the database.

Once the data have been encapsulated as an object in the database, the only way to change or to query them is to send a request known as *message*, to carry out one of its operations. The types of querying possible depend on the operations that have been used to define the objects. The response of the object to a message will depend on its state, and the same message may bring about a different reaction when received by different objects or in a different context; this is termed *polymorphism*.

Data used in object-oriented databases need to be clearly definable as unique entities. Given that, these databases (as with their network and hierarchical counterparts) provide very efficient structures of organizing hierarchical, interrelated data, establishing the database is obviously time-consuming as the objects may be defined more explicitly and the various links need to be established. Once this is finished, the database provides a very efficient structure for querying especially with reference to specific objects.

Choosing the Most Appropriate Database Structure
It should be apparent that the four basic databases structures-hierarchical, network, object-oriented, and relational-all have something to offer for spatial information systems. Hierarchical systems allow large databases to be divided up easily into manageable chunks, but they are inflexible for building new search paths and they may contain much redundant data. Network systems contain little redundant data, and provide fast, directed, and inflexible links between related entities. Object-oriented systems permit relations, functionality, persistence, and interdependence to be built into one system at the expense of the programming tools being more complex and heavier demands on computing power. Relational systems are open, flexible, and adaptable, but may suffer from large data volumes, redundancy, and long search times. Consequently it is not surprising that these techniques are often used together in spatial information systems to complement each other, rather than assigning all the work to one of them.

A hierarchical approach is often useful for dividing spatial data into manageable themes, or into manageable areas, so that continuous, seamless mapping becomes possible. The network approach is ideal for topologically linked vector lines and polygons. The relational approach is good for retrieving objects on the basis of their attributes, or for creating new attributes and attribute values from existing data. Object orientated structuring is useful when entities share attributes or interact in special ways.

SPATIAL DATA ANALYSIS 10

10.1 INTRODUCTION

After creation of a database in a GIS using any DBMS, the data is to be analyzed for specific purposes. This chapter explains how the data is analyzed within a GIS to extract information so that it can be used for decision-making. The one of major concerns are to investigate the patterns in a spatial data and to identify possible relationships between patterns and other attributes within the study region.

There are a wide range of functions available for data analysis in most GIS packages, including measurement techniques, query on attributes on proximity analysis, overlay operations, and analysis of models of surfaces and networks. This chapter introduces the range of data analysis functions available and their practical application is GIS.

10.2 DATA ANALYSIS TERMINOLOGY

Since different GIS software packages often use different terminology to describe the same function, it is necessary to explain some of the commonly used terminology in this chapter.

Entity: An individual point, line, or area in a GIS database is called an entity.
Attribute: Attributes are characteristics of an entity. In a vector GIS, attributes, such as name of a road, average rainfall in an area, are stored in a database. In a raster GIS, an attribute is the value of a cell in the raster grid in the form of a numerical code, such as '1' for metalled road, '2' for unmetalled road.
Feature: A feature is an object in the real world to be encoded in a GIS database.
Data layer: A data layer is a data set for the area of interest in a GIS, normally containing data of only one entity type (*i.e.*, points, lines, or areas).
Image: A data layer in a raster GIS is called as image.
Cell: An individual pixel in a raster is a cell.
Function or operation: Function or operation is a data analysis procedure performed by a GIS.
Algorithm: A sequence of actions to be performed by computer to solve a problem is known as an algorithm.

10.3 MEASUREMENT OF LENGTH, PERIMETER AND AREA

Measurement of length, perimeter, and area is a common task in GIS. It is important to note that all measurements from a GIS will be approximate due the reason that vector data are made up of straight line segments, and all raster entities are approximated using a grid cell representation.

In a raster GIS, the Euclidean distance or the shortest path between two points A and B can be measured by any one of the methods given below.

(i) By drawing a straight line between A and B, and computing its length as hypotenuse of a the right angled triangle ABC by Pythagorean geometry (Fig. 10.1),

i.e. $AB = \sqrt{AC^2 + CB^2} = \sqrt{4^2 + 4^2} = 5.7$ units.

(ii) By measuring distances along raster cell sides Aa, ab, bc, ... gB, and adding up them, i.e.,

AB = (Aa)+(ab)+ .... + (fg)+(gB),
or = 1+1+1+1+1+1+1+1 = 8 units

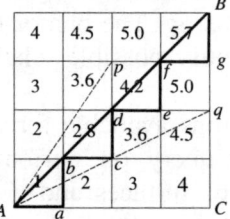

Fig. 10.1 Distance measurement in raster GIS

(iii) By forming concentric equidistant zones around the starting point A. Thus,

$$AB = \sqrt{4^2 + 4^2} = 5.66 \approx 5.7 \text{ units}$$
$$AP = \sqrt{2^2 + 3^2} = 3.61 \approx 3.6 \text{ units}$$
$$AQ = \sqrt{4^2 + 2^2} = 4.47 \approx 4.5 \text{ units}$$

The measurement of perimeter in raster GIS is done in terms of the number of cell sides that form the boundary of the feature. The perimeter of the feature shown in Fig. 10.2 are given by

Perimeter = side-1 + side-2 + ... + side-13 + side-14
= 14 units

and area enclosed by the boundary of the feature is the number of cells within the boundary, i.e.,

Area = 7 units

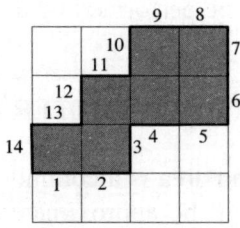

Fig. 10.2 Measurement of perimeter and area

If length of one side of a cell is known in metres, say 10 m, then in the previous examples,

$AB = 5.7 \times 10 = 57$ m
Perimeter $= 14 \times 10 = 140$ m
Area $= 7 \times (10 \times 10) = 700$ m^2

In the raster GIS, the area and perimeter measurements are affected by location of origin and orientation of the raster grid, and these problems are solved by proper selection of grids with north-south alignment and use of consistent origins.

Measurement of length, perimeter, and area in vector GIS is easier and accurate compared to raster GIS. In Fig. 10.3, the length of the line AB is calculated as below:

$$AB = \sqrt{(x_b - x_a)^2 + (y_b - y_a)^2} \qquad \ldots(10.1)$$

where (x_a, y_a) and (x_b, y_b) are the coordinates of A and B, respectively.

The area of a feature is calculated by totaling the areas of simple geometric figures formed by subdividing the feature or directly by the following formula (Chandra, 2002a).

$$\text{Area of } ABCDEA = \frac{1}{2}\left[x_a(y_b - y_e) + x_b(y_c - y_a) + x_c(y_d - y_b) + x_d(y_e - y_c) + x_e(y_a - y_d)\right] \qquad \ldots(10.2)$$

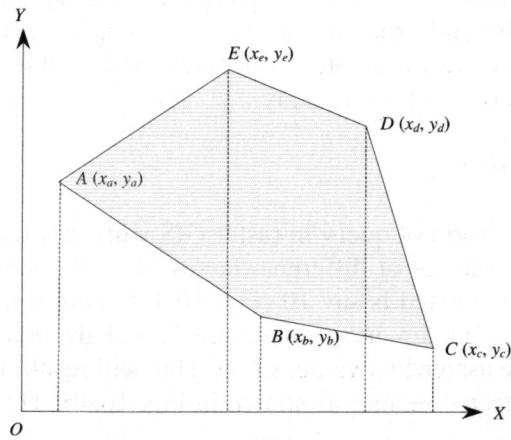

Fig. 10.3 Area by Coordinates (*Source: Chandra, 2002a*)

The perimeter is determined by summing up lengths of all the lines. The calculated length, perimeter, and area can be stored permanently in vector GIS database as attributes and to use them when needed thus avoiding repetitive calculations.

10.4 QUERIES

Performing query on the database is an essential part of GIS. Some queries may require the data that is the result of data analysis. Generally the queries may be in the form.

(i) What is the shortest distance between two points?
(ii) What is the route that will take minimum time to travel between two points?
(iii) What is the total area covered by forest?
(iv) How many police stations are with in a distance of 500 m from a particular point?
(v) What is the distance of a nearest hotel from a particular point?

Queries help in GIS analysis for checking the quality of data and the results obtained. The queries may be classified as

(i) spatial query, and
(ii) aspatial query.

The queries, which require spatial analysis, fall under the class of *spatial query*, while aspatial *queries* use only attribute data of features involving no spatial analysis. A query such as how many hospitals with heart care facility are located in a given area is an aspatial query as it can be performed by database software alone and it does not involve analysis of the spatial component of data.

It is possible to combine the individual queries to identify the entities in a database to satisfy two or more spatial and aspatial criteria, for example where are the hospitals which have more than 10 private wards. Boolean operators AND, NOT, OR, and XOR are often used to combine different data sets by overlay.

10.5 RECLASSIFICATION

Reclassification is another form of query in raster GIS with only difference that it results in a new image in which the features of different classes have different codes. If cell value of a particular class of feature selected is say 10 (Fig. 10.4a), then in reclassification all the cells with value 10 may be assigned a new value, say 1, and the remaining cells with various classes of features may be assigned a value, say 0. This will result in a new image which only highlights a particular selected feature as shown in Fig. 10.4b. This new image is a Boolean image.

It is also possible to create a new image by reclassifying the features assigning new cell values that indicate weightage of different classes. An example is given in Table 10.1. The weightage may be based upon certain criteria.

10.6 BUFFERING AND NEIGHBOURHOOD FUNCTIONS

Buffering function, also known as *proximity function*, in GIS is one of the *neighbourhood functions* and it is used to create a zone of interest around an entity, or set of entities.

Buffering allows a spatial entity to influence its neighbours, or the neighbours to influence the character of an entity. Other neighbourhood function that include *data filtering*, involves the recalculation of cells in a raster image based on characteristics of neighbours. If a point is buffered, a circular zone is created (Fig. 10.5a), and buffering lines and areas creates new areas (Fig. 10.5b and c).

11	11	11	11	11	9	9	9	9	9
11	11	10	10	10	9	9	9	9	9
11	11	10	10	10	11	11	11	10	10
12	12	12	10	10	10	11	11	11	10
12	12	12	12	10	11	11	11	11	10
12	8	8	8	8	10	11	11	11	10
12	8	8	9	9	9	9	11	11	11
12	12	10	10	10	10	9	11	11	11
12	12	10	10	10	10	10	11	11	11
12	12	10	10	10	10	10	11	11	11

(a) Old cell values

0	0	0	0	0	0	0	0	0	0
0	0	1	1	1	0	0	0	0	0
0	0	1	1	1	0	0	0	1	1
0	0	0	1	1	1	0	0	0	1
0	0	0	0	1	0	0	0	0	1
0	0	0	0	0	1	0	0	0	1
0	0	0	0	0	0	0	0	0	0
0	0	1	1	1	1	0	0	0	0
0	0	1	1	1	1	1	0	0	0
0	0	1	1	1	1	1	0	0	0

(b) New cell values

Fig. 10.4 Reclassification

Table 10.1 Reclassification of land use data by assigning weights

Land use	Old cell value	New cell values (Boolean)	New cell values (weight)
Wetland	8	0	1
Water	9	0	2
Agriculture	10	1	5
Forest	11	0	4
Industrial	12	0	3

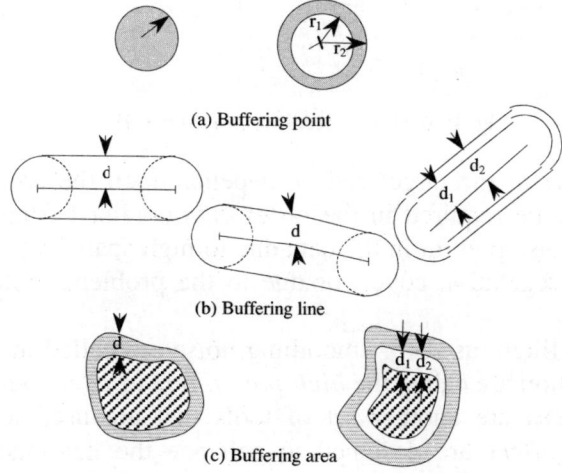

(a) Buffering point

(b) Buffering line

(c) Buffering area

Fig. 10.5 Different types of buffering function

Buffering function is used to answer the questions such as which hotels fall within a distance of 500 m from venue of a conference, or which hotels are within 500 m of a main road. An alternative approach would be to calculate distances of hotels and then identify those, which are less than the specified distance. There could be several approaches to get the answer of a query, but the one that is less time consuming, should be adopted.

Different GIS softwares use different types of computational operation to answer a query. Fig. 10.5 illustrates only the common and basic type of buffer operations. Buffer zones may be fixed or varying according to feature attributes. In a road network, major roads may have wider buffer zones while the minor roads may have narrower buffer zones.

In vector GIS, buffer zones are often created with the use of a single command or option while raster GIS utilize proximity calculation [Sec 10.3, (*iii*)], resulting in a new raster data layer where the attribute of each cell is a measure of distance.

In raster GIS, neighbourhood function alters the values of individual cells on the basis of adjacency. Filtering is one of the functions of neighbourhood, used for processing of remote sensing data. Filtering changes the value of a cell, which depends on the attributes of the cells in neighbourhood. A filter comprises of a group of cells around a target cell, and its size and shape are decided by the operator. A filter may have shape of squares or circles, and its size determines the number of neighbouring cells used in the filtering process. The filtering process involves passing of a filter of predetermined size and shape across the raster data set to recalculate the value of the target cell lying at the centre of the filter. In Fig. 10.6, a 3×3 square filter has been applied to recalculate the value of the cell *dc* which has a value of 4.

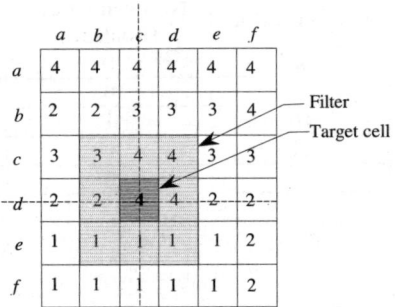

Fig. 10.6 Filter operation in raster GIS

The recalculated value of the target cell *dc* depends upon the criteria used as given in Table 10.2. Filtering may be required in the raster data obtained from a classified satellite images to smoothen the noise present in the data due to high spatial variability in a particular class of feature such as vegetation cover, or due to the problems with the data collection devices.

In signal processing, filters used for smoothing noise are called the *low-pass filters* and those used for edge detection are called the *high-pass filters* or *edge-detecting filters*.

Texture transformations are another set of tools that are used to identify the spatial pattern in data. *Texture filters* are designed to enhance the heterogeneity of an area. A common algorithm for a texture filter is to calculate the standard deviation of the cell values in a 3×3 neighbourhood. If the attribute values in this neighbourhood are all similar, the standard deviation is small, and it is said that this neighbourhood has *low texture* or *low variability* and in case there are many different attributes in a neighbourhood, we have *high texture*. Texture filters are also used to find the boundaries between delimited areas, since texture within a homogeneous area must be zero.

Table 10.2 Filtering operation criteria

Criteria	Target cell	Original value	Recalculated value
Minimum filter	dc	4	1
Maximum filter	dc	4	4
Mean filter	dc	4	2.67
Model filter [1]	dc	4	4
Diversity filter [2]	dc	4	4

1 = Most frequently occurring class; 2 = Number of different classes

There are other spatial transformations useful particularly while dealing with elevation data. A *slope transformation* turns a data layer of elevation into one of slope, by calculating the local first derivative. A companion to slope is aspect. An aspect calculation is used to determine the direction that a slope faces. A mathematical way of explaining aspect is to calculate the horizontal component of the vector perpendicular to the surface. Aspect is usually classified into bins of fixed size, so that the resulting data layer is not continuous, but ordinal. A frequent choice is to classify slope into eight categories, each representing an eighth of a circle (or 45°) range in aspect.

The characteristics of a data set along a specified line can also be determined by the use of another neighbourhood function. Such a process is commonly presented in the form of graphs as profiles or cross-sections of data set. Profiles are used to determine the slope along a line between two points for construction of highway, canal, or to determine the intervisibility between the two points.

10.7 DATA INTEGRATION - MAP OVERLAY

Integrating two or more different thematic map layers of the same geographic area is a common operation in GIS analysis. The technique of *map overlay* has many applications, such as for visual comparison by overlaying a map showing only hospitals on a road network map, to answer the query that 'where are the hospitals located'. In this case no new data are produced. This technique is also used for the overlay of vector data on a raster background image, which is a scanned topographic map.

When overlays involve merging of data from two or more data layers of same area, new spatial data sets are created in the form of a new output data layer. To answer certain types of query, such as which hospitals are within 500 m from a highway, buffering operation and an overlay function are required, which cannot be done without creating a new data layer by merging.

GIS vector data and raster data have different approach of performing map overlays, and are discussed in the following sections.

Vector Overlay

In a vector-based system, the analysis is based on a polygon intersection algorithm in which new polygons are created as needed, and redundant boundaries are eliminated (Fig. 10.7). Vector map overlay relies heavily on the two associated disciplines of geometry and topology. The data layers being overlaid need to be topologically correct boundaries so that the lines meet at nodes and polygon boundaries are closed. To create topology for a new data layer produced as a result of the overlay process, the intersection of lines and polygons from the input layers need to be calculated using geometry.

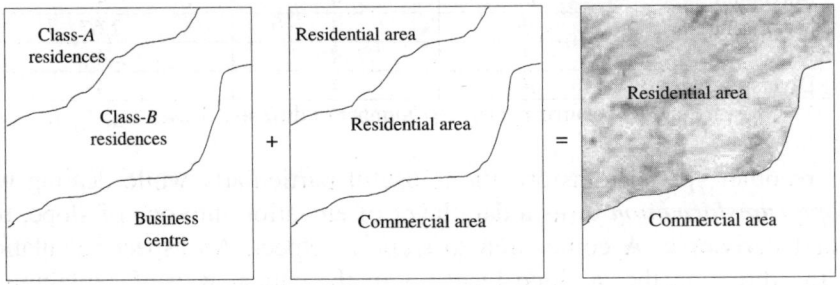

Fig. 10.7 Merging and redefinition of features.

There are three main types of vector overlay (Fig. 10.8), which are

   (*i*) point-in-polygon,
   (*ii*) line-in-polygon, and
   (*iii*) polygon-on-polygon.

Point-in-polygon

When it is desired to locate a point with respect to polygons, *i.e.*, in which polygon a point lies, point-in polygon overlay is used. Fig. 10.8a shows two data layers containing wells as points and soil types as polygons. The point-in-polygon overlay on these layers results an output layer indicating that in which soil category the wells lie. This information is given as an output map of a new set of wells created with additional attributes describing the soil type.

Line-in-polygon

To answer the queries like whether a road lies on sandy or clay type of soil, line-in-polygon overlay is used. Fig. 10.8b shows data layers of roads and soil types. When the two layers are integrated, the roads are split into smaller segments, depending on which part of a road falls in which type of soil category. A database record of for each new road segment is created in the output map. The output layer is more complicated then the two input layers as topological information is required to be retained, and therefore, line-in-polygon is more complex.

Polygon-on-polygon

This overlay as shown in Fig. 10.8c is used to answer the queries such as, (*i*) where are the different types of soil or areas lying within urban area, (*ii*) where is the urban area boundary and where clay types of soils within urban area boundary are, or (*iii*) where are sandy soils within urban area. The query of type (*i*) involves Boolean (OR) operator, type (*ii*) involves

(NOT) and the operation is referred to as *cookie cutting*, and type (*iii*) involves (AND). In mathematical terms these operations are respectively referred to as UNION, IDENTITY, and INTERESECT.

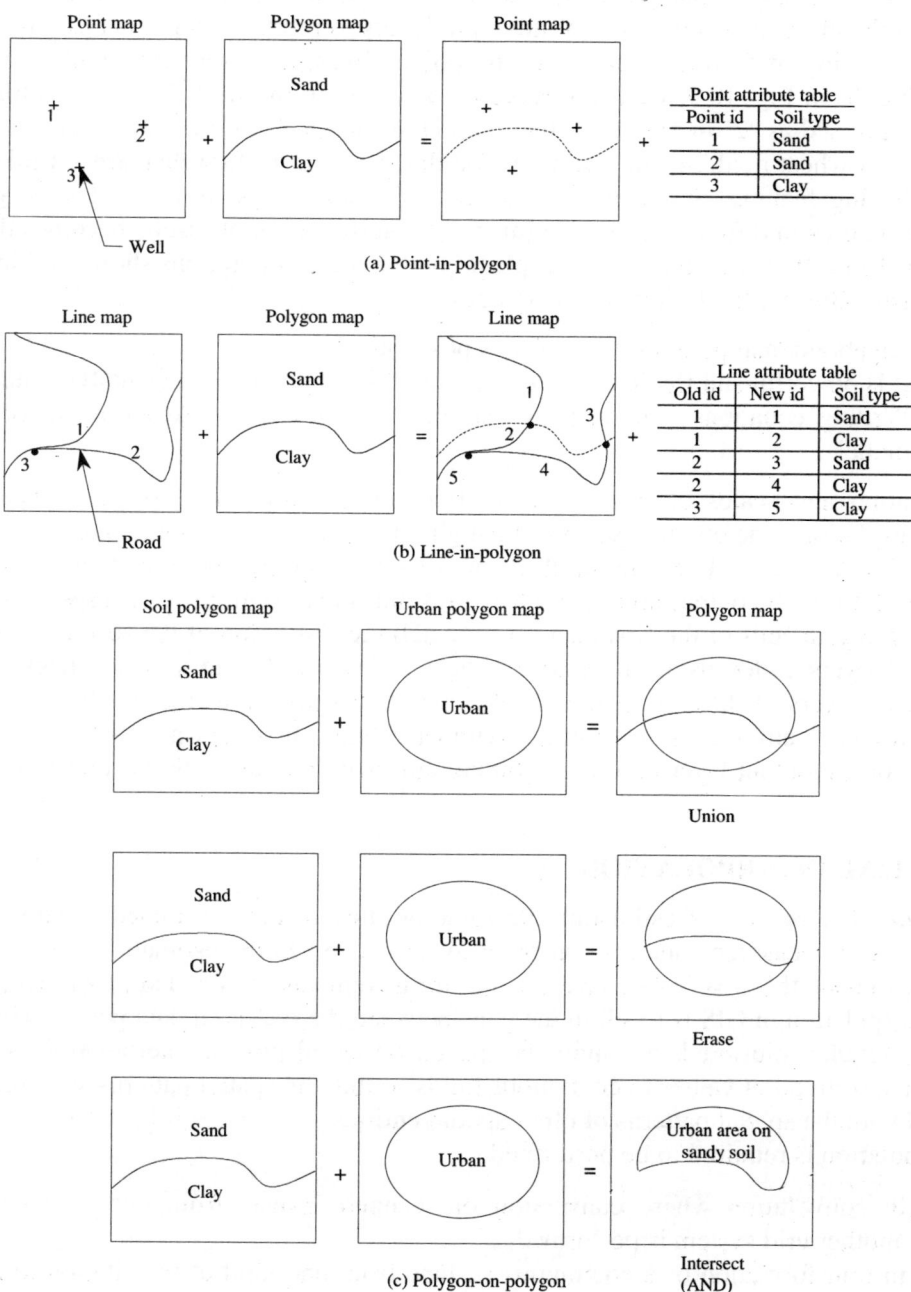

Fig. 10.8 Vector overlay operations

Raster overlay

In a raster-based system, cells in input data represent the raster data structure. A single cell represents a line, a string of cells represents a line and a group of cells represents an area. Raster overlays employ mathematical operations addition, subtraction, multiplication or division on the individual cell values of the input layers to produce output data. This requires appropriate coding of features represented by points, lines, and areas the input data layers. For example, as in Fig. 10.9, wells are represented as '1' in the well map layer while sewer lines are expressed as '2' in the sewer line map layer. In the land use map layer, the coding may be '3' for wheat field, '4' for forest, '5' for clay soil and '6' for urban areas while for all the cells having features of no interest as 0, in all data layers. If the codes assigned to different land uses in different layers are same, the interpretation of results becomes different as given in Table 10.3, which gives the explanation to various operations shown in Fig. 10.9.

The raster GIS has the following advantages:

(*i*) Algebraic manipulation of images is possible,
(*ii*) It is powerful and flexible for integration of data and organizing analysis, and
(*iii*) Writing equations with map as variables is possible for development of spatial models.

Resolution and scales of measurement affect raster overlays. If the two data are at different resolutions, the output layer has the higher resolution out of the two resolutions. For example, if a data set at a resolution of 10 m is to be combined with another data set at a resolution of 40 m, the output layer will have a resolution of 10 m. In such cases, it would be better to aggregate cells of the first data set to match the resolution of the second data set. In the case of layers coded using nominal, ordinal, interval, and ratio scales (Chapter 9), the result of two layers of different coding scales may be of no use as the numbers in the two layers do not have any logical relationship with each other. It is very necessary to assess the usefulness of the output layer before combining any two data layers that it makes some real sense.

10.8 SPATIAL INTERPOLATION

Interpolation is a process of estimating the value of attributes at unsampled locations within area covered by measurements made at point locations. When the estimation is done for the sites lying outside the area covered by existing measurements, it is called *extrapolation*. The role of interpolation in GIS is to fill in the gaps between the observed data points, which may be having regular, clustered, or randomly spaced observed points. Interpolation is used to convert data from point values to continuous fields so that the spatial patterns sampled can be compared with the spatial patterns of other spatial entities.

Interpolation is required to be performed

(*i*) in convolution where conversion of scanned images from one grid system to another grid system is performed,
(*ii*) in transformation of a continuous surface from one kind of tessellation to another, and
(*iii*) in conversion of data from sets of sample points to a discretised, continuous surface.

Table 10.3 Explanation of the operations shown in Fig 10.9.

(a) Operation equivalent to vector point-in-polygon (Fig. 10.9a)

Well	(i) Well station = 1
	(ii) Feature of no interest = 0
Wheat	(i) Wheat = 3
	(ii) No feature of no interest = 0
Output map	(i) Wheat field with well = 4
	(ii) Wheat field = 3
	(iii) Well = 1
	(iv) Neither well nor wheat field = 0

(b) Operation equivalent to vector line-in-polygon. (Fig. 10.9b)

Sewer line map	(i) Sewer line = 2
	(ii) Feature of no interest = 0
Urban map	(i) Urban = 6
	(ii) Feature of no interest = 0
Output map	(i) Sewer line in urban area = 8
	(ii) Sewer line not in urban area = 2
	(iii) Urban area with no sewer line = 6
	(iv) Neither urban area nor sewer line = 0

(c) Operation equivalent to vector polygon on polygon, (Fig. 10.9c)

Soil map	(i) Clay soil = 5
	(ii) Feature of no interest = 0
Forest map	(i) Forest = 4
	(ii) Feature of no interest = 0
Output map	(i) Forest on clay soil = 9
	(ii) Forest not no clay soil = 4
	(iii) Clay soil not in forest = 5
	(iv) Neither forest nor clay soil = 0

(d) Operation: Addition (Fig. 10.9d)

Soil map	(i) Clay soil = 1
	(ii) Feature of no interest = 0
Forest map	(i) Forest = 1
	(ii) Feature of no interest = 0
Output map	(i) Forest on clay soil = 2
	(ii) Forest or clay soil = 1
	(iii) Neither forest nor clay soil = 0

(e) Operation: Multiplication (Fig. 10.9e)

Soil map	(i) Clay soil = 1
	(ii) Feature of no interest = 0
Forest map	(i) Forest = 1
	(ii) Feature of no interest = 0
Output map	(i) Forest on clay soil = 1
	(ii) Other areas = 0

198 *Remote Sensing and Geographical Information System*

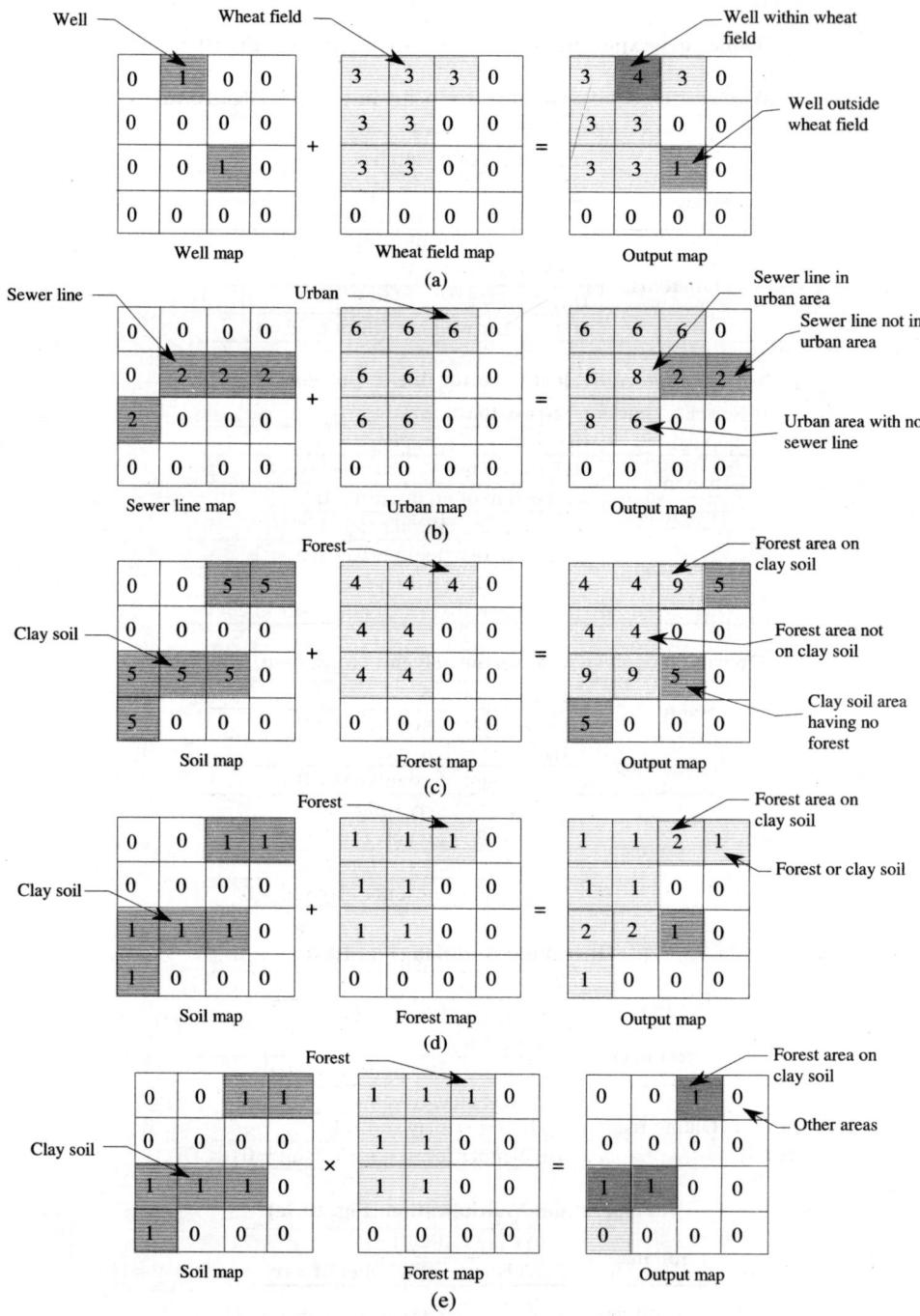

Fig. 10.9 Different types of raster overlay operations

Interpolation is most commonly used process for generating contours using DEM or DTM which contain spot levels obtained by levelling or photogrammetry. Most GIS packages often provide a number of interpolation methods to generate continuous surfaces for use as map overlays or display in their own right.

Classification of Interpolation Methods
Interpolation methods can be broadly classified as under.

Local or global: When a single mathematical function is applied uniformly to all observed data points, the interpolation method is said to be global, and when applied repeatedly to small subsets of the entire data set of observed points, the method is said to be local. The regional surfaces so formed are combined to create a surface of the whole area. In contrast to the local methods, the global method usually generates smooth surfaces.

Exact or approximate: The surface generated by exact interpolation methods passes through the observed data points keeping their values unaltered. The interpolation methods that fall under approximate methods do not pass the generated surfaces through the observed data points and they may alter or smoothen the values of observed data points to obtain a general trend for generating the surface. The exact methods of interpolation are employed when the observed data points have high degree of reliability whereas the approximate methods are used when there is uncertainty with the observed data points.

Gradual or abrupt: Gradual and abrupt methods of interpolation are distinguished by the continuity of the surface generated. As indicated by the names of the methods, a smooth surface is produced between the sample points by gradual methods to represent rolling terrain, and surfaces with abrupt changes by abrupt methods as required to represent vertical cliffs, ridges, and valleys, and therefore, terrain model requires both the methods to depict a terrain having such features.

Deterministic or stochastic: When a suitable mathematical model is available to generate a surface representing the terrain, deterministic methods are used. Since rarely such case exists in the real world, stochastic methods are used to take care of the randomness in the interpolated surfaces.

There are various methods of interpolation available but whichever method is used, the data derived are only an estimate of the real-world values, and therefore, before using a particular method, its limitations must be understood. Most commonly and frequently used methods are Thiessen polygons, TIN, and spatial moving average, and these are briefly discussed in the following sections.

Thiessen Polygon Method
Thiessen polygons, named after the climatologist A.H. Thiessen, are used in exact method of interpolation. In this method, global characteristics of the data set do not influence over the interpolation process, which is of local nature. It is also an abrupt method as sharp boundaries are present between the polygons.

The sampled points in the data set form the vertices (sometimes referred to as nodes) in the chain of triangles formed by the sampled points as shown in Fig. 10.12. *Delaunay triangulation* is the most commonly used method of creating a network of triangles from sampled points over a particular geographic space, and it is closely associated with *proximal regions* that are also called *nearest-neighbourhood regions, Thiessen polygons,* or *Voronoi polygons*. Proximal regions are created as the result of subdividing a particular geographic space encompassing a set of points into a set of convex polygons. Considering three points A, B, and C in a plane passing through them, the plane is partitioned into three proximal regions V_1, V_2, and V_3 using the perpendicular bisectors of AB, BC, and AC, respectively, and thus having one *anchor point* (the original point) in each region. A proximal region can be defined as a polygon within which every point is closer to its own anchor point than the anchor points of all other regions. For example, any point located in V_1 region is closer to B than A and C.

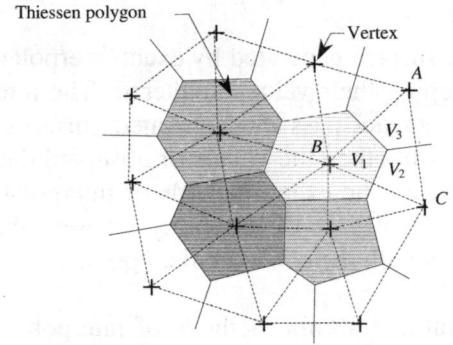

Fig. 10.12 Delaunay triangulation

Thiessen polygons are used in interpolation if there are only a few scattered sampled points yet the analyst wants to characterize regions on the basis of these points assuming that in each polygon, the influence of the enclosed sampled point is absolute. Mostly Thiessen polygons are involved in the determination of the influence of point data representing such as shopping centres, industries, or other economically based activities. By modifying the position of perpendicular lines on the basis of size or any other magnitude of each point, the Thiessen polygons become even more representative of the actual influence of shopping or industries on the surrounding space making the analysis better. Thiessen polygons have also been used to detect spatial patterns in vegetation.

Triangulated Irregular Network (TIN)

Triangulated irregular network is a method of constructing a surface from a set of irregularly spaced data points. In the TIN data model, the terrain is recorded as a continuous surface made up of a mosaic of non-overlapping triangular facets formed by connecting selectively sampled points of elevation data using a consistent method of triangle construction (Fig. 10.13). To make the interpolation simpler mathematically, most TIN models assume planer triangular facets. The sides of a triangular facet are called the *edge*, and the ends of an edge are called the *vertices* or nodes. In a TIN model, edges depict terrain features, for example, ridges, river channels, and breaks, while the vertices describe nodal topographic features, such as peaks, pits, and passes.

The TIN data model is distinct from the DEM data model in the following respect:

(*i*) In a TIN, every data point has (x, y, h) where h is the elevation, while in DEM (x, y) are implicit.
(*ii*) A TIN may include explicit topographical relationships between points and their proximal triangles.

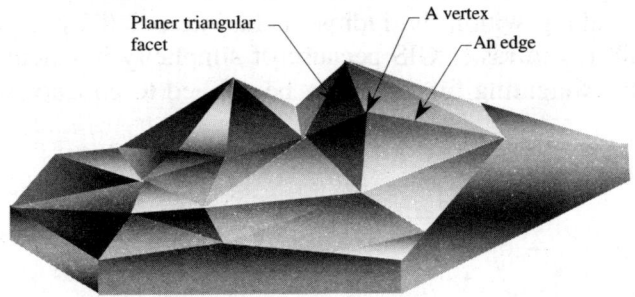

Fig. 10.13 Triangulated Irregular Network

In TIN data models, topological relationships play a significant role as using the method of triangulation, the totally unstructured collected elevation points are turned into a properly organized geographic database suited for terrain modeling applications. Since a given set of data points can be triangulated in many ways as shown in Fig. 10.14, different contour maps will be generated by different triangulation network of data points. Thus, a necessary procedure must be adopted to ensure the creation of an unique TIN for a given set of data points.

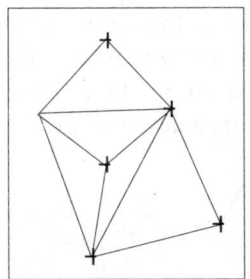 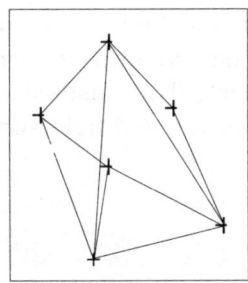 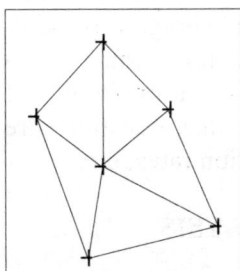 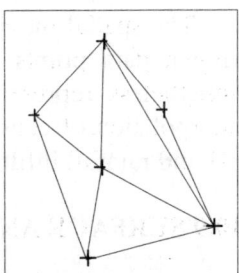

Fig. 10.14 Different triangulation network using same data points

The TIN model is an exact interpolation method based on local data points. It uses linear equation and trigonometry to calculate the interpolated values at point other than the data points within the boundary of the TIN.

Spatial Moving Average

The spatial moving average involves calculating a value at a point based on the range of values attached to neighbouring points falling within the range defined by the user. This user-defined neighbourhood filter is passed systematically across the region of interest, calculating new interpolated values as the filter moves over the dataset. This interpolation method produces a surface that suppresses the values of known data points to represent the global pattern in the data.

The spatial moving average is perhaps the common interpolation method used in GIS. It is an approximate point interpolation method since it does not honour the values of known data points. Generally, a *circular filter* is used since known data points in all directions have an equal chance of falling within the radius of the interpolated points, (Fig. 10.15a), but a *square filter* is preferred in raster GIS because of simplicity in calculation (Fig. 10.15b). If necessary, a distance weighting function may be applied to enhance the influence of closer data points.

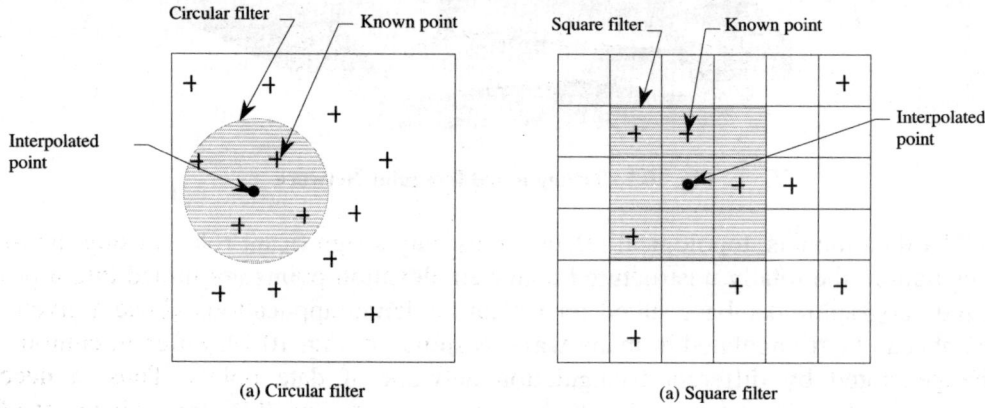

Fig. 10.15 Using filters in spatial moving average method

The spatial moving average method is most suited when it is found that the values of the known data points are not certain or exact, due to measurement errors but where they nonetheless represent variation in global pattern. This method has been applied for the interpolation of census data, questionnaire responses, and field survey measurements of soil pH and rainfall infiltration rates, etc.

10.9 SURFACE ANALYSIS

To generate a surface as close to a real-world surface, sufficient numbers of points at suitable locations are required. If the data points are not sufficient in number or at suitable locations, interpolation is used to fill the gaps and create the surface for analysis. In practice, however, a surface-fitting algorithm is commonly used to improve the results of terrain modeling. A local approach of interpolation that takes into account only two points in the neighbourhood of interpolated point lacks the consideration of the characteristics of terrain continuity and smoothness. The global approach of interpolation removes such shortcomings in the local approach by utilizing all or most of the data points to characterize the surface at a point, thus allowing estimation to be made on the trend of the surface.

A widely used global surface-fitting method is *trend surface analysis* where the terrain surface is approximated by a polynomial expansion of the coordinates of the sampled points. Some of the commonly used polynomials are

Linear	$z = a + bx + cy$	…(10.3)
Quadratic	$z = a + bx + cy + dx^2 + exy + fy^2$	…(10.4)
Cubic	$z = a + bx + cy + dx^2 + exy + fy^2 + gx^3 + hx^2y + ixy^2 + jy^3$	…(10.5)

where z is the estimated height at a point having coordinates (x, y), and $a, b, c,…, j$ are polynomial coefficients. The coefficient values are determined by solving a set of simultaneous equations formed using the known values at data points, by the method of least squares.

Finally, an optimum local interpolation method known as *kriging*, named after the pioneering work of Danie Krige, is now widely used in DTM software packages as it is flexible and can handle any type of data. The various approaches of kriging, such as *simple kriging*, *ordinary kriging*, *universal kriging*, *block kriging*, and *Co-kriging*, employ different criteria to estimate values at points. Kriging is a generic name for a family of least-square linear regression algorithms that are used to estimate the value of a continuous attribute (such as terrain height) at any unsampled location using only attribute data available over the study area. Kriging treats the continuous attribute to be interpolated as a regionalized variable. It is possible to map one attribute using other attributes, if available and spatially correlated, using co-kriging approach. This applies when the data for one particular attribute are more difficult or too expensive to obtain. Therefore, the other less expensive set of attribute data becomes surrogate for the more expensive set.

Creating New DEM Data Sets

New digital terrain data are sometimes required when the existing data are not of a resolution or quality compatible to a certain application. The production of new DEMs may also be warranted because of the availability of new data source. With the aid of interpolation using a surface-fitting algorithm, an economical approach is to refine an existing DEM by incorporating the new data set instead of creating a completely new data set, and create a DEM of a higher resolution from a lower resolution data. This requires a good understanding of the nature and requirements of both the old and new DEM data sets. For situations that require the use of surface fitting, the choice of method depends on the geographic extent of the new DEM. Local or global approaches may be used for DEMs covering a small area while for continual or national DEMs global approaches such as kriging may be computationally impractical due to millions of data point to be considered. In the development of a countrywide DEM, of Australia, an iterative finite difference interpolation technique developed by Hutchinson (1993) was used. This technique has the computational efficiency of local interpolation method without sacrificing the major advantages of global methods.

Extraction of Topographic Attributes and Landscape Features

Topographic attributes which are defined as numerical description of the terrain, and may be classified as (Moore, *et al*, 1991)

(*i*) Primary, and
(*ii*) Secondary (also referred to as compound).

Primary attributes are those topographic parameters that may be directly calculated from digital terrain, such as elevation and slope etc. *Secondary attributes* are those parameters of topography that are formed by combining the primary attributes with other environmental indices that characterize the spatial variability of specific processes occurring in the landscape, such as soil water content distribution.

The following primary attributes used to describe landforms, can be easily estimated from DEM:

(*i*) Elevation (the altitude or height of terrain points)
(*ii*) Slope (the gradient of the land)
(*iii*) Aspect (The azimuth of the maximum local slope)
(*iv*) Specific catchment area (the upslope area draining across a unit width of contour)
(*v*) Flow path length (the maximum distance of water flow to a point in a catchment)
(*vi*) Profile curvature value (the curvature of slope profile)
(*vii*) Plan curvature (the curvature of contour).

Further, parameters pertaining to geomorphological analysis may also be extracted, including local relief, drainage density, and statistics of slope and convexity. The extraction of topographic attributes and geomorphometric parameters is usual and intermediate process in digital terrain modeling.

DEMs and TINs are used to extract a wide variety of landscape features, conventionally associated with drainage and hydrological applications. However, landscape features are now extracted increasingly for studies in different branches of science and technology in a broader context of using GIS. Numerous algorithms available to extract landscape features can be categorized on the basics of primary landscape features they extract. These are

(*i*) Surface-specific points,
(*ii*) Linear or network features, and
(*iii*) Areal features.

Surface-specific points: Surface-specific points are topographic features such as peaks, pits, passes, and saddles. A straightforward process to extract these features from DEM is to compare elevation differences in a local neighbourhood. In a DEM, local elevation maxima represents peaks and local elevation minima represents pits, passes, and saddles.

Linear or network features: At present, the most commonly used method to extract linear topographic features from DEMs, is a hydrological approach (Mark, 1984). In this method, the drainage area of each DEM elevation grid is first determined by climbing recursively through the DEM (Fig. 10.16a and b) resulting into a matrix, called the drainage area transform that contains the drainage areas for all the grids in the DEM shown in Fig. 10.16c (Lo and Yeung, 2002).

The drainage area transform is then used to trace the channel pixels as identified by those grids with large drainage areas. Channels are recursively followed upstream until there is no more point that exceeds a minimum threshold (Fig. 10.16d). From the drainage network, ridges may be delineated, and non-continuity, if any, in the channels and ridges, is removed by interpolation. Thinning process may be required if channels and ridges consist of multiple adjacent lines to make them continuous line of one grid width.

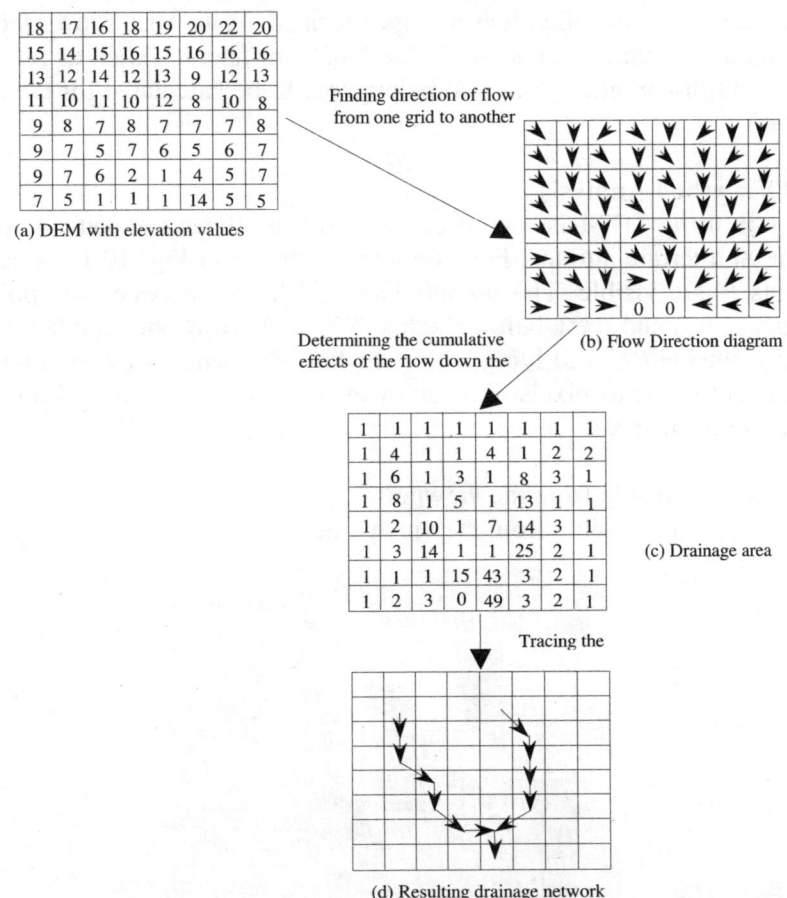

Fig. 10.16 Hydrological approach of delineating drainage network from a DEM

Areal features: The delineated drainage networks and ridges are used for the delineation of areal hydrological features such as catchments, by using variants of the recursive DEM climbing algorithms to identify those grids belonging to a specific channel or ridge.

Calculating Slope and Aspect

For many GIS applications, slope or steepness and aspect of a unit of terrain are required. In raster DEMs, slope and aspect are calculated using a 3 × 3 window that is passed over a database to determine a 'best fit tilted plane' for the cell at the centre of the window. This approach is used to determine the coefficients in polynomial given by Eq. (10.1), and then slope and aspect can be calculated using the following formulae:

Slope $\quad S = b^2 + c^2$ ...(10.6)

Aspect $\quad A = \tan^{-1}(\dfrac{c}{b})$ ...(10.7)

In the vector GIS, slope and aspect are calculated at the time of generation of TIN using a series of linear equations for individual triangles formed. Plate 10.1 shows the examples of shaded slope and aspect images generated. The landform features such as rate of change of slope and convexity or curvature may also be required to be calculated for landform analysis and classification.

Visibility and Viewshed Analysis

Visibility analysis using DTM is identification of areas of terrain that are visible from a particular point on a terrain surface. For example, as shown in Fig. 10.17, P is a point from where the points P' are visible. The points visible P'' falling between the points P and P', called as the *viewpoints* and lying below the line PP', will always be visible from P'. But the points P''' falling outside PP'' and lying below the line $P'P''$ will always be invisible from P'. In raster GIS the points are as pixels. For each viewpoint in a DEM, a visibility matrix V can be constructed. In this matrix if

V_{ij} = 1, point V_{ij} *is* visible from the viewpoint
$\quad\;\,$ = 0, point V_{ij} is not visible from the viewpoint.

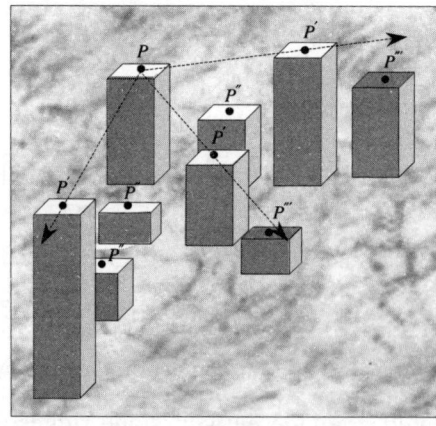

Fig. 10.17 Visibility analysis using a DTM

The area in which all the pixels are visible from the viewpoint is called as *visibility region* of the viewpoint. The determination of intervisibility between two pixels is called as *line of sight analysis*. The work associated with the computation and interpretation of the visibility matrices and visibility regions, is referred to as *viewshed analysis*.

The method of viewshed analysis is similar in both raster and vector GIS, and in each case the result of analysis would be presented as a viewshed map. Some GIS packages offer more sophisticated visibility analysis allowing the user to specify the height of the viewing position as an offset above the terrain surface height at that location, making the computations easier for locating the top of tall structures, such as tall chimneys, or pylons. Some GIS provide facility of incorporating barrier features, or 'walk' or 'fly' through a terrain model. Other data such as satellite can also be draped onto the surface of a DTM for better views and analysis.

10.10 NETWORK ANALYSIS

In the context of GIS, a network is defined as a set of interconnected linear features through which resources can flow. Common examples of networks include highways, railways, city streets, canals, rivers, transportation routes, such as for garbage collection, mail delivery, and school buses, and utility distribution systems, for example, telephone, electricity, water supply, and sewage. There are many spatial problems that require the use of network analysis for their solution. These include, for example (Lo and Yeung, 2002),

(*i*) to find the shortest path (in terms of physical distance or least cost) that can be followed to visit a series of features in a network, known as *pathfinding*,

(*ii*) to assign one or more portions of a network to be served by a facility or business location, called as *allocation*,

(*iii*) to find all portions of the network that are connected with the movement of a particular feature (*e.g.*, city transport), called as *tracing*,

(*iv*) to depict the accessibility of a location and the interactions that occur between different locations (based on a technique known as *gravity modeling*). This is widely used in economics, geography, engineering, and urban planning, known as *spatial interaction*,

(*v*) to generate a distance matrix between different pairs of locations in the network, known as *distance matrix calculation*, and

(*vi*) to determine simultaneously the locations of existing and planned facilities, as well as the allocation of demand to these facilities, known as *location-allocation modeling*.

Some of the classic network applications are discussed below.

Shortest Path Problem

The shortest path which is the shortest distance (or least-time path) between two points on a network, is determined by proximity analysis in a raster GIS. Impediments to travel are added to a raster grid by increasing the value of cells that are barriers to travel. Then the result is obtained as least-cost route. Vector GIS network analysis of finding shortest path is more flexible and provides a thorough impediment analysis such as restrictions and congestion in traffic routes.

Route Tracing
Route tracing through the network analysis is required to identify the routes for unidirectional flow of resources with special reference to stream networks and services, such as sewerage systems and cable TV networks. Other applications are determination of streams contributing to a reservoir, customers serviced by a particular sewer main, or clients affected by a broken cable. The key concept in route tracing is connectivity of network links at network nodes. Knowledge of direction of flow is an important factor for route tracing. Therefore, each link in the network must be associated with direction of flow that can be defined at the time of digitizing process by keeping the directions of digitization and flow same. Now tracing the links downstream or upstream of a point on the network, is performed by moving in the direction of flow or against it.

Salesman Travel Problem
A salesman may be required to visit a specific set of clients in a day, for which he would like to know the best (usually the quickest) route that he can follow to finish his job. Similar problems are collection of garbage and distribution of mails. Such problems are solved by analyzing the order of stops and paths between them. Getting solutions of such problems, is a complex task, and to simplify such tasks, the ordering of the stops can be determined by calculating the minimum path between each stop and every other stop in the list based on impedance met in the network. A trial and error method (also referred to as heuristic) is then applied to order the visits minimizing the total impedance from first stop to the last.

Location-allocation Modeling
An important application of network analysis is allocation of resources. This is done by modeling of supply and demand in which help of movement of goods, people, and information or services through the network are required to match the demand with the supply. Allocation of resources is usually done by allocating links in the network to the nearest center of supply taking into account impedance values. The maximum catchment area of a particular supply centre can also be determined on the basis of the demand located along adjacent links in the network. A situation may arise by imposing limitation on supply and demand, in which some parts of the network may not be serviced despite a demand being present in that part. Such problems can be solved by reducing the supply to some parts of the network or by identifying the optimum location for a new centre to meet the shortfall in supply relative to demand.

10.11 DIGITAL TERRAIN VISUALIZATION

Visualization of DTM in the context of GIS has two primary objectives:
    (*i*) to communicate geographic information, and
    (*ii*) to provide a means for data exploration and hypothesis refinement,

and therefore, it is an integral component of DTM. Visualization means representing relevant terrain information about a given geographic area that result from the analysis of the characteristics of its topography and related spatial phenomena. There are numerous well-developed techniques, and some are being developed. These techniques have been classified according to the dimension of the graphical display, and have been presented below.

Two-dimensional Visualization of Terrain
The most conventional and commonly used method of digital terrain visualization is using contours. This method is a quantitative approach of representing three-dimensional terrain in two dimensions. A major drawback of this approach is that an inexperienced map user finds it difficult to visualize relief in three-dimensional terrain from two-dimensional representation. As a result, different methods have been developed to create a three-dimensional impression of relief using contours.

Shaded contours method: In this method contours drawn by varying their widths according to illumination brightness. This method has not found much use in GIS.

Analytical or automated shading method: Hill shading method of two-dimensional terrain visualization is based on the principles of applying illumination to the surface normal vector for the DEM or TIN (Fig. 10.18). Similar principle but different in algorithm, is employed for hill shading using DEM (Fig 10.19). Hill shading is referred to as a qualitative method of two-dimensional terrain visualization, and therefore, numerical values of elevation cannot be measured from the display.

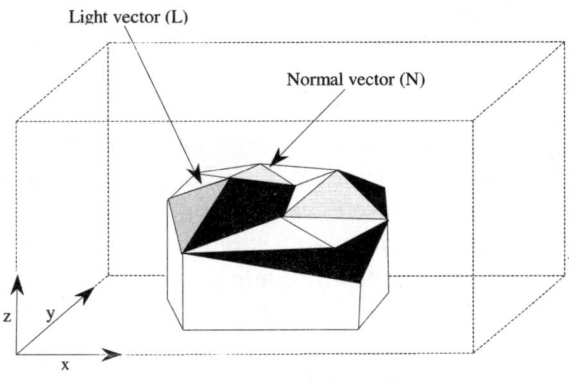

$L \times N = 0.0 <$ illumination values < 1.0

Fig. 10.18 The concept of hill shading (*Source: Lo and Yeung, 2002*)

Two-and-a-Half-Dimensional Visualization of Terrain
It is possible to simulate three-dimensional terrain from a two-dimensional display through the use of clever graphics. However, technically a two-and-a-half-dimensional `display is basically an isometric model. In an isometric model, the principal axes x, y, and z are equally inclined to each other, and the scale along these axes is also same. Therefore, two-and-a-half-

dimensional visualization though being relatively easy to implement from programming point of view, it does not create the impression of visual depth. Classic uses of the two-and-a-half-dimensional approach are wire frame diagrams (Plate 10.2). A more realistic perspective view of the terrain is created by draping other kinds of images on two-and-a-half-dimensional models (Plate 10.3). However, with the development of powerful computer graphics software and hardware, three-dimensional terrain visualization has become possible, superceding the two-and-a-half dimensional visualization.

Fig. 10.19 Hill shade relief image (*Source: Lo and Yeung, 2002*)

Three-Dimensional Visualization of Terrain

In a three-dimensional model, x, y, z data points are used to form a solid structure that may be visualized in a perspective view. This three-dimensional model is a solid model and can be viewed as an analog for the physical space as perceived by an observer. Three-dimensional modeling is one of the most exciting developments in GIS technology as it allows the full specification of three-dimensional operations on the objects and phenomena within the constraints of the geometrical model used (Lo, *et al*, 2002). It is conceptually complex and computationally intensive. It includes several sophisticated computer graphics techniques such as geometric transformation, depth cueing, hidden edge and surface removal, antialiasing, shading, ray tracing, shadows, surface texture detail, and atmospheric attenuation.

Multidimensional Visualization of Terrain

Multidimensional visualization of terrain makes use of techniques to present the terrain in sequences like the frames of a movie. Time as fourth dimension is included by animation techniques to visualize the temporal changes in the terrain from a static view point. More advanced animation techniques are being used to change the viewpoint continuously to introduce the fly-by, or fly-through effect of visualization. Digital terrain data of multidimensional nature are best managed in virtual environments.

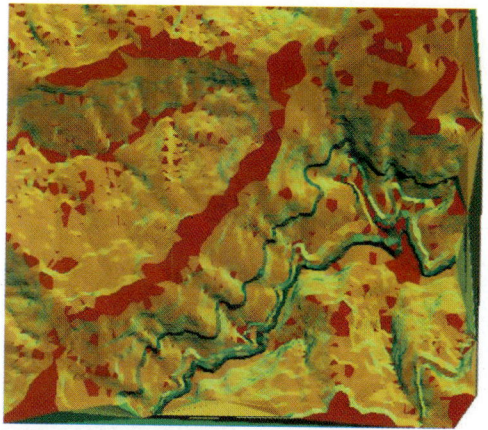

(a) Digital Elevation Model

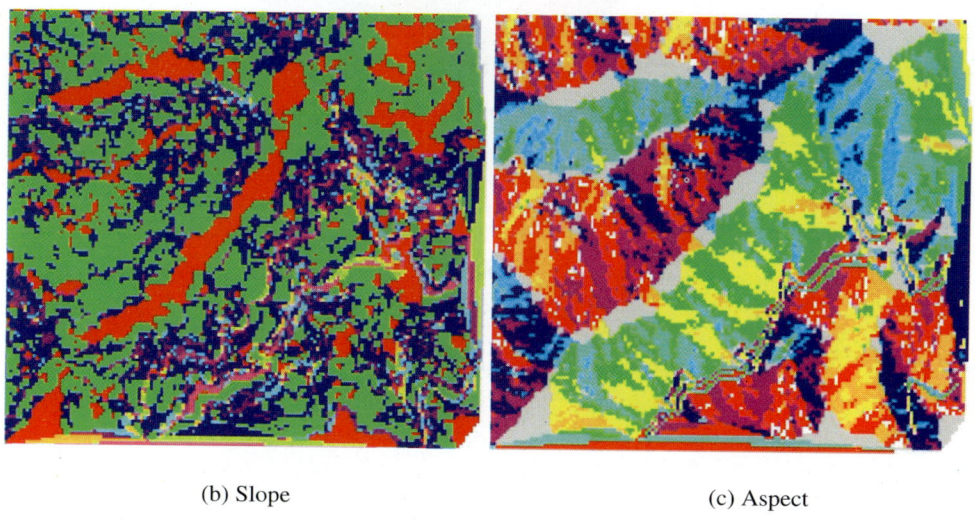

(b) Slope (c) Aspect

Plate 10.1 DEM and its derived product

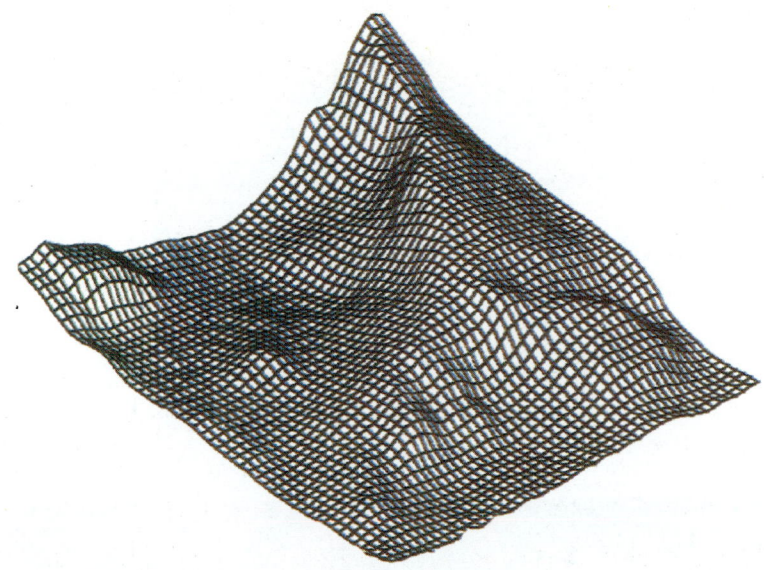

Plate 10.2 A wire frame diagram to show relief (*Source: Heywood, et al, 2000*)

Plate 10.3 The wire frame model in Fig. 10.21 draped with orthorectified aerial photograph
(*Source: Heywood, et al, 2000*)

GIS APPLICATION 11

11.1 INTRODUCTION

GIS application involves practical aspects of designing and managing a GIS for a project. Good project design and management are essential for producing a useful and effective GIS application. This chapter, therefore, aims to provide a framework for the development of GIS application. Design techniques help in identifying the nature and scope of a problem, defining the system requirements, quantifying the amount and type of necessary data, and indicating the data model needed and the analysis required. Delivering a completed project on time and ensuring its quality are a part of management techniques.

A general approach for project design and management has been outlined in this chapter followed by some case studies. It may be noted that each project is unique in nature, and requires a specific design and management approach depending on the technology available to the users, and the organizational culture in which the GIS must reside. In general, a GIS application may include:

- (*i*)　problem identification,
- (*ii*)　designing a data model
- (*iii*)　project management
- (*iv*)　identifying implementation problems, and
- (*v*)　project evaluation.

It would be out of scope of this book to discuss the above points in detail, and therefore, the discussion would be limited to only giving a direction to appreciate the steps involved in designing and managing a GIS project.

11.2 PROBLEM IDENTIFICATION

The first step towards the GIS application is to identify the problem that has to be addressed by the GIS. There can be two approaches as under:

- (*i*)　Creating a rich picture
- (*ii*)　Developing a root definition.

Rich Picture: A schematic view of the problem to be addressed in a project is referred to as *rich picture*. This schematic view presents the main components of the problems in a project, as well as any interactions that exist, and helps to organize the ideas that arise during discussions between, for example, a property dealer and a property buyer. A rich picture may adopt symbols to indicate certain activities. For example, crossed swords are used to express conflicts, eyes to represent external observers, speech bubbles to indicate personal or group opinions. A consensus view of the problem emerging out of participation of all concerned can be well represented by a rich picture.

Root Definition: Different users may have different views of the problem. The *root definition* is a view of a problem from a specific perspective. The system developer must arrive at a common root definition considering all view points. This helps others to evaluate and understand the way GIS is being constructed. In designing a GIS application, it should be ensured that it addresses to a range of needs and solutions.

Soft Systems Approach: The soft systems approach, originally developed by Checkland (1981) to problem identification, is a method of addressing unstructured problems. A problem is said to be a structured problem when a definite location is identified for a particular problem, whereas in unstructured problem, instead of definite location, a neighbourhood is known. Soft systems approach depends on the context or *world view* from which the problem is being considered.

11.3 DESIGNING A DATA MODEL

Data modeling has different meaning in different context. In the context of the design and management of a project, it can be viewed as consisting of a *conceptual model* and a *physical model*. In a conceptual data model, elements of spatial form and process are included in the rich picture. The detailing to represent the conceptual model within the computer is done through physical data model. It includes details about the spatial data model, the data structure, and the analysis scheme. GIS application requires a well-planned data modeling to meet out the expectations of GIS users.

One way of creating a conceptual data model is to identify clearly the elements of data model: the entities, their states, and their interrelationships, and present them in the form of flowcharts using standard symbols that illustrate different aspects of the model. The advantage of using a flow chart is that it illustrates all the stages involved modeling of the problem from data requirements to output requirements.

To create physical data model, additional details are required that describe modeling of spatial entities, their associated attributes, and interrelationships of entities in the computer. In other words, in physical data model emphasis is on developing a model of the relationships between the entities, and it is referred to as an *analysis of scheme*. Cartographic modeling is one of the most frequently used techniques of designing an analysis scheme. Cartographic modeling is an acceptable methodology for processing of spatial information. It is a generic way of expressing and organizing the methods by which spatial variables and spatial operations are selected, and used to develop a GIS data model. Cartographic modeling involves geographic data processing methodology that views maps as variables in algebraic equations. In map algebra the symbols are assigned numeric attributes of map elements or even whole maps instead of x, y, z as algebra. Using mathematical operations such as addition, subtraction, multiplication, and division, new numbers are generated by interaction of numbers assigned to symbols in an equation, producing new maps by use of specific spatial operations. For example, if *a* is road transport map, *b* is rail transport map, and *c* is the communication map, then the equation from map algebra is

$$a + b = c \quad\quad\quad \ldots\ldots (11.1)$$

where + sign indicates the spatial overlay operation 'union', performed on *a* and *b* to produce *c*.

The various stages in the development of a cartographic model consist of

(*i*) identifying the required map layers or spatial data sets,
(*ii*) using natural language to explain the process of moving from the data available to a solution, drawing a flow chart representing graphically the process, and
(*iii*) annotating the flowchart with the commands necessary to perform these operation within the GIS being used (Burrough, 1986).

11.4 PROJECT MANAGEMENT

A good project management is an essential prerequisite for the success of a GIS project. After constructing the data model, the GIS must be implemented, and in many cases integrated into the wider information strategy of an organization. An Information Technology (IT) project can be managed by different approaches. The commonly used approaches are the system life cycle and prototyping.

The *systems life cycle* is a linear approach used to manage the development and implementation an information technology system. It provides a very structured framework for the management of a GIS project. In this approach, also referred to as *waterfall model*, (Skidmore and Wroe, 1988), the outputs from the first stage of the process are used in the second and the output from the second stage in the third stage, and so on. Various stages may be feasibility study, system investigation and system analysis, system design and implementation, review, and maintenance.

Though the system life cycle approach is a popular management tool for information technology projects, it has a number of limitations. Designers who use the system life cycle approach often fail to address the context of the business for which the system is being developed. The time scale and linear nature of the system life cycle approach fails to allow for change in the scope and character of the problem. This approach does not put the user at the centre of the system design, and also it is often considered to favour hierarchical and centralized systems of information provision.

In the *prototype approach* to manage information technology projects, the basic requirements of the system are defined by the user using the rich picture and root definition techniques. These basic ideas are utilized by the system designer to create a prototype structure fulfilling the needs identified by the user. This developed system is then experimented by the user to find out if it fulfills the requirements as expected. The system designer may improve the system on the recommendations of the user and potential users to make it of wider value.

The prototype approach also has some problems such as Difficulty in managing the prototyping, changing of resources implications following the development of the first prototype system, and knowing when to stop the development.

11.5 IMPLEMENTATION PROBLEMS

GIS design and management will always have problems, which cannot be predicted by any mean. These problems, for example, may be non-availability of data in the format required by the GIS software, lack of knowledge about the GIS being used imposing technical and conceptual problems in implementation, users changing their requirements.

When the GIS data available is not in the desired format, the GIS designer has two options. First, the designer should look for a supplier who can provide the data in the desired format, or the second option is to convert the available data into the desired format. Many times, it is found that a better and easier approach is to employ an independent expert to undertake application development or specific analysis when it is inevitable that applications will be limited by technical and conceptual knowledge of the users about which spatial data model, data structure, or analytical operation is most appropriate for the project. The solution to the problem of changing needs of users is to get frequent-feedback from the end-users of the GIS.

11.6 PROJECT EVALUATION

Project evaluation requires that the output produced by the system is usable, valid and meets the objective of the project. Testing the GIS and validating output are a crucial part of the design process. If the results are in the form of predicted values, then validation becomes difficult. GIS can be made more economical and appropriate by taking prototyping approach in which frequent testing and evaluation take place automatically.

To determine that the developed GIS applications meet the objectives defined at the beginning of design process, the following approaches may be adopted:

(i) Feedback can be obtained from all the parties involved in the process of design and development of the GIS about achieving the goals for which it was designed.
(ii) GIS output can be checked against reality, and
(iii) The adaptations and changes that had to be made at the time of moving from the rich picture through the GIS data model to the implementation can be evaluated.

11.7 CASE STUDIES

In order to highlight the utility of GIS in various fields, some case studies have been taken up for discussion. Basic purpose of providing these case studies is to give the reader an idea regarding the different types of data required for a GIS project, the methodology to be adopted, the procedure of analysis, and the presentation of output. The case studies discussed are

- (*i*) Site suitability for urban planning,
- (*ii*) Road accident analysis,
- (*iii*) Irrigation water management,
- (*iv*) Tourism information system,
- (*v*) Worldwide earthquake information system,
- (*vi*) Drainage problem in tea plantation area, and
- (*vii*) Knowledge base system for military use.

It may be noted that here the intention is to provide a framework to undertake GIS analysis, the methodology and analysis procedure may vary with individual analyst and the problem at hand.

GIS for Site Suitability in Urban Planning

Urbanization, which is expansion of urban areas due to urban development and migration of rural population to urban areas, is a dynamic phenomenon. For proper urban planning satisfying various needs, accurate and timely data are required. In order to extract and utilize the required information for Dehradun city, for which the study was taken up, facing fast development and urbanization are taking place after its becoming capital city of Uttaranchal State. Remote sensing technique was combined with the GIS technology for subsequent use to analyze the urban growth and its direction of expansion, and to find suitable sites for further urban development (Soni, 2003). The data and software used for this study are given in Table 11.1 and 11.2.

Table 11.1 Data used (*Source: Soni, 2003*)

S.No.	Data	Year	Scale
1	Guide map of Dehradun	1945	1:20,000
2	Guide map of Dehradun	1965	1:20,000
3	Topo sheet 53J/3	1984	1:50,000
4	Topo sheet 53F/15	1965	1:50,000
5	IRS LISS II	1988	
6	IRS 1D PAN	1997	
7	IKONOS	2001	

Table 11.2 Softwares used (*Source: Soni, 2003*)

S.No.	Software	Use
1.	ERDAS Imagine 8.5	For Registration and onscreen digitization.
2.	Arc GIS	For finding change, suitability analysis and preparing maps.
3.	Adobe Photoshop	For scanning and mosaicing maps.
4.	MS Excel	For numerical analysis on attribute tables.

The methodology adopted in this study is presented in the form of flowcharts (Figs. 11.1 and 11.2). All the guide and topographical maps were converted into the desired land use classes. The classes identified are urban, agriculture, forest, vacant/scrub, river, main roads

218 *Remote Sensing and Geographical Information System*

and other roads and railway line. The same classes were identified also on the satellite images for the year 1988, 1997 and 2001 and land use maps were prepared (Plate 7.6). Subsequently these land use maps were scanned and mosaiced using Adobe Photoshop. The maps were registered to one another using ERDAS Imagine 8.5 GIS software and then digitized. MS Excel was used for preparing the attribute table. Arc GIS software was used for finding change in land use, performing suitability analysis and preparing final output maps.

With the help of these land use maps, urban expansion of Dehradun city for the period 1945-2001 is shown in Plate 7.7, while Plate 11.1 shows the direction of change in land use during 1945-2001. Table 11.3 shows the change in land use for the above period.

In order to identify the sites suitable for urban growth certain factors have to considered which are

(*i*) Land should be vacant or having low usage value presently,
(*ii*) Land should be at an appreciable distance from river so that flood is not a hazard,

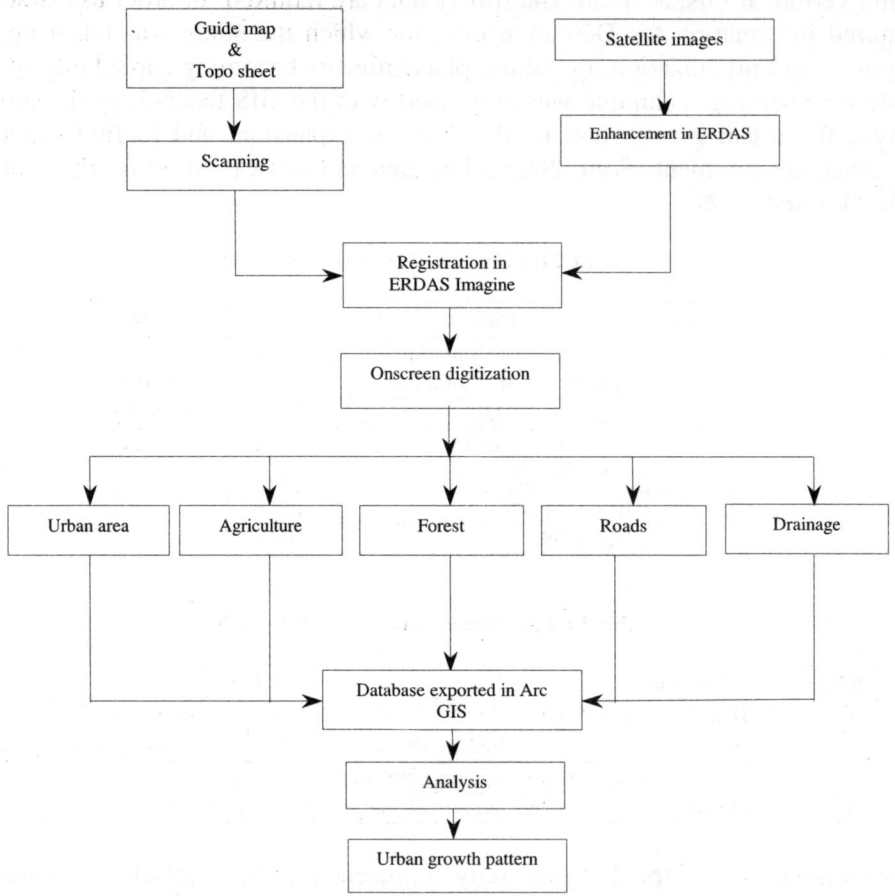

Fig. 11.1 Flowchart for study of urban growth (*Source: Soni, 2003*)

(*iii*) Land should have proper accessibility,
(*iv*) Land should be more or less flat, and
(*v*) Land should have good supply of water.

On the basis of these factors, site suitability map is generated by assigning weight to different thematic maps. Here thematic maps of land cover, accessibility, slope, flood hazard and ground water have been given weights as 10, 9, 6, 4, and 4, respectively. The thematic maps were converted in raster format to determine score for each pixel in the study area; and finally combined to produce a composite suitability map (Plate 11.2) by simple addition of rescored maps with weight system.

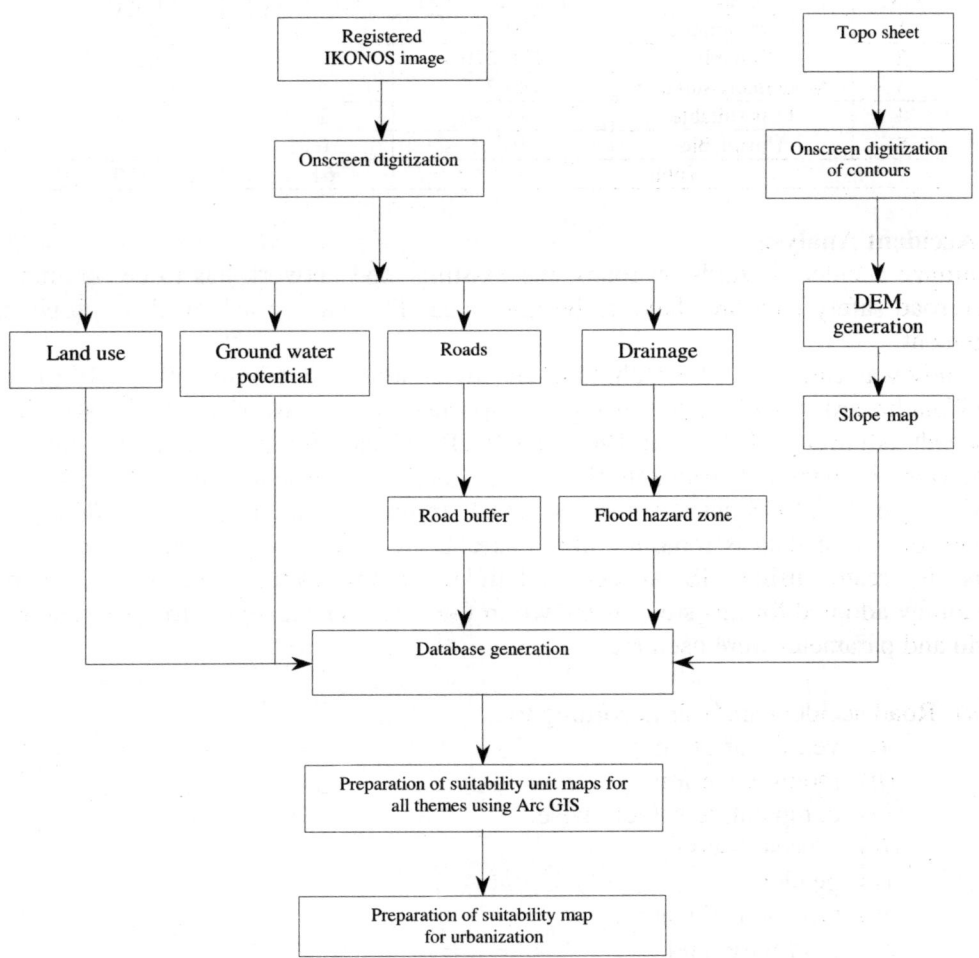

Fig. 11.2 Flowchart for site suitability analysis (*Source: Soni, 2003*)

Thus suitability score = 10*(Land use score) + 9*(Accessibility score) + 6*(Slope score) + 4*(Flood hazard score) + 4*(Ground water score).

The range of suitability score has been equally divided into five parts with high scores indicating larger suitability. From the suitability map, existing urban area and river has been removed in order to find the suitable sites for further development. Table 11.4 shows suitability classes for further urban development.

It is found that out of the 61.67 km^2 available for further development, it is found that near 85% of the area is either very suitable or suitable for urban expansion.

Table 11.4 Suitability classes for urban development (*Source: Soni, 2003*)

S.N.	Class	Suitability score	Area (km^2)	Area (%)
1	Very suitable	270-315	22.41	36.35
2	Suitable	225-270	30.21	48.97
3	Moderately suitable	180-225	5.94	9.63
4	Less suitable	135-180	2.90	4.71
5	Unsuitable	90-135	0.21	0.34
	Total		61.67	100

Road Accident Analysis

To minimize accident hazards on roads, the existing road network has to be optimized and also the road safety measures have to be improved. This can be achieved by proper traffic management.

A study was carried out for Dehradun city by creating a database within GIS to analyze the accident hazards (Uraon, 2003) using the toposheet of the area of the year 1984, the guide map of Dehradun city of the year 1965 and IKONOS data for the year 2001. Initially, the road network is extracted from the 1984 guide map and subsequently updated using high resolution (1m) IKONOS data for the year 2001. In order to ascertain the accident trends, a minimum of 5-year data is required, and was collected from traffic police records. A GIS database is created using MS Access, and linked to the spatial road network map. The methodology adopted for this study is shown in Fig. 11.3. The analyses for different accident scenario and parameter were used are

(*a*) Road accident analysis according to
    (*i*) yearly variation,
    (*ii*) monthly variation,
    (*iii*) comparative vehicle-wise,
    (*iv*) time slot-wise
    (*v*) gender
    (*vi*) type of accident
    (*vii*) road-wise, and
(*b*) Identification of black spots.

Two typical histograms depicting the type of vehicles involved in accidents and the number of accidents, which took place in different periods on particular road of the city, is shown in Plate 11.3 and 11.4. The accident-prone locations were identified with the help of Accident Rate (AR) method and Accident security Index (ASI) method (Plate 11.5). By identifying the accident-prone locations with their accident severity, remedial measures can be planned by the district administration to minimize the accidents in different parts of the city.

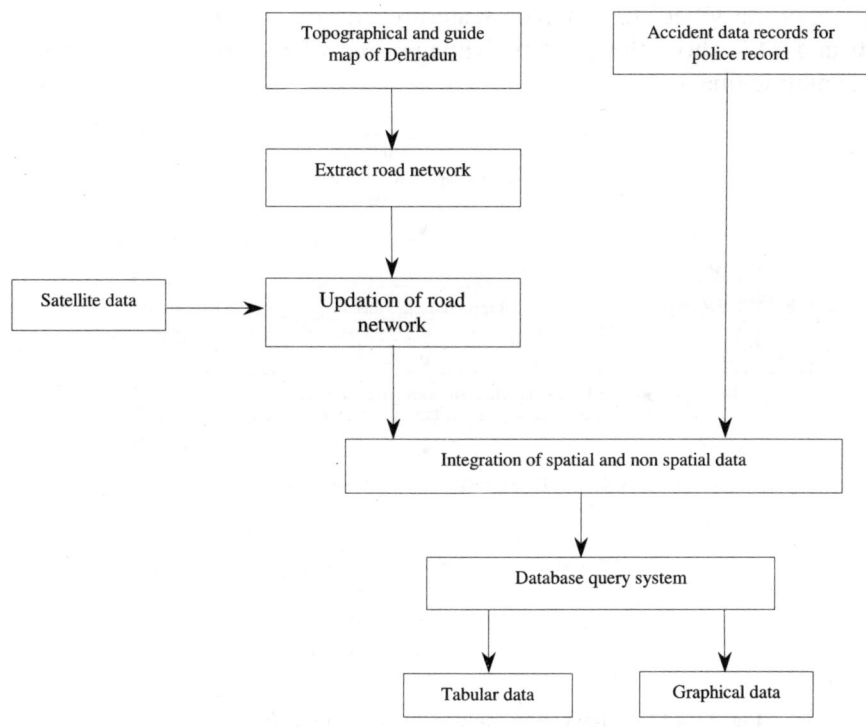

Fig. 11.3 Flowchart of methodology used (*Source: Uraon, 2003*)

Irrigation Water Management

Availability of water for irrigation purposes for any area is vital for crop production in that region and needs to be properly and efficiently managed for proper utilization of water. To evaluate the irrigation performance, integrated use of satellite remote sensing and GIS assisted by ground information has been found to be efficient technique in spatial and time domain for identification of major crops and their conditions, and determination of their areal extent and yield.

A study based on crop classification using synthetically generated images for Bahadurabad area of Haridwar district in Uttaranchal, was carried out to evaluate the crop water requirements using hydrological data. The classification technique adopted was multispectral classification. Figs. 11.4 to 11.6 present the flowcharts showing the methodologies adopted for the study, image classification, and GIS application (Sirisha, 2003).

222 Remote Sensing and Geographical Information System

The data used in the study were extracted from topographical map, IRS LISS-III image, IRS PAN image of the area. The synthetic images were created in order to highlight different features and to reduce the redundancy by image transformation techniques, such as PCA, NDVI, Tasseled Cap, Ratio, and subtraction. The hydrological data was obtained from the local administration.

It was found by analyzing the satellite data that the major crops of the area are sugarcane and wheat. Other classified data include waterlogged area, dry sand, urban area, and water. Plates 11.6 and 11.7 show the classified image area of the irrigated command using raw image and synthetic image.

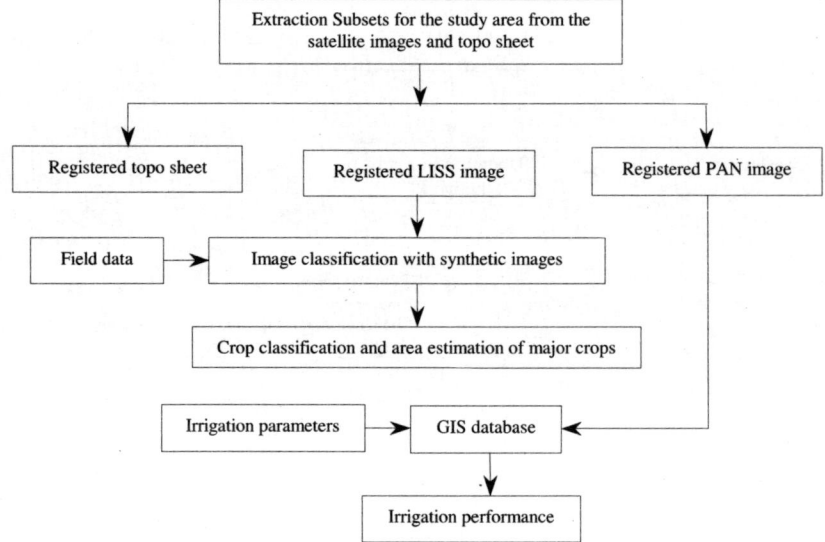

Fig. 11.4 Flowchart showing methodology (*Source: Sirisha, 2003*)

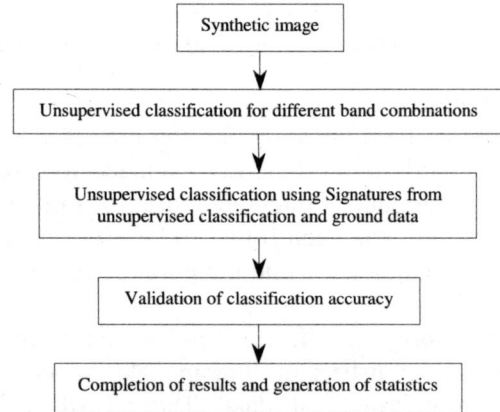

Fig. 11.5 Classification approach (*Source: Sirisha, 2003*)

Irrigation requirements of crops were determined by considering the factors such as evapotranspiration, Net Irrigation Requirement (NIR), Field Irrigation Requirement (FIR), Gross Irrigation Requirement (GIR), and month total volume of water required, by organizing them in GIS environment.

Gross irrigation requirement for different crops in different command areas were determined. Plate 11.9 shows the requirement in one of the command areas, and Plate 11.10 shows the variation in demand and supply of irrigation water in the same command area. Finally, irrigation efficiencies for all command areas were calculated by considering irrigation demand and supply. The study has indicated that by considering peak monthly demand, performances of minor irrigation canals in the area are within acceptable limits except that of Kherli minor.

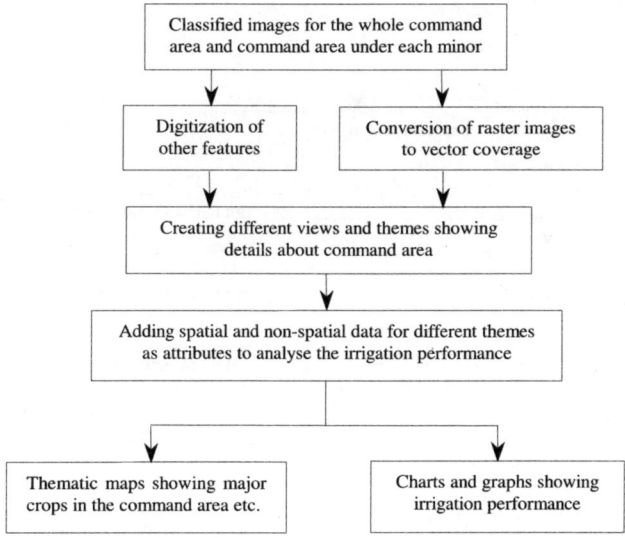

Fig. 11.6 GIS approach adopted. (*Source: Sirisha, 2003*)

Tourism Information System

Information plays a vital role to tourists in planning their travel from one place to another, and success of tourism industry depends on providing timely information to the tourists.

State of Uttaranchal being one of the most beautiful and visited places in India due to its religious value and natural beauty, a DBMS was generated to handle the large volume of data related to tourist places and display them in the form maps by linking the data to the maps with the help of GIS, as it can provide with the tools required to generate a better understanding, and can serve the needs of tourists (Venkateshvarlu, 2003).

For the development of tourist information system for Uttaranchal state, the data used were topographical map, atlas of Uttaranchal published by NATMO, Kolkata, and tourist guide map, data/information from Internet, field data, and available books and literatures on tourism in Uttaranchal. The methodology adopted in this work, is given in Fig. 11.7. In order to provide all data at a common scale and to carry out measurements, the administrative map, physiographic, drainage, land use, and transport maps were converted into digital form, and then registered and georeferenced with each other using ERDAS Imagine software.

Plate 11.11 shows an overall view of "Tourism GIS' developed in this work for Uttaranchal state. The 'Tourism GIS' can perform the following tasks:

(*i*) **Find** a particular feature,

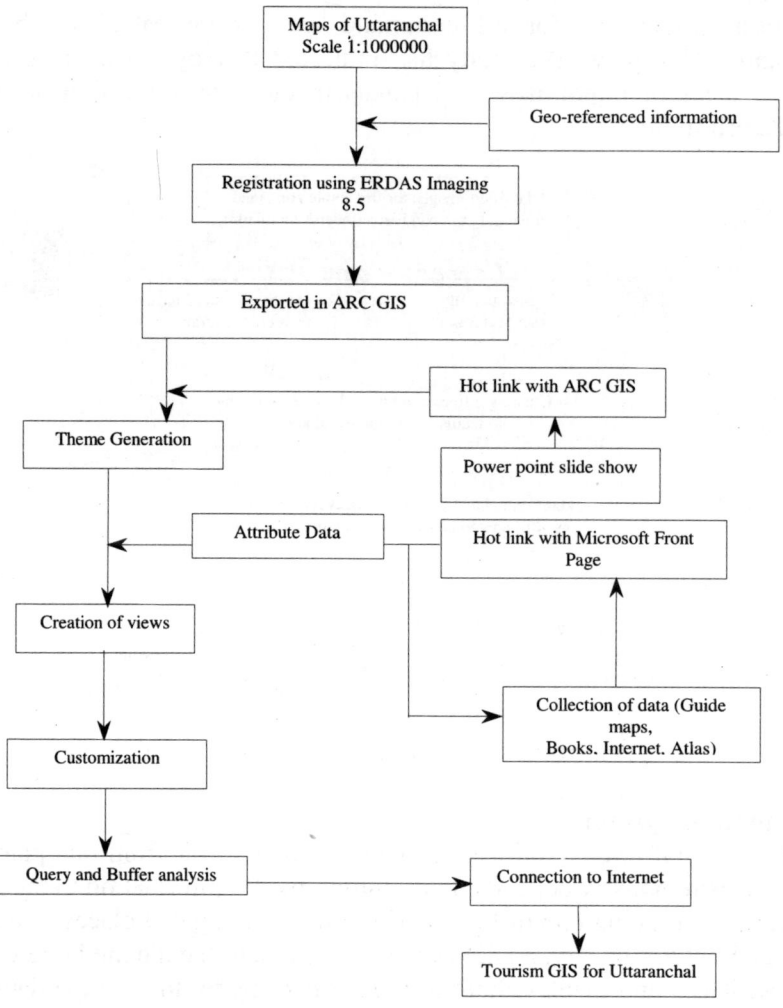

Fig. 11.7 Flow chart of methodology adopted (*Source: Venkateswarlu, 2003*)

(*ii*) **Pan** to drag the window to view a certain area
(*iii*) **Zoom** to enlarge or minimize the view,
(*iv*) **Measurement** to measure distances between points of interest,
(*v*) **Buffering** to locate places of interest within a specified distance,
(*vi*) **Hyper link** for adding the attributes to the location, and
(*vii*) **Identify** tool for displaying the window identify results showing the attribute data of any feature in an active theme.

The developed *Tourism GIS* is user friendly with Graphical User Interface (GUI).

Worldwide Earthquake Information System

One of the most frightening and destructive phenomena of nature is the occurrence of an earthquake. There is a need to have knowledge regarding the trends in earthquake occurrence worldwide. The catalogues having the information about the earthquake parameters and source parameters are prepared by collecting seismic data, and are used for seismic hazard estimation. Since the manual interpretation and analysis of worldwide seismic data are tedious and time consuming, the implementation of integrated GIS technology can provide an approach for rapid evaluation of complex inventor database under a variety of earthquake scenario, and allowing the user to view the results interactively.

Jaganniwas (2004) has made an attempt to compile seismicity data on GIS platform for use by seismologist and earthquake engineers. The data collected and used are Digitized World Map, GSHAP Map, Tectonic Plates Map, Indian Seismic Zoning Map, and Earthquake Catalogue. The softwares employed for the study are Arc-GIS 8.3, ERDAS 8.6, Visual Basic 6.0, and MS Access. The methodology adopted is presented in Fig. 11.8. The GIS Based Earthquake Information System developed, allow the user with the options of Add Layer, Check Box, Zoom, Selection, Queries, Graphs, Tools, and About, within an user interface window (Plate 11.12). This figure also shows a selected circular area for which some information may be required, such as seismic hazard data. Plate 11.13 shows that how the queries are answered.

A GIS based user interface system for querying on earthquake catalogue will be of great help to the earthquake engineers and seismologists in understanding the behaviour pattern of earthquakes in spatial and temporal domain.

Drainage Problem in Tea Plantation Area

Drainage, defined as *removal of excess water from crop root zone*, is a critical element for sustained, efficient, and productive agriculture. Drainage problems in tea plantations differ widely because of its varied nature of physical conditions. The topography, soil, and source of water for any given area vary so greatly that it requires a complete and thorough evaluation of the factors such as permeability of the soil, consumptive use of water, location of water table, and surface topography.

Sub-surface drainage problems mainly arise due to poor permeability of the sub-soil, seepage from high lands, and presence of springs or artesian at shallow depth below ground, and presence of hard clay pan in the sub-soil. Since tea crop requires moisture at adequate levels at all times of growth, any variation either excess or lack, has a direct impact on the

226 *Remote Sensing and Geographical Information System*

tea yield, and thus water logging in the area is going to greatly influence the productivity of tea. A study was carried out to perceive the complexities involved in water logging areas of Helem Tea State of Tezpur direct in Assam (Naidu, 2000)

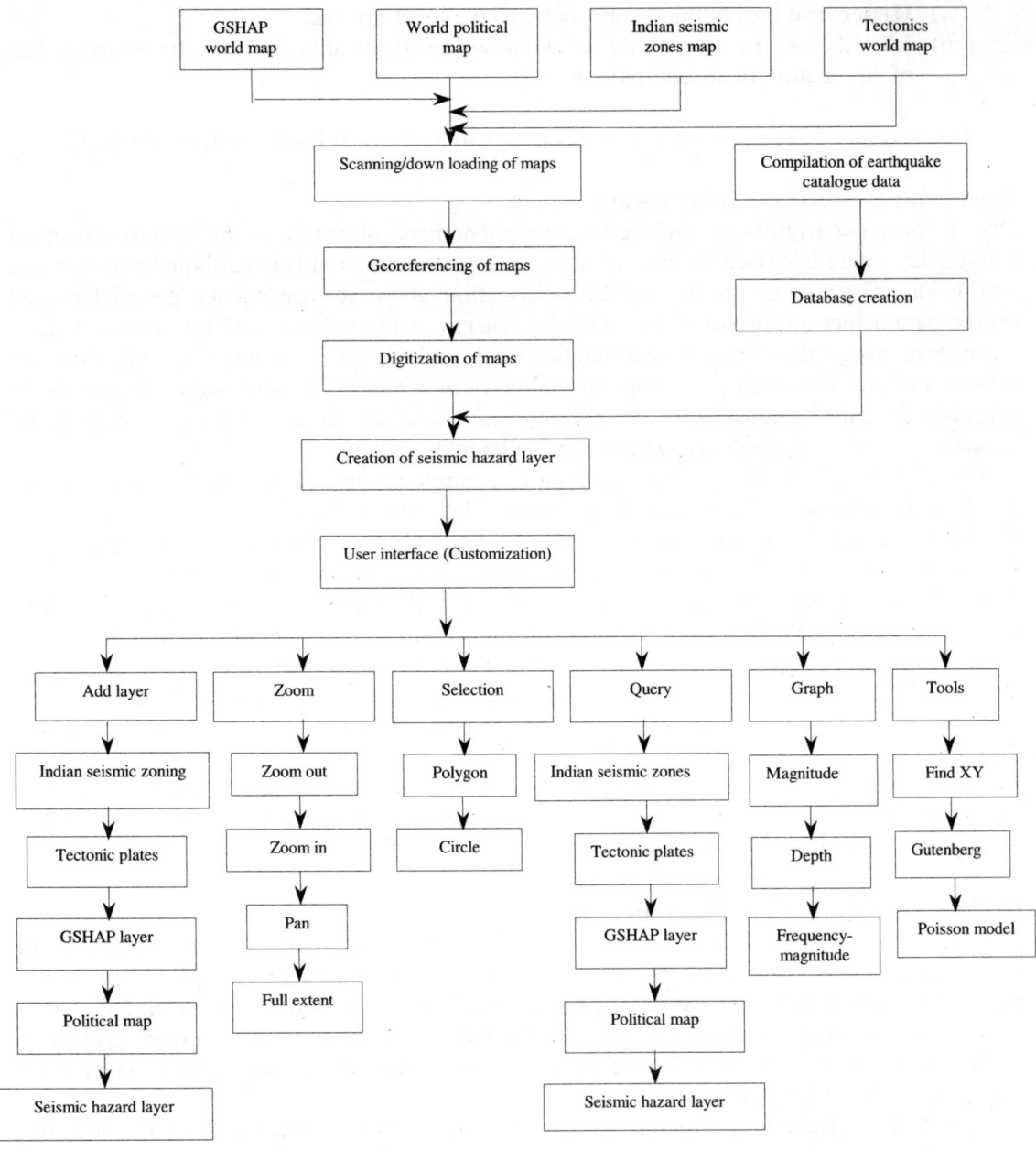

Fig. 11.8 Flow chart for the development of worldwide earthquake system (*Source: Jagannivas, 2004*)

The piezometric data for the tea garden area indicates serious built up water table in the root zone of tea plants all throughout the rainy season. The runoff from the high intensity rainfall during rainy season is drained through the main drains, and discharged into Balom River, and all these drainages having insufficient water carrying capacity leading to flooding conditions in the tea garden. To obtain the inputs required for hydraulic design in a GIS environment, this study was taken up, involving the following steps.

    (*i*) Generation of DEM of the tea area,
   (*ii*) Terrain analysis and land form analysis through slope map,
  (*iii*) Determination flow direction over land surface using slope vector map,
  (*iv*) Visualization of variation of ground level at desired sections using profiles,
   (*v*) Classification of water logged areas,
  (*vi*) Calculation of surface and sub-surface runoff,
 (*vii*) Design of drains, and
(*viii*) Checking the adequacy of the river Balom.

The methodology adopted for this study is given in Fig. 11.9. The GIS packages used in the study was MGE (Modular GIS Environment). For hydraulic design, a hydraulic simulation program HEC-6 was used.

The study involves classification of water logging in the area during different time periods of the year (Plate 11.14). The water table conditions were also obtained for different time periods of the year (Plate 11.15) by drawing contours. Finally, surface flow and subsurface runoff were analysed (Plate 11.16 and 11.17).

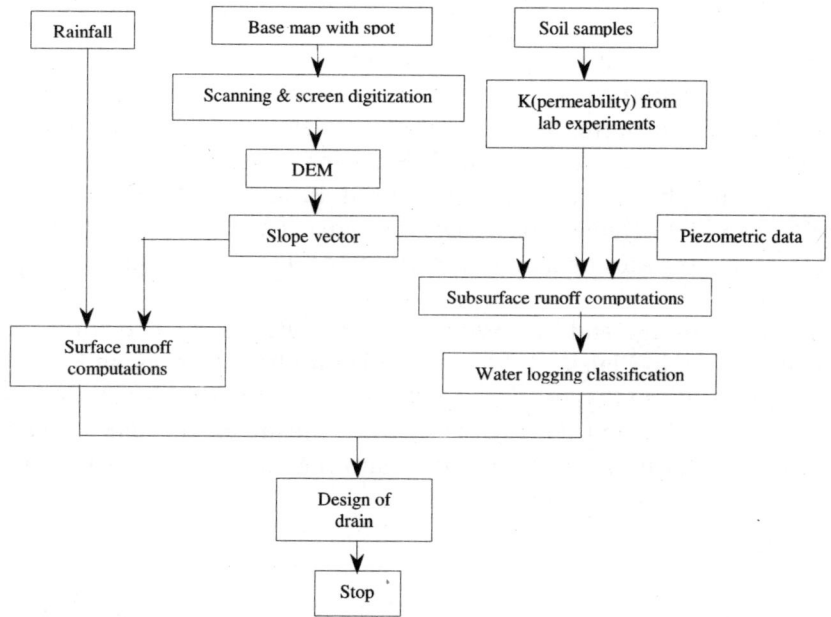

Fig. 11.9 Flow chart of study of drainage problem in a tea plantation areas (*Source: Naidu, 2000*)

The results of the study reveal that the problem of water logging is due to various reasons. The existing drainage systems being inadequate require modifications to augment the drainage in the area to alleviate the problems of water logging problem.

Knowledge Based System for Military Use

History of warfare shows weaker forces have achieved victory by using the terrain conditions to their advantage, and hence, ground conditions dictate the outcome of a war to a large extent. Correct and timely analysis of terrain is essential for today's fast paced battlefield. Conventional method of studying paper topographical maps is being replaced by use of maps in digital form to get terrain information. Satellite remote sensing data and GIS are increasingly being used to derive terrain information from digital images that can be used along with knowledge Base (KB) derived from experts for developing a Decision Support System which makes the process of selection of suitable sites for various military uses more accurate and faster.

Baijal (2002) has developed a knowledge-based system for some military tasks in the area lying south of Saharanpur district of Uttar Pradesh. The system developed has the capability to:

(*i*) select wet and dry bridging sites,
(*ii*) select likely ferry sites,
(*iii*) select potential helipads,
(*iv*) identify tactically important roads, prepare vehicle mobility maps, and
(*v*) investigate the behaviour of various parameters for a particular military use by assigning different confidence levels to the parameters.

The data used in this study are primarily extracted from the topographical maps of the area. However, some hypothetical data have also been used for creation thematic layers. Using ERDAS Imagine software, thematic layers of water bodies, roads, power lines, DEM, slope map, and land use map were prepared. For generating the thematic layers of soil and water depth (river only), hypothetical data were used. Plate 11.18 shows some of the thematic maps prepared. Plate 11.19 shows a raster DEM of area. The DN values of the pixels in raster images for different categories of land use, soils, slope, etc, were assigned as given in Table 11.5.

To identify a wet bridge and dry bridge sites the rules in the text form for knowledge-based system are shown in Plates 11.20, and to identify helipad sites, in Plate 11.21. On the basis of this, site for different types of bridges, ferry site and overall mobility in the area can be identified (Plate 11.22 and 11.23). The study illustrates the use of GIS along with knowledge base provides useful information regard the terrain features which can be useful for planning present day war strategies.

Table 11.5 DN values of the pixels in raster images (*Source: Baijal, 2002*)

Feature	DN Value assigned
River	1
Road (Category 1)	
Sand	
Water depth (Adequate)	
Slope (<5%)	
Power Line	
Built up areas	
Canal	2
Road (Category 2)	
Silt	
Water depth (inadequate)	
Slope (5% to 10%)	
Lakes	3
Clay	
Slope (> 10%)	
Cultivated Land	4

GIS Application 231

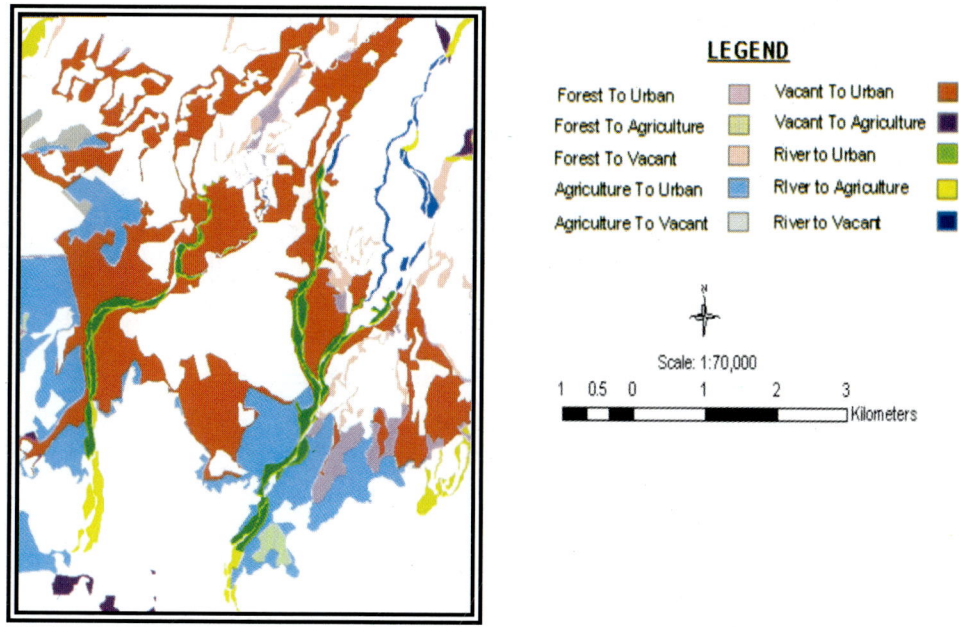

Plate 11.1 Land use change direction

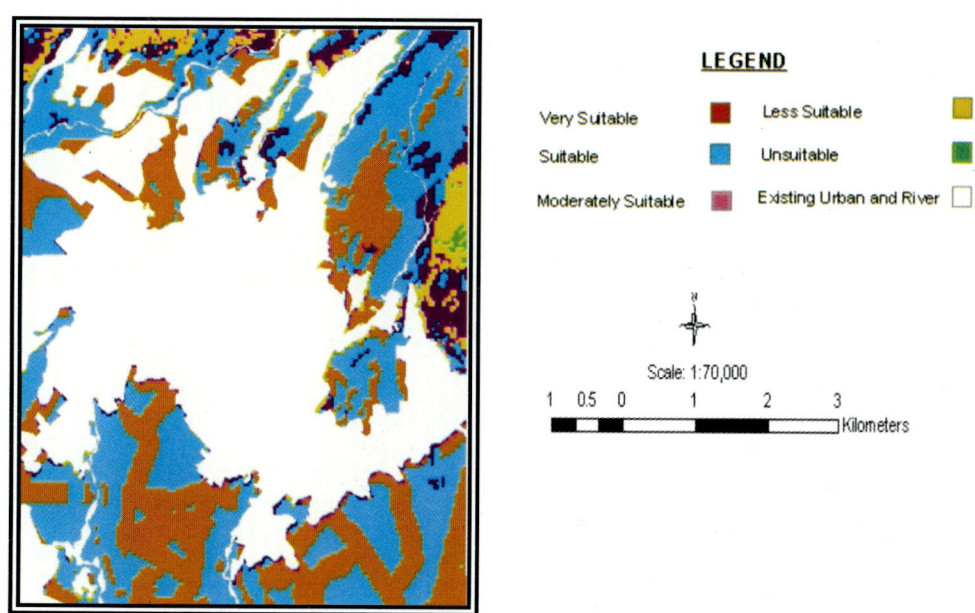

Plate 11.2 Suitability map for future urban growth

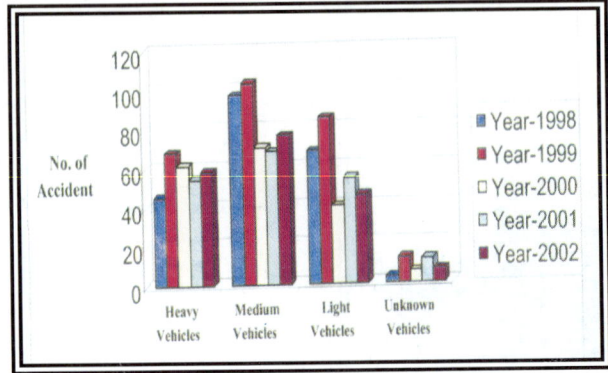

Plate 11.3 Type of vehicle involved

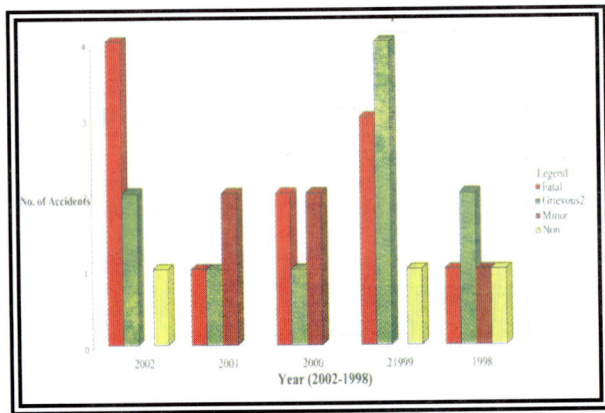

Plate 11.4 Accidents on Raipur road

Plate 11.5 Accident prone location map

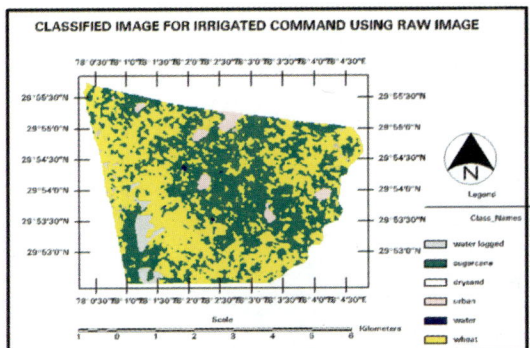

Plate 11.6 Classified image of irrigated Command

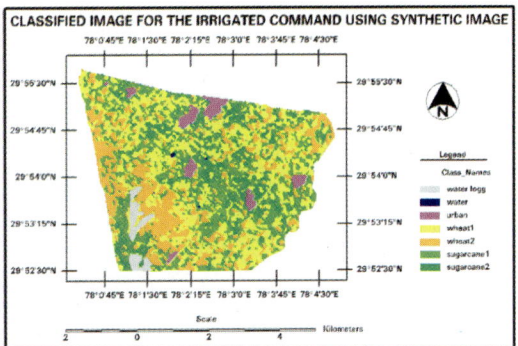

Plate 11.7 Shows the major crops in the irrigated command

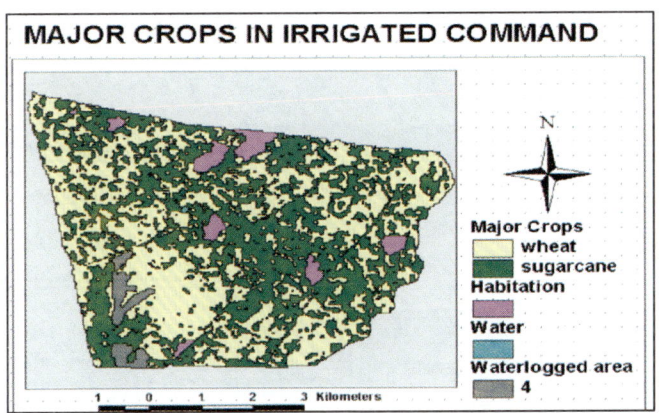

Plate 11.8 Major crops in irrigated command

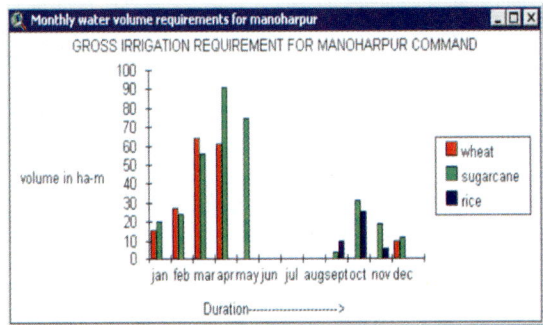

Plate 11.9 Gross irrigation requirement for different crops

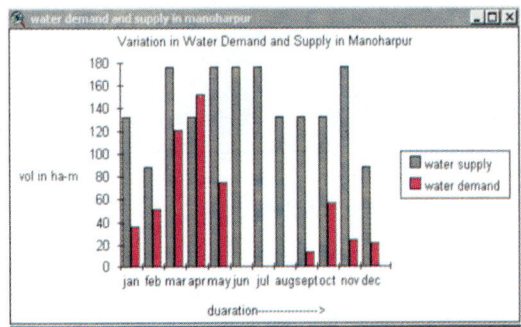

Plate 11.10 Variation in demand and supply of irrigation water

234 *Remote Sensing and Geographical Information System*

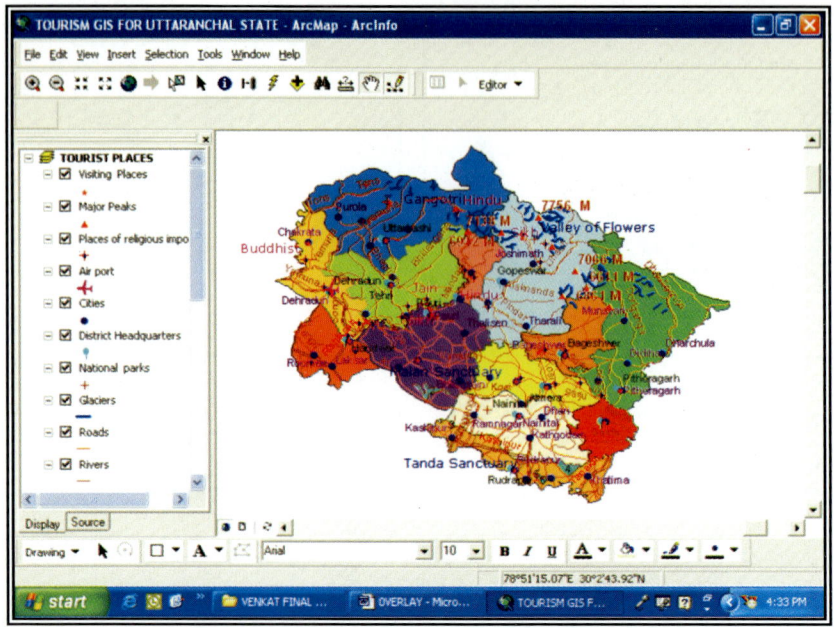

Plate 11.11 Overall view of 'Tourism GIS' of Uttaranchal

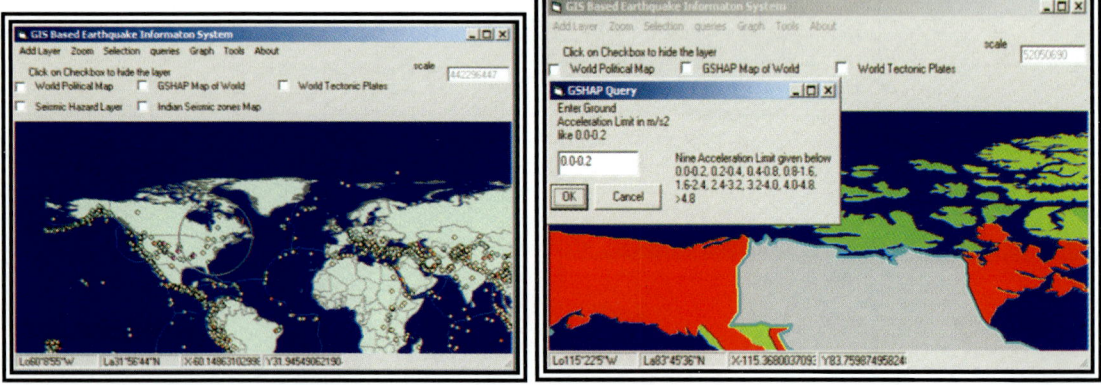

Plate 11.12 Selection of area creating circular zone   Plate 11.13 Window showing Query in GSHAP layer

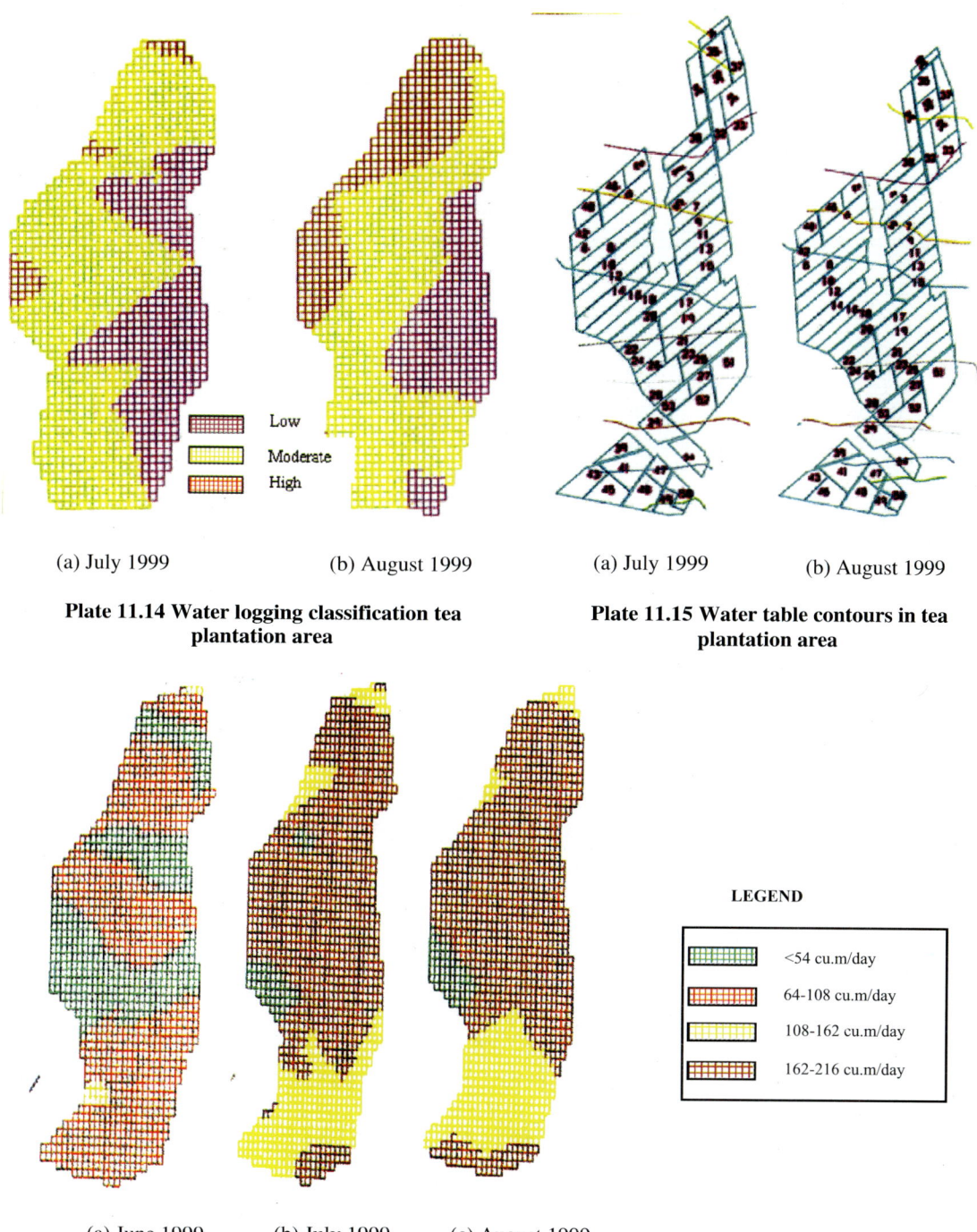

(a) July 1999   (b) August 1999   (a) July 1999   (b) August 1999

Plate 11.14 Water logging classification tea plantation area

Plate 11.15 Water table contours in tea plantation area

(a) June 1999   (b) July 1999   (c) August 1999

Plate 11.16 Surface flow in the tea plantation area

LEGEND

- <54 cu.m/day
- 64-108 cu.m/day
- 108-162 cu.m/day
- 162-216 cu.m/day

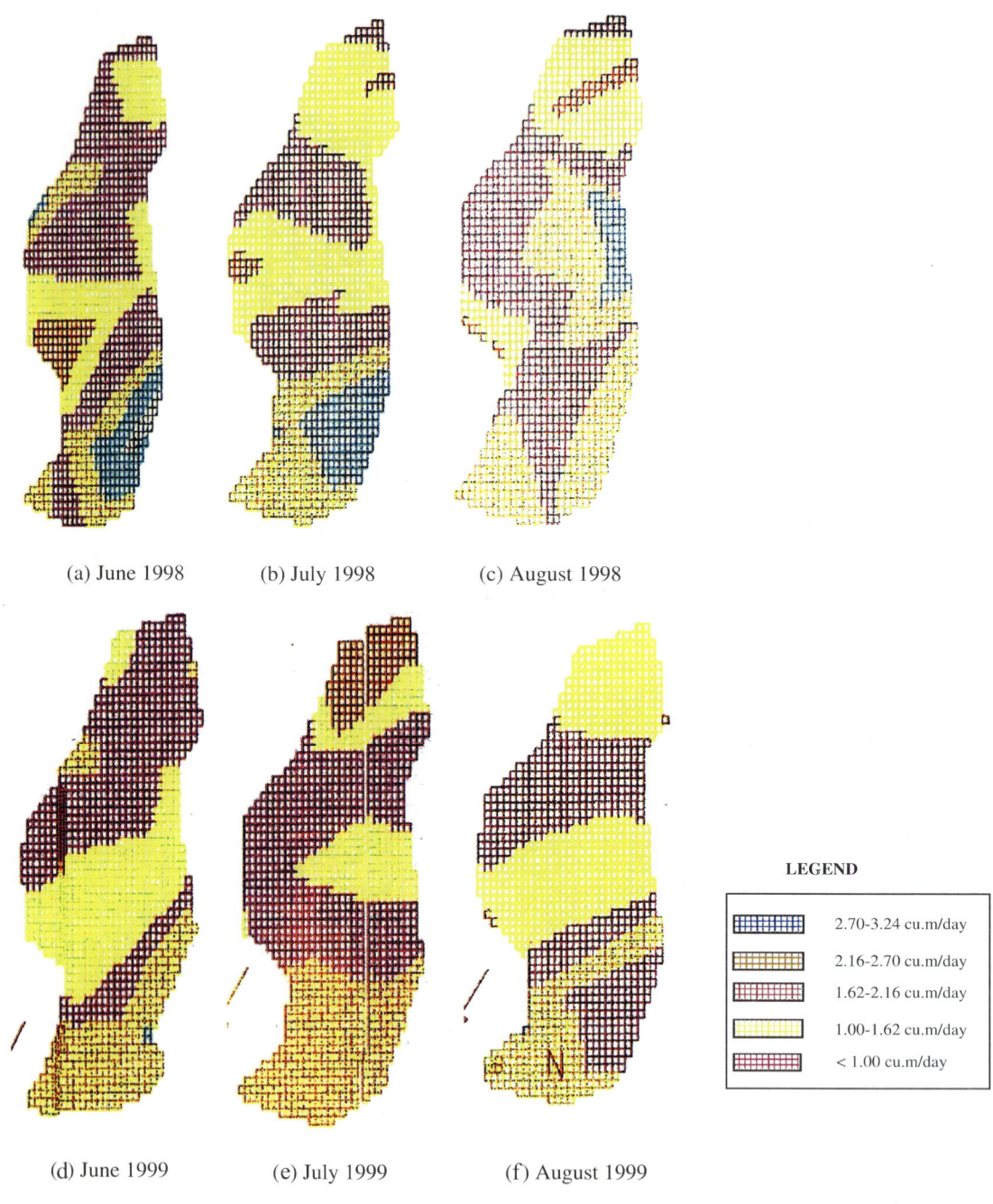

11.17 Subsurface runoff in the tea plantation area.

GIS Application 237

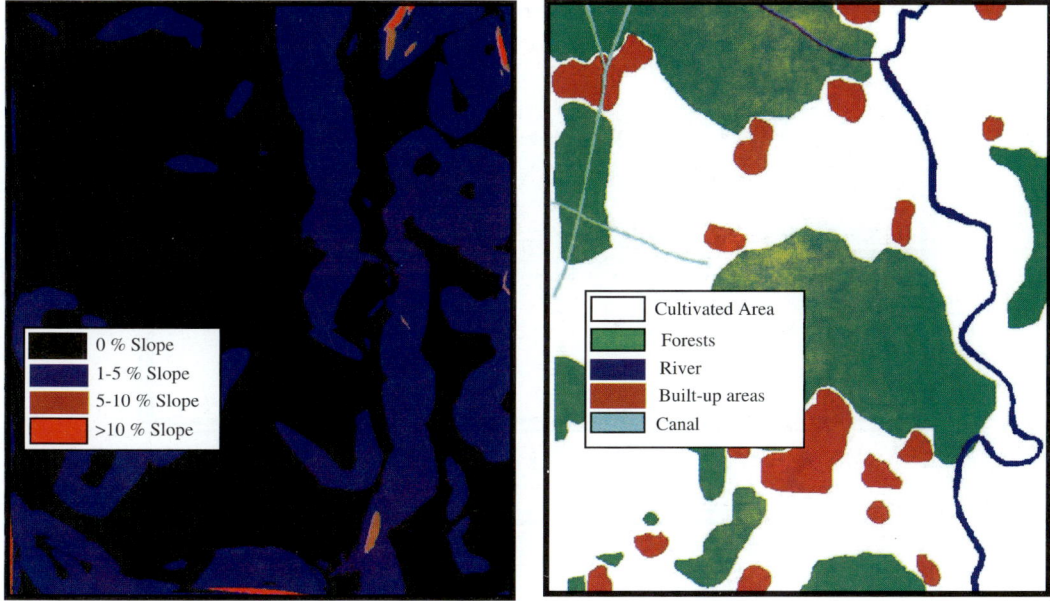

(a) Slopes (b) Land use

Plate 11.18 Thematic maps

Plate 11.19 Raster DEM of study area

238 *Remote Sensing and Geographical Information System*

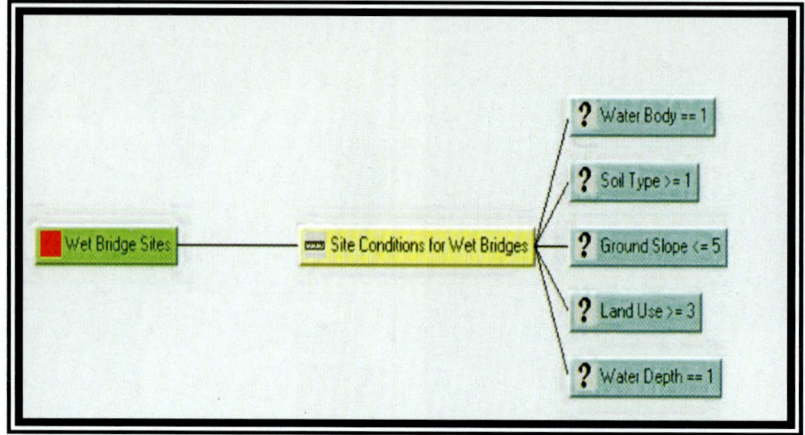

(a) Wet bridge

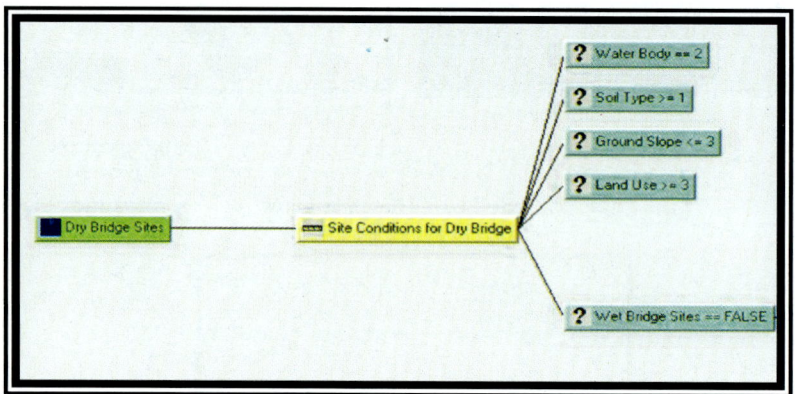

(b) Dry bridge

Plate 11.20 Knowledge base for selection of bridge sites

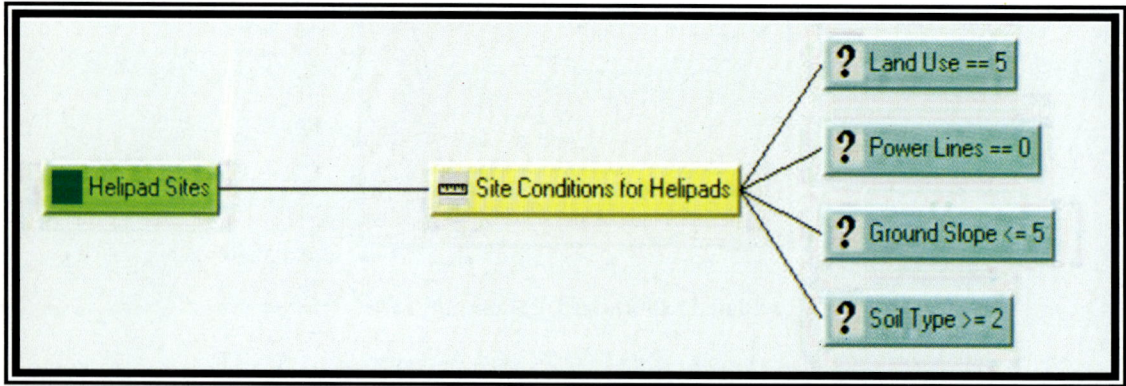

Plate 11.21 Knowledge base for selection of helipad sites

GIS Application 239

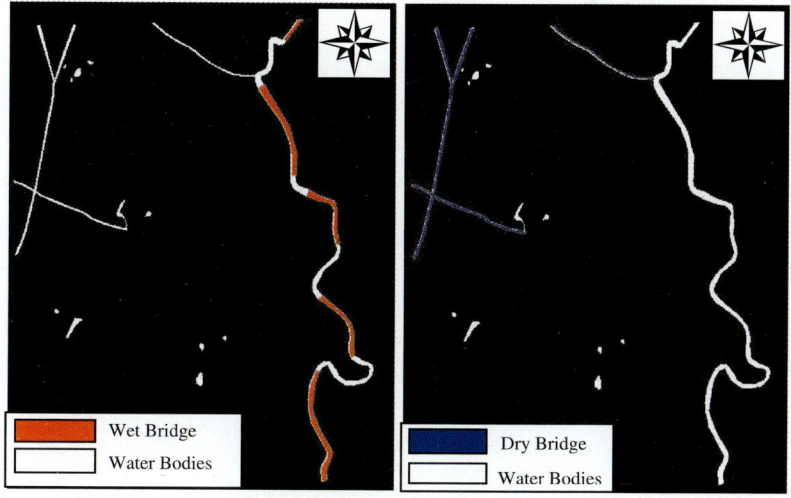

(a) Suitable sites for Wet Bridge   (b) Suitable sites for Dry Bridge

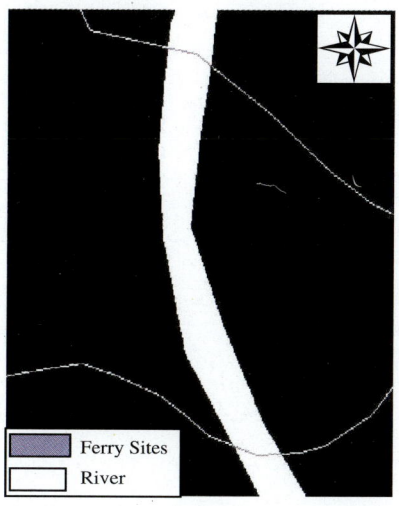

(c) Suitable ferry sites

Plate 11.22 Suitability maps for selection of site

FUTURE TRENDS 12

12.1 INTRODUCTION

With the launching of satellites from different countries with varying spatial and spectral resolution, the remote sensing analyst now has recourse to a large volume of data. These data can be used for a large number of different types of application. However, one problem, which has always been attracting the attention of analyst for research, is the problem of mixed pixel in image classification briefly discussed in chapter 2. A concerntrated effort in this direction has been given by analysts in accounting for mixed pixel in classification process. A brief discussion of the same has been given in the following section.

GIS has now found its applicability and utility in many areas, it is now becoming an important tool for decision and policy makers and researchers. In this chapter, a brief introduction to the recent advances in the field of GIS has been presented. The readers may note that applications and utility of remote sensing data and its analysis, and GIS will continue to grow and thus newer technologies and applications will keep on adding to it with time.

12.2 ADVANCES IN REMOTE SENSING

The problem of mixed pixels in classification of remote sensing data has been a reason for worry for the remote sensing analysts with the advent of satellite data. With coarse resolution data, the occurrence of mixed pixels had been intense, and it was thought that this aspect will reduce with increase in spatial resolution. However, this problem has remained same in magnitude with increase in spatial resolution. With coarse resolution, the chances of two or more classes contributing to a mixed pixel were high but the number of such pixels was small. With improved spatial resolution, the number of classes within a pixel has reduced but the number of mixed pixels has increased. In a way, the problem of mixed pixels remained, may be its direction of impact has changed.

Whatever is the nature and origin of mixed pixels, these will always create problems in image classification. Let us consider a simple land area consisting of two classes, namely, water and land (Fig. 12.1). It can be clearly seen that two pixels belong to only one class, *i.e.*, pixel 1 has water and pixel 4 has land, and they are called as pure pixels, whereas pixels 2 and 3 has varying composition of land and water, and are called mixed pixels.

A mixed pixel displays a composite spectral response that may be dissimilar to the spectral response of each of its component classes, and therefore, the pixel may not be allocated to any of its constituent classes. Therefore, an error is likely to occur in the classification of the image. Convention statistical based image classification (also known as hard classification) which assumes that the pixels contain pure information, would identify the pixel to one and only one class. Thus pixel 2 may be classified as water and pixel 3 as land (Fig. 12.1b). Depending upon the proportion of mixed information, it may result into a loss of pertinent information present in a pixel and subsequently in an image.

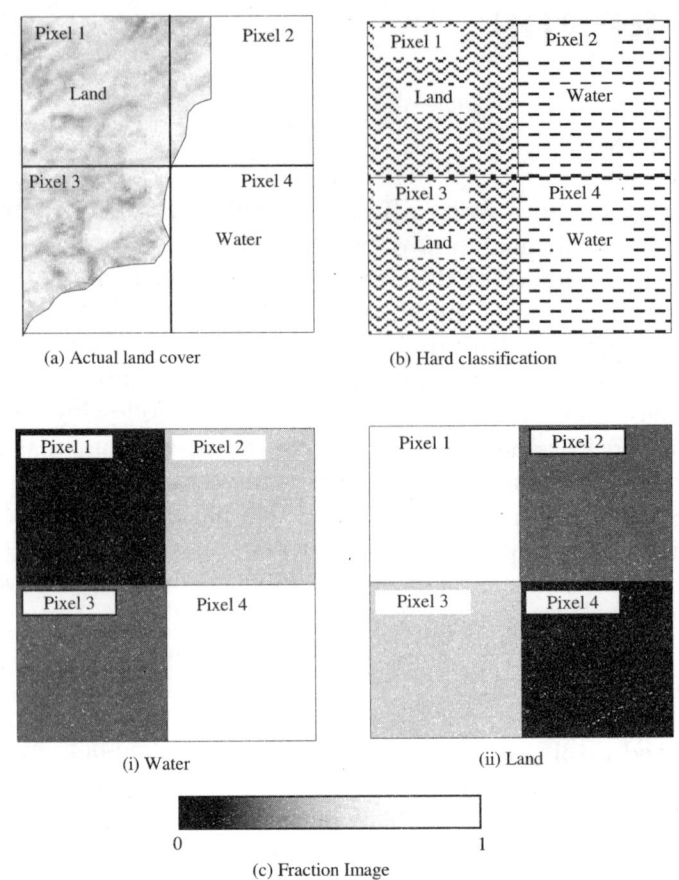

Fig. 12.1 The concept and analysis of mixed pixel

Hard classifiers/classification methods available may tend to over or under estimate the actual area covered by different land cover/land use classes on the ground. The land cover/land use maps so derived from hard classification may be used as input for other remote sensing or GIS applications. Therefore, the mixed pixels have to be accommodated in the classification process in some way, by making use of subpixel or soft classification methods based on certain heuristic and logical reason has to be adopted. The output from these methods is a set of class membership values for each pixel known as soft, fuzzy or sub-pixel classification outputs which represent the probability fraction or proportion images (Fig. 12.1c). These soft outputs strongly relate to actual extents of the classes on ground.

Soft classification methods are either based on spectral mixture analysis, fuzzy set theory, or artificial neural network. Some of the most commonly used methods have been discussed below.

Linear Mixture Model (LMM)

The Linear Mixture Model (LMM) is widely used for the decomposition of the class proportion of mixed pixels (Settle and Drake, 1993). The method assumes that the spectral response of a pixel is a linear sum of the mean spectral responses of the various land cover classes weighted by their relative proportion on the ground (Thomas, et al, 1996) and the model can be mathematically expressed as

$$x_i = \sum_{j=1}^{c} f_j M_{ij} + e_i \qquad \ldots(12.1)$$

where M_{ij} is the end member spectra representing the mean class spectral response of jth land cover class in the ith band, f_j are the proportions of jth land cover class in a pixel, and e_i is the error term for i^{th} band, which expresses the difference between the observed spectral response and the model derived spectral response of the pixel. It may be noted that class proportions of a mixed pixel are not negative and that the sum of all the class proportions is equal to one, and can be expressed as

$$\sum_{j=1}^{c} f_j = 1 \qquad \ldots(12.2)$$

and $f_j \geq 0$ for all j land cover classes.

The end member spectra matrix M represents the spectral responses of classes, and may be calculated by taking the average spectral response of that class having pure pixels, or estimated from laboratory and field measurements of the classes, or by performing principal component analysis.

Fuzzy Set Based Method

The use of fuzzy set based classification methods in remote sensing has received growing interests in situations where the geographical phenomena are inherently fuzzy (Zhang and Foody, 1998). Although, a number of fuzzy set based methods may be adopted, mathematical details of only two unsupervised methods, most commonly used, namely Fuzzy c-Means (FCM) and Possibilistic c-Means (PCM) are discussed below.

Fuzzy c-Means (FCM): FCM is an iterative clustering method that may be employed to partition pixels of remote sensing images into different class membership values. The key is to represent the similarity that a pixel shares with each cluster with a function (membership function) whose value lies between zero and one. Each pixel will have membership in every cluster. Memberships close to unity signify a high degree of similarity between the pixel and that cluster (Bezdek, et al, 1984). The net effect of such a function of clustering is to produce fuzzy c-partitions (U) of a given data. A fuzzy c-partition of the data is the one which characterizes the membership of each pixel in all the clusters by a membership function that ranges from zero to one. Additionally, the sum of the memberships for each pixel must be unity. This is achieved by minimizing the generalized least-square error objective function,

$$J_m(U,V) = \sum_{i=1}^{N} \sum_{j=1}^{c} (\mu_{ij})^m \|x_i - v_j\|_A^2 \qquad \ldots(12.3)$$

subject to constraints

$$\left. \begin{array}{l} \sum_{j=1}^{c} \mu_{ij} = 1 \text{ for all } i \\[6pt] \sum_{j=1}^{c} \mu_{ij} > 1 \text{ for all } j \\[6pt] 0 \leq \mu_{ij} \leq 1 \text{ for all } i, j \end{array} \right\} \qquad \ldots(12.4)$$

where x_i is the vector denoting spectral response of a pixel i (*i.e.*, a vector of spectral response of a pixel in various bands of a multi-spectral remote sensing image), V is the collection of vector of cluster centres, v_j, μ_{ij} are class membership values of a pixel (members of fuzzy c-partition matrix), c and N are number of clusters and pixels respectively, m is a weighting exponent ($1 < m < \infty$), which controls the degree of fuzziness. $\|x_i - v_j\|_A^2$ is the squared distance (d_{ij}) between x_i and v_j, and is given by

$$d_{ij}^2 = \|x_i - v_j\|_A^2 = (x_i - v_j)^T A(x_i - v_j) \qquad \ldots(12.5)$$

where A is the weight matrix.

Two parameters of J_m are m and A. Weighted exponent m controls the relative weights placed on each of the squared error d_{ij}^2. At $m = 1$, partitions that minimize J_m become increasingly hard and as m tends to reach ∞, partitions that minimize J_m become increasingly soft or completely fuzzy. Consequently, increasing m tends to degrade the membership towards the fuzziest state. Each choice of m will generate one FCM solution keeping all other parameters fixed. No theoretical or computational evidence provides an optimal value of m. If a test set is available for the process under investigation, the best strategy for selecting m at present seems to be experimental dataset as proposed by Bezdek, *et al*, 1984.

Another parameter in J_m that deserves a special mention is the weight matrix A. This matrix controls the shape of the optimal clusters. Amongst a number of A-norms, three norms namely Euclidean, Diagonal, and Mahalonobis norm, each induced by specific weight matrix are widely used. The formulations of each norm are given as (Bezdek, 1981)

$A = I$	Euclidean Norm	...(12.6)
$A = D_j^{-1}$	Diagonal Norm	...(12.7)
$A = C_j^{-1}$	Mahalonobis Norm	...(12.8)

where I is the identity matrix, D_j is the diagonal matrix having diagonal elements as the eigen values of the variance covariance matrix. C_j is given by,

$$C_j = \sum_{i=1}^{N}(x_i - c_j)(x_i - c_j)^T \qquad ...(12.9)$$

where $$c_j = \sum_{i=1}^{N} \frac{x_i}{N} \qquad ...(12.10)$$

When the diagonal norm is used, each dimension is effectively scaled *via* the Eigen values. The Euclidean norm is the only choice for which extensive experience with the data is available. The type of norm may have to be decided before carrying out classification using fuzzy c-means clustering method.

Now that all parameters in Eq. (12.3) have been defined, the fuzzy c-partition is obtained through an iterative process of optimization using the following two equations. The cluster centres are updated by

$$v_j = \frac{\sum_{i=1}^{N} \mu_{ij}^m x_i}{\sum_{i=1}^{N} \mu_{ij}^m} \qquad ...(12.11)$$

and the class membership matrix μ_{ij} is obtained by

$$\mu_{ij} = \frac{1}{\sum_{k=1}^{c}\left(\frac{d_{ij}^2}{d_{ik}^2}\right)^{1/(m-1)}} \qquad ...(12.12)$$

where $$d_{ik}^2 = \sum_{j=1}^{c} d_{ij}^2 \qquad ...(12.13)$$

The class membership values of a pixel denote the class proportions, which in turn may represent the soft classified output for a pixel. The general procedure of FCM algorithm may be formalized in the following steps:

Step 1. Fix values of *m* and *c,* and select the type of A-norm and maximum number of iterations.

Step 2. Choose an initial membership matrix U^i with its elements selected as random start, random non-fuzzy start, and uniform start. In random start, each membership value is assigned a number between 0 and 1 to all classes. In random non-fuzzy start, a value of 1 is assigned to a randomly chosen class and 0 to all other classes. In uniform start, a membership value of $(1/c \pm a)$ is assigned to all classes where *a* is a small random number.

Repeat
Step 3. Input the image file.
Step 4. Computer the vector of cluster centres, v_j using Eq. (12.11).
Step 5. Compute the distance based on the selected *A-norm* using Eq. (12.6).
Step 6. Update *U* matrix for the next iteration U^{i+1} using Eq. (12.12).
Until $(\|U^{i+1} - U^i\| < \varepsilon)$, where ε is the user defined limiting error.
Step 7. Write final *U* to file, which represent class proportions.

Possibilistic c-Means (PCM): The main motivation behind the use of PCM relates to the relaxation of the probabilistic constraint of FCM [*i.e.* Eq. (12.5)]. Therefore, the formulation of PCM is based on a modified FCM objective function whereby an additional term called as *regularizing term* is also included i.e. the second term in Eq. 12.14. This forces μ_{ij} to be as large as possible such that pixels with high degree of typicality (*i.e.* the membership value of a pixel in a cluster representing the possibility of the pixel belonging to that cluster) with respect to a cluster may have high value of μ_{ij} and the pixels with less typicality may have low μ_{ij} values in all clusters. Thus, similar to FCM, PCM clustering is also an iterative process where the class membership values are obtained by minimizing the generalized least-square error objective function (Krishnapuram and Keller, 1993), given by

$$J_m(U,V) = \sum_{i=1}^{N}\sum_{j=1}^{c}(\mu_{ij})^m \|x_i - v_j\|_A^2 + \sum_{j=1}^{c}\eta_j \sum_{i=1}^{N}(i - \mu_{ij})^m \qquad \ldots(12.14)$$

subject to constraints

$$\left.\begin{array}{l} \max \mu_{ij} > 0 \text{ for all } i \\ \sum_{i=1}^{N}\mu_{ij} > 0 \text{ for all } j \\ 0 \leq \mu_{ij} \leq 1 \text{ for all } i, j \end{array}\right\} \qquad \ldots(12.15)$$

where η_j is a parameter that depends on the distribution of pixels in the cluster *j* and is assumed to be proportional to the mean value of the intra cluster distance. For clusters with similar distributions, η_j may be set to the same value for each cluster (Massone, et. al, 2000). Generally, η_j depends on the shape and average size of the cluster j and its value may be computed as

$$\eta_j = K \frac{\sum_{i=1}^{N} \mu_{ij}^m d_{ij}^2}{\sum_{i=1}^{N} \mu_{ij}^m} \qquad \ldots(12.16)$$

where K is a constant and is generally kept as 1. The class memberships, μ_{ij} are obtained from

$$\mu_{ij} = \frac{1}{1 + \left(\frac{d_{ij}^2}{\eta_j}\right)^{1/(m-1)}} \qquad \ldots(12.17)$$

The cluster centres are updated in the similar fashion as in FCM. The general procedure of PCM algorithm may be formalized in the following steps:

Step 1. Fix values of m and c, and select the type of A-norm and maximum number of iterations.
Step 2. Chose an initial membership matrix, U^i as described in FCM.
Step 3. Estimate η_j using Eq. (12.16).

Repeat

Step 4. Input the image file.
Step 5. Compute the vector of cluster centres v_j using Eq. (12.11).
Step 6. Compute the distance based on the selected A-Norm using Eq. (12.5).
Step 7. Update U matrix for the next iteration U^{i+1} using Eq. (12.17).

Until $(\|U^{i+1} - U^i\| < \varepsilon)$, where ε is the user defined limiting error.

Step 8. Estimate the final value of η_j Eq. 12.16 from final U^{i+1} matrix.
Step 9. Compute elements of U matrix using final value of η_j using Eq. 12.16
Step 10. Write final U to the file, which represents class proportions when scaled from 0 to 1.

It may be mentioned that the value of m and its interpretation is different in the PCM than in FCM (Krishnapuram and Keller, 1996). The weighting exponent m in PCM determines the rate of decay of the membership values. It is true that in both PCM and FCM, as m approaches 1, partitions that minimize J_m become increasingly hard whereas when m approaches ∞, partitions that minimize J_m become increasingly soft or complete fuzzy. However, in FCM, as m increases, it represents increase in sharing of pixel in all clusters, whereas in PCM, it represents increased possibility of all pixels in the data set completely belonging to a given cluster.

Neural Network Based Methods
Another set of soft classification methods comes from artificial neural network theory. Artificial neural networks have the capability to generalize the relation between the evidence (*e.g.,* remote sensing data) and the conclusion (*e.g.*, landcover classification) without developing any mathematical models. Thus, unlike statistical parametric methods, they do not assume that the data follows a distribution. The neural network contains interconnected layers each containing a number of units, symbolizing the biological concept of a neuron. The interconnections carry weights, which are adjusted in an interactive learning process to provide neural network solution (Schalkoff, 1997). The learning process may be supervised or unsupervised depending on whether training data are required or not. Accordingly, a number of supervised an unsupervised neural network algorithms have been developed.

Typically, a supervised neural network such as an MLP consists of three layers; an input layer, a hidden layer and an output layer (Fig. 12.2). The input layer receives the data (*i.e.*, the multi-spectral remote sensing image data). As a result, the units in the input layer equal the number of bands used for the classification. Unlike input layer, hidden and output layers process the data. The output layer produces the neural network results. The number of units in the output layer is generally equal to the number of classes to be mapped. Therefore, the number of units in the input and output layers are fixed by the application designed. Selection of the number of hidden layers and their units is a critical step for the successful operation of the neural network. Using too few units in the hidden layer may result into inaccurate classification as the network may not be powerful enough to process the data. On the other hand, by using a large number of hidden units, the computational time becomes large. It may also result into the network being over-trained. The optimum number of units in the hidden layer is often determined by trial and error, though some empirical relations do exist (Kavzoglu and Mather, 2003).

Back Propagation Neural Network (BPNN)
The back propagation learning algorithm is a generalized least squares algorithm that adjusts the connection weights between units to minimize the mean square error between the network output and the target output. The target output is known from reference data. Data provided to input unit are multiplied by the connection weights and are summed to derive the net input to the unit in the hidden layer as shown in Fig. 12.2, and it is given by

$$net_s = \sum_i x_i W_{is} \qquad \qquad \ldots(12.18)$$

where, x_i is a vector of magnitude of the ith input (*i.e.*, spectral response of pixel), W_{is} is matrix of the connection weights between i^{th} input layer unit and s^{th} hidden layer unit. Each unit in s^{th} hidden layer computes a weighted sum of its inputs, and passes the sum *via* an activation function to the units in the jth output layer through weight vector W_{sj}. There is a range of activation functions to transform the data from hidden layer unit to an output layer unit. These include pure linear, tangent hyperbolic, sigmoid functions, etc. Although, the use of these functions may lead to difference in accuracy of classification (Ozkan and Erbek, 2003), sigmoid function has been widely used, and may be defined as

$$O_s = 1/[1 + \exp^{-\lambda net_s}] \qquad \qquad ...(12.19)$$

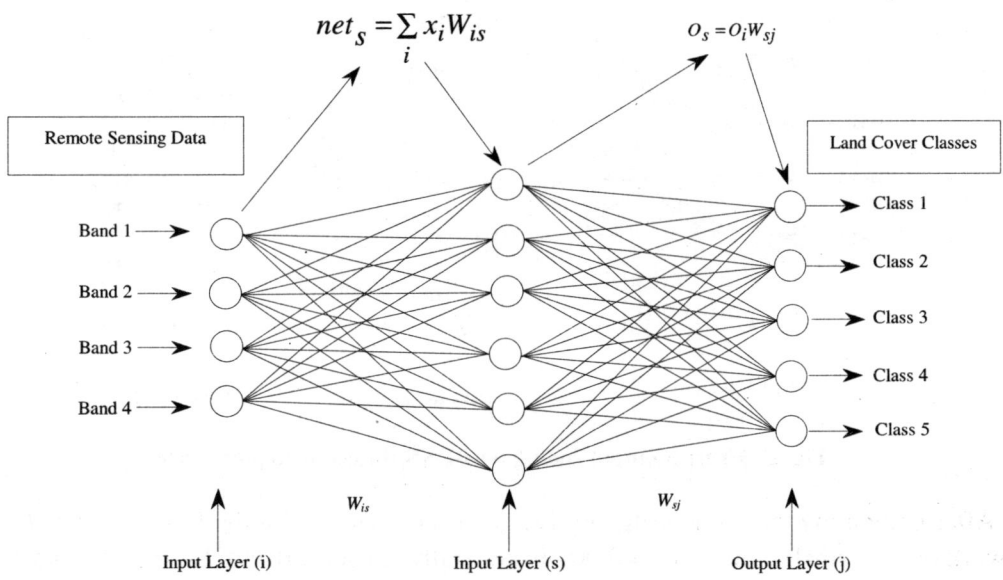

Fig. 12.2 Typical supervised neural network architecture

where O_s is the output from the sth hidden layer unit, and λ is a gain parameter that controls the connection weights between the hidden layer unit and the output layer unit. Outputs from the hidden units are multiplied with the connection weights, and are summed to produce the output of the jth unit in the output layer.

$$O_j = O_s W_{sj} \qquad \qquad ...(12.20)$$

where O_j is the network output for the jth output unit (*i.e.*, the land cover class) and W_{sj} is the weight of the connection between s^{th} hidden layer unit and j^{th} output layer unit.

An error function E, determined from a sample of target (known) outputs and network outputs, is minimized iteratively. The process continues until E converges to some minimum value, and the adjusted weights are obtained. E is given by

$$E = 0.50 \sum_{j=1}^{c} (T_j - O_j)^2 \qquad \ldots (12.21)$$

where T_j is the target output vector, O_j is the network output vector, and c is the number of classes. The target vector is determined from the known class allocations of the training pixels, which are coded in binary form. For example, in Fig. 12.3, a pixel belonging to class 3 shall be coded as 0 0 1 0 0 at the five output units. The collection of known class allocations of all pixels will form the target vector.

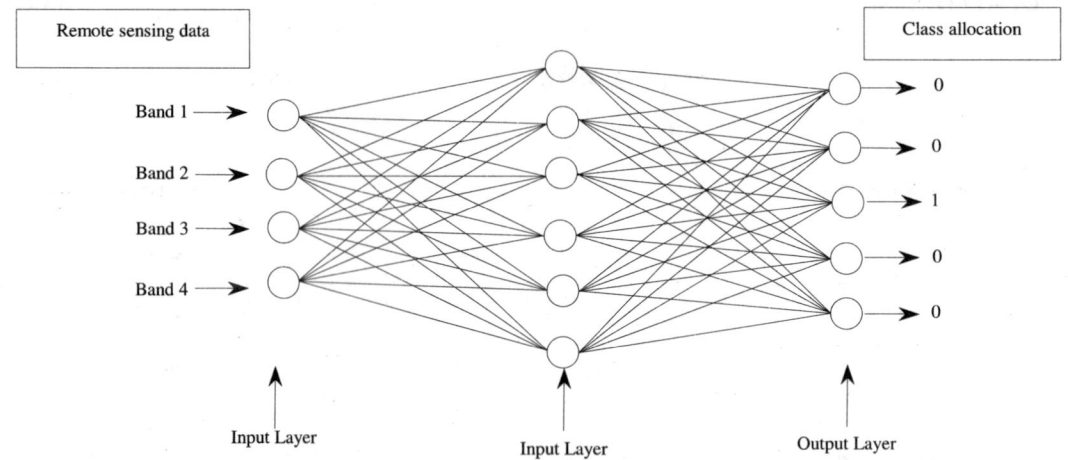

Fig. 12.3 Target output coding for BPNN trained with pure pixels

After computing the error of the network, it is compared with the limiting error E_L of the network (a very small number $E_L \approx 0.001$ is generally chosen). If $E \leq E_L$, the network training is stopped otherwise E is back propagated to the units in the hidden and the input layers. The connection weights are updated after several iterations. The number of iterations may vary from one data set to the other, and is generally determined by trial and error. The process of back propagation and weight adjustment is explained below

First the error vector at each unit of the output layer is computed as

$$E_j = O_j (1-O_j) E \qquad \ldots (12.22)$$

and the error vector for each unit at the hidden layer is computed as

$$E_s = O_s (1-O_s) \sum_{j=1}^{c} O_j E_{sj} \qquad \ldots (12.23)$$

Thereafter, the net error in connection weights between output layer, and hidden layer is computed as

$$E_{sj} = O_s E_j \qquad \qquad \ldots(12.24)$$

and error in connection weights between hidden layer and input layer as

$$E_{is} = F_m x_i E_s \qquad \qquad \ldots(12.25)$$

where F_m is the momentum factor which controls the momentum of the connections between the hidden layer unit and the input layer unit.

Once the error vectors are computed, the weight updation takes place for the next iteration. The weights between output layer and hidden layer are updated as

$$(W_{sj})_{new} = (W_{sj})_{old} + E_{sj} \qquad \qquad \ldots(12.26)$$

and the weights between input layer and hidden layer are updated as

$$(W_{is})_{new} = (W_{is})_{old} + E_{is} \qquad \qquad \ldots(12.27)$$

The gain parameter (λ) is also updated as,

$$(\lambda)_{new} = (\lambda)_{old} + L_R E_s \qquad \qquad \ldots(12.28)$$

where L_R is the learning rate which controls the time of the learning process.

The next iteration starts with new set of weights and parameters, and the process is repeated till the convergence is achieved and the adjusted weights are obtained. At this stage, the network is assumed to be trained. The adjusted weights are then used to determine the outputs of the unknown pixels of the image using Eqs. (12.18), (12.19), and (12.20). The network outputs are called as *activation levels*. For hard classification, the pixel is allocated to the class with the highest activation level whereas for soft classification, these activation levels are scaled from 0 to 1 for a pixel to produce soft outputs (Foody, 1996). The general procedure of BPNN algorithm may be formalized in the following steps.

Step 1. Initialize the network parameters such as limiting error E_L, learning rate L_R, momentum factor F_m, gain parameter λ, number of iterations, and randomly choose initial weights between the interconnected layers.

Step 2. Set the number of units in all layers (*i.e.*, input, hidden, and output layers).

For $i = 1$ to N_t, do;

Step 3. Define the target output vector of the selected pixel.
Step 4. Compute net_s for the selected pixel using Eq. (12.18).
Step 5. Compute the value of O_s using Eq. (12.19).
Step 6. Compute the network output O_j using Eq. (12.20).
Step 7. Compute the network error E using Eq. (12.21).

Step 8. Compare the network error by the limiting error.
Step 9. If $E \leq E_L$ go to step 14, else continue
Step 10. Update the weight vector between input layer and hidden layer using Eqs. (12.22), (12.24), and (12.26).
Step 11. Update the weight vector between hidden layer and output layer using Eqs. (12.23), (12.25), and (12.27).
Step 12. Adjust the gain parameter λ using Eq. (12.28).
Step 13. Return to Step 4, until ($E \leq E_L$).

End

Step 14. Save the final weight vectors and gain parameter to a file.
Step 15. Apply steps 5 and 6 using the set of final weights to classify the whole image.

Competitive Learning Neural Network followed by Learning Vector Quantizers (CLNN-LVQ)

Competitive learning can be described as a process in which output layer units compete among themselves to get authority to produce output for given pixels (Mannan and Ray, 2003). It may be viewed as a procedure that learns to group the input data in clusters on the basis of the characteristics of input data. The heart of competitive learning lies in the 'winner-takes-all-strategy'. A competitive learning neural network (CLNN) contains two layers; input layer consisting of n units and output layer consisting of c units (Fig. 12.4). The input layer unit encodes the input data and their values are supplied through weighted connections to output layer unit. W_{ij} refers to the weight of the connection from i^{th} input unit to the jth output unit. The output layer units compete among themselves to acquire ability to activate in response to the input data. Therefore, the output layer is sometimes called as *competitive layer*. An appropriate weight update rule is used by the learning algorithm to determine the connection weights from input layer units to the competitive layer units.

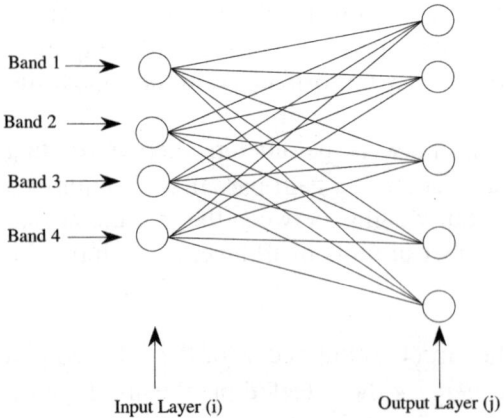

Fig. 12.4 Typical CLNN network architecture

Given that the overall purpose of the network is to assign a unique output layer unit for each input pixel, a reasonable criterion is that each output unit should represent the input data to be associated with that unit. The jth unit in the output layer is completely described by the vector of weights of connections from input units {*i.e.*, $W_j = (W_{1j}, W_{2j}, \ldots\ldots\ldots\ldots W_{nj})$}. Each such weight vector has the same dimensionality as each input data vector. A distance measure may be used to estimate the similarity between an input vector to the weight vector associated with a unit and is given as (Mehrotra, *et al*, 1997),

$$D_j = \sqrt{\sum_{i=1}^{n}(x_i - W_{ij})^2} \qquad \ldots(12.29)$$

where D_j is the Euclidean distance between input vector x_i (*i.e.*, the spectral response of the pixel) and the weight vector between the ith unit in the input layer, and the jth unit in the output layer W_{ij}. The weight update rule is formulated in such a way that the weight vector associated with each unit is as close as possible to all input pixels for which the unit is the winner of the competition. The winner is judged on the basis of minimum distance between the input layer unit and the weight vector to the input unit i.

The connection weights of the winner unit are modified so that the weight vector at the position of the winner unit moves closer to the input unit, while the other weights remain unchanged. This is the winner-takes-all phase. The weight update rule can be described as

$$(W_{ij})_{new} = (W_{ij})_{old} + \sum_{i=1}^{n} L_R(t)\{x_j - (W_{ij})_{old}\} \qquad \ldots(12.30)$$

where $(W_{ij})_{old}$ is the weight at the position of the winning unit, $L_R(t)$ is the learning rate of the network, a predetermined constant which decreases with the time of unsupervised learning in each iteration as

$$L_R(t+1) = \frac{L_R(t)}{2} \qquad \ldots(12.31)$$

Thus, the process of weight update brings the winning unit's weight vector closer to those training pixels that fall within its zone of attraction. The learning stops when the error vector given by Eq. (12.32) is minimized.

$$E_L \geq \|(W_{ij})_{old} - (W_{ij})_{new}\| \qquad \ldots(12.32)$$

where, E_L is the user defined limiting error of the network.

Once the error has been minimized, adjusted weights are obtained. These adjusted weights reflect the cluster centres similar to the conventional hard *k*-means clustering algorithm. The distance between each pixel and the adjusted weights, are computed to group the pixels into user defined clusters.

The cluster information of each pixel from CLNN is then used as desired output for LVQ algorithm to further refine the cluster information. An LVQ illustrates that how the unsupervised learning mechanism (*i.e.*, CLNN) can be adapted to solve the supervised learning tasks in which the cluster information is known for every pixel in the image. LVQ, thus, uses the cluster information from CLNN as training data as it is done in a supervised learning process.

LVQ processing is also based on winner-takes-all-strategy, but here since the desired output is known, the weights of the winning unit are adjusted in such a fashion that if the output of the winner unit matches with the desired output, it is moved towards that input vector.

$$(W_{ij})_{new} = (W_{ij})_{old} + L_R(t)\{(W_{ij})_{old} - x_j\} \qquad \ldots(12.33)$$

Otherwise it moves far away from input vector.

$$(W_{ij})_{new} = (W_{ij})_{old} + L_R(t) * (W_{ij})_{old} - x_j) \qquad \ldots(12.34)$$

At the start of LVQ classification process, weights are initialized by small random values, which are then updated in the process of learning. The weight update rule uses a learning rate that is a function of time t, and is given as

$$L_R(t) = 1/t \qquad \text{or} \qquad L_R(t) = a(1-(t/A)) \qquad \ldots(12.35)$$

where a and A are positive constants ($A > 1$).

The updating of learning rate allows the network to converge to a state in which the weight vector is stable and has little change with any further input pixel vector presentations. The network error is determined from the difference between the old and the new weights, which is then compared to the limiting error, after each iteration, until the error is minimized. After network convergence, the final weights are obtained, which are treated as the final cluster centres. The distances of each pixel from each cluster centre are determined and the pixel is allocated to the class having minimum distance to produce hard classification. Soft classified outputs may be derived by scaling these distances from 0 to 1 to denote the class memberships of each pixel. For applying the CLNN-LVQ classifier, the steps given below may be followed

Step 1. Randomly set a weight matrix $W_{i,j}$.
Step 2. Set an initial value of the learning parameter, limiting error and number of iterations.

For input pixel x_i, do:
Step 3. Compute distances D_j.
Step 4. Define the position of the winning unit (*i.e.*, the class which has the minimum value of D_j) named as Min.
Step 5. Update the weight of the position of the winning unit using Eq. (12.30).

Step 6. Update the learning rate using Eq. (12.31).
Step 7. If $E_L \geq \|(W_{ij})_{old} - (W_{ij})_{new}\|$ continue, else go back to step 4.

End

Step 8. Calculate the distances and allocate the pixel to class having minimum distance, and save the output to a file to be used in training of LVQ.
Step 9. Use the initial values of weight matrix, and the learning rate from step 1 and 2, respectively.

For input pixel x_i, do:
Step 10. Compute the distance D_j using Eq. (12.29).
Step 11. Define the position of the winning unit (*i.e.* the class which has the minimum value of D_j) and that position is named as Min.
Step 12. Update the weight of the position of the winning unit using either Eq. (12.33) or Eq. (12.34), depending of, if the position of winning unit matches with the output of CLNN or not respectively.
Step 13. Update the learning rate using Eq. (12.35)
Step 14. If $E_L \geq \|(W_{ij})_{old} - (W_{ij})_{new}\|$ continue, else go back to step 14.

End

Step 15. Compute the distances between a pixel and final weight using Eq. (12.29)
Step 16. For hard classification, allocate the pixel to the class having the minimum distance.
Step 17. For soft classification, scale the distances to produce membership values.

12.3 CLASSIFICATION ACCURACY ASSESSMENT

No classification is considered complete unless an assessment of accuracy has been performed (Jensen, 1986). The accuracy assessment is a critical step in any mapping process, and thus is an essential component that allows a degree of confidence to be attached to maps for their effective use. Traditionally, the accuracy of classification has been assessed using error matrix based measures. Here, each pixel in the image is assumed pure, containing one class per pixel on the ground. Thus, in essence, the continuum of variation found in the landscape is divided into a finite set of classes such that pixels representing these classes became pure, and the error matrix based measures may be used. However, these classes become less separable as the class mixture increases, and therefore, the error matrix based measures may be inappropriate. Alternate accuracy measures are, therefore, sought to evaluate the accuracy of soft classification which represents the class mixture in a meaningful way.

When the class allocation is soft (*i.e.*, pixels with varying class membership values both in the classified image and the reference data), other measures are sought. These include the Euclidean and the L_1 distance (Foody and Arora, 1996), the cross-entropy (Foody, 1995a), and correlation coefficients (Maselli, *et al*, 1996). All these measures may be treated as indirect methods of assessing the accuracy of soft classification because the accuracy evaluation is interpretative rather than a representation of actual value as denoted by the traditional error matrix based measures. Recently, the concept of fuzzy error matrix (FERM) has been put forth to assess the accuracy of soft classification (Binaghi, et al, 1999).

In this section, some soft classification accuracy measures are introduced, and have been categorized as, distance measures and fuzzy error matrix based measures.

Distance Measures

An easy approach to assess the accuracy of soft classification is to measure the distance between the soft classification output and the soft reference data. A number of distance measures may be used.

The Euclidean distance D_i may be defined as the square root of sum of the squares of the difference in the class proportions of a pixel of the soft classification output ($^1\mu_{ij}$) and the soft reference data ($^2\mu_{ij}$) and may be derived for each pixel from (Foody and Arora, 1996)

$$D_i = \sqrt{\sum_{j=1}^{c}(^1\mu_{ij} - {}^2\mu_{ij})^2} \qquad \ldots(12.36)$$

Mean of the Euclidean distance computed over all the testing pixels may then be given as

$$\overline{D} = (1/N_s)\sum_{i=1}^{N_s} D_i \qquad \ldots(12.37)$$

where N_s is the total number of testing pixels.

The L_1 distance (L_1) may be defined as the sum of the absolute differences in the class proportions of a pixel of the soft classification output and the soft reference data, and may be derived for each pixel from (Foody and Arora, 1996),

$$L_1 = \sum_{j=1}^{c} \left| {}^1\mu_{ij} - {}^2\mu_{ij} \right| \qquad \ldots(12.38)$$

Mean value for L_1 distance over all the testing pixels may be computed from

$$\overline{L} = (1/N_s)\sum_{i=1}^{N_s} L_i \qquad \ldots(12.39)$$

The cross entropy $d(^1\mu, {}^2\mu)_i$, is also a distance measure to evaluate the degree of similarity between the class proportions of a pixel of the soft classification output and the soft reference data, and may be derived for each pixel from (Foody, 1995)

$$d(^1\mu, {}^2\mu)_i = -\sum_{j=1}^{c}(^1\mu_{ij})\log_2(^2\mu_{ij}) + \sum_{j=1}^{c}(^1\mu_{ij})\log_2(^1\mu_{ij}) \qquad \ldots(12.40)$$

Mean value of the cross entropy may be given as

$$\overline{d(^1\mu, {}^2\mu)} = (1/N_s)\sum_{i=1}^{N_s} d(^1\mu, {}^2\mu)_i \qquad \ldots(12.41)$$

Root mean square error ($RMSE_j$) for each class may be computed from

$$RMSE_j = \sqrt{\frac{\sum_{i=1}^{N_s}(^1\mu_{ij} - {}^2\mu_{ij})^2}{N_s - 1}} \qquad \ldots(12.42)$$

The mean value of all these measures indicates the accuracy of the whole classification. Lower the value of each of these measures, higher will be the accuracy of classification.

The correlation coefficient CC may also be used to indicate the accuracy on individual class basis estimated from a soft classification output and a soft reference data. The higher the correlation coefficient, the higher is the classification accuracy of a class. The formulation of CC is given by (Maselli, et al, 1996)

$$CC = \frac{Cov(^1\mu_{ij}, {}^2\mu_{ij})}{\sigma_{^1\mu_{ij}} \sigma_{^2\mu_{ij}}} \qquad \ldots(12.43)$$

where $Cov(^1\mu_{ij}, {}^2\mu_{ij})$ is the covariance between the two distributions (*i.e.* the soft classified output and the soft reference data) and $\sigma_{^1\mu_{ij}}, \sigma_{^2\mu_{ij}}$ are the standard deviations of both the distributions.

Fuzzy Error Matrix Based Measures

Recently, the concept of fuzzy error matrix (FERM) being used to assess the accuracy of soft classification (Binaghi, et al, 1999). The layout of a fuzzy error matrix is similar to the traditional error matrix that is used for assessing the accuracy of hard classification. The exception is that elements of a fuzzy error matrix can be any non-negative real numbers instead of non-negative integer numbers. The elements of the fuzzy error matrix represent class proportions corresponding to reference data (*i.e.*, soft reference data) and classified outputs (*i.e.*, soft classified image), respectively.

Let R_n and C_m be the sets of the reference and classification data assigned to class n and m, respectively, where the values of n and m are bounded by one and the number of classes c.

Note here that R_n and C_m are fuzzy sets and $\{R_n\}$ and $\{C_m\}$ form the two fuzzy partitions of the testing sample data set X where x denotes a testing sample in X. The membership functions of R_n and C_m are given by

$$\mu_{R_n} : X \to [0,1] \text{ and } \mu_{C_m} : X \to [0,1] \qquad \ldots(12.44)$$

where [0,1] denotes the interval of real numbers from 0 to 1 inclusive. Here, $\mu_{R_n}(x)$ and $\mu_{C_m}(x)$ is the class membership (or class proportion) of the testing sample x in R_n and C_m, respectively. Since, in the context of fuzzy classification, these membership functions also represent the proportion of a class in the testing sample, the orthogonality or sum-normalization is often required, which for the fuzzy reference data may be represented as

$$\sum_{j=1}^{c} \mu_{R_i}(x) = 1 \qquad \ldots(12.45)$$

The procedure used to construct the fuzzy error matrix M employs fuzzy min operator to determine the element $M(m,n)$ in which the degree of membership in the fuzzy intersection $C_m \cap R_n$ is computed as

$$M(m,n) = |C_m \cap R_n| = \sum_{x \in X} \min(\mu_{C_m}, \mu_{R_n}) \qquad \ldots(12.46)$$

The layout of a typical FERM is shown in Table 12.1

Table 12.1 A fuzzy error matrix

Soft classification	Soft reference data				Total grades
	Class 1	Class 2	...	Class c	
Class 1	$M_{(1,1)}$	$M_{(1,2)}$	...	$M_{(1,c)}$	C_1
Class 2	$M_{(2,1)}$	$M_{(2,2)}$	...	$M_{(2,c)}$	C_2
...	⋮	⋮	⋮	⋮	
Class c	$M_{(c,1)}$	$M_{(c,2)}$	...	$M_{(c,c)}$	C_c
Total grades	R_1	R_2	...	R_c	

Definition of terms: $M_{(m,n)}$ is the member of FERM in the mth class in the soft classified output and n^{th} class of the soft reference data, C_m is the sum of class proportions of class m in the classified output and R_n is the sum of class proportions of class n from the reference data.

From FERM, overall accuracy OA may be calculated from

$$OA = \frac{\sum_{j=1}^{c} M_{(i,j)}}{\sum_{j=1}^{c} R_j} \qquad \ldots(12.47)$$

Similarly, User's and Producer's accuracy of class j (UA_j) and (PA_j) may be computed as

$$UA_j = \frac{M_{(j,j)}}{C_j} \quad \text{and} \quad PA_j = \frac{M_{(j,j)}}{R_j} \qquad \ldots(12.48)$$

The average User's and Producer's accuracy $(AA_u$ and $AA_p)$ may then computed from

$$AA_u = \frac{\sum_{j=1}^{c} UA_j}{c} \quad \text{and} \quad AA_p = \frac{\sum_{j=1}^{c} PA_j}{c} \qquad \ldots(12.49)$$

For clear understanding of fuzzy error matrix, consider a simple example, where a pixel has been decomposed into three classes. The corresponding class proportions from soft reference data and soft classification output for that pixel are given by

$\mu_{R_1}(x) = 0.3,$ $\qquad \mu_{R_2}(x) = 0.2,$ $\qquad \mu_{R_3}(x) = 0.5$ and
$\mu_{C_1}(x) = 0.7,$ $\qquad \mu_{C_2}(x) = 0.1,$ $\qquad \mu_{C_3}(x) = 0.2$, respectively

The fuzzy error matrix for the above pixel is then created using fuzzy min operator, and is displayed in Table 12.2. The fuzzy error matrix for the whole classification is accumulated by summing up the elements of fuzzy error matrix of each testing pixel (*i.e.*, testing sample).

Table 12.2 Fuzzy error matrix for an individual testing sample

Soft classification	Soft reference data			Total grades
Class 1	0.3	0.2	0.5	0.7
Class 2	0.1	0.1	0.1	0.1
Class 3	0.2	0.2	0.2	0.2
Total grades	0.3	0.2	0.5	1.0

For this example, the overall accuracy (OA) is 60%, user's accuracy is 42.90%, 100.00%, and 100.00% for classes 1, 2, and 3 respectively, producer's accuracy is 100.00%, 50.00%, and 40.00% for classes 1, 2, and 3 respectively, average user's accuracy is (80.97%), and average produce's accuracy is (63.33%).

Thus, fuzzy error matrix can be used to derive a number of accuracy measures similar to the ones obtained from traditional error matrix for hard classification. Therefore, the FERM based measures appear more appropriate for the assessment of accuracy of soft classification than the distance measures and correlation coefficients.

12.4 ADVANCES IN GIS

GIS technology is changing at very fast rate, and it has moved from mainframe computer to work station and from desktop based PC system to palm top system. With the recent advancement in broadband and wireless communication technology along with Internet facilities, a new area in GIS known as Internet/Web GIS has now provided a new dimension to GIS analyst. A subsequent advance in wireless technology, another area of GIS development, is Mobile GIS, based on mobile computing and mobile Internet (Fangxiong & Zhiyong, 2003). It is not conventional GIS modified to operate on a smaller computer, but an extension of Web GIS to mobile Internet including wireless Internet/Intranet and mobile communication network.

12.5 INTERNET GIS

Internet is a collection of interconnected network of computing machines worldwide, where networks operate using a standard set of addresses allowing millions of computer connectivity on a global scale. Internet based Worldwide Web (WWW) has emerged as an alternative means of accessing, viewing, and distributing spatial information. In Internet GIS, results are achieved by integrating front end query capabilities supported through standard Internet and WWW interfaces, and protocols with the capabilities of a commercial DBMS and GIS software residing in the background. User defined queries are translated into corresponding SQL commands, and passed to the *back-end* GIS database for handling. The resulting response is returned back through a getaway or bitmap format suitable for fast transmission and quick viewing across the Internet. Online Internet GIS can perform two major Internet GIS applications, *i.e.*, client side and server side base on common getaway interfaces (CGI) or gateway scripts.

In a client-side Internet GIS application, the client (Web browser) is enhanced to support GIS functionality while in a server-side GIS application, a Web browser is used only to generate server requests and display the results. Client-side GIS applications are implemented typically by enhancing the Web browser with a Java applet, Active X, or plug-ins. Some client-side applications even require users to install a complete client application. In either case, client-side applications require software of some kind (other than a browser) to be transferred to the user (Fig. 12.5). An example of a client-side Internet GIS application is one that runs as a Java applet. The code for the applet is transferred to the Web browser as binary instructions that provide a graphical user interface (GUI) for the GIS application. Vector-based data is then transferred to the client enabling the complex GIS functions on the client. This architecture should not be confused with a similar server-side GIS architecture that implements a Java applet to create a GUI for the GIS application. In this case, the applet is simply an interface for an image. The complex GIS calculations and data remain on the server.

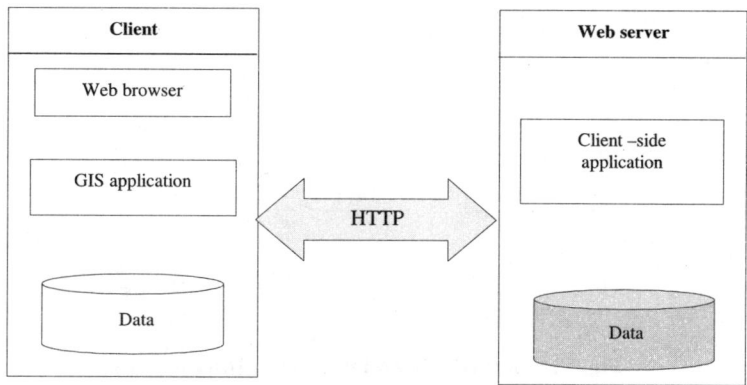

Fig. 12.5 Client-side architecture of Internet GIS

An example of a server-side Internet GIS application is one typical of the mapping applications found on the major Internet portals. In these applications, users send a request to a server (i.e. an address), and the server processes the request and sends the results back as an image embedded in an HTML page via standard HTTP. The response is a standard Web page that a generic browser can view. In server-side Internet GIS applications, all the complex and proprietary software, in addition to the spatial and tabular data remain on the server (Fig. 12.6). This architecture has several advantages because the application and data are centralized on a server. These advantages include simplified development, deployment, and maintenance. A comparison of the advantages and disadvantages of server-side and client-side GIS applications is given in Table 12.3.

The Web is a stateless environment. A Web server receives a request from a client, processes the request, and sends a response with no knowledge about the client's state unless state maintenance is used. This is similar to a common software architecture known as pipe-filter architecture. Once the architecture of the system is selected, there are two additional architectural choices pertaining to state maintenance for a server-side application. One can choose a pipe-filter approach and accomplish state maintenance on the client side, or could choose an object-oriented approach and accomplish state maintenance on the server side. Using a pipe-filter approach, state maintenance on the client side is accomplished by storing all state maintenance variables (extents, layers, command, input variables, etc.) on the client side. This is done by one of two methods, or the combination of both; the use of browser cookies, and/or the use of hidden input tags (or parameter tags) in the HTML page.

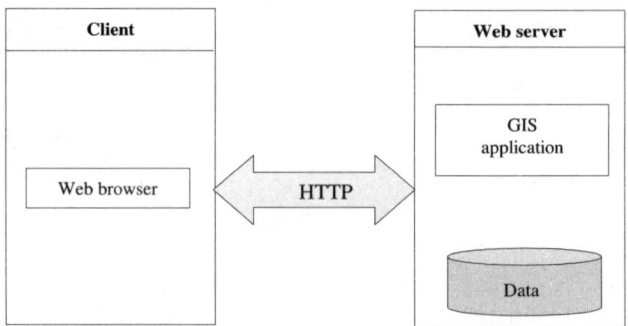

Fig. 12.6 Server-side architecture of Internet GIS

Table 12.3 Comparison of server side and client side

		Advantages	Disadvantages
Server side Internet GIS		Simpler to develop	Primitive Graphical User Interface
		Easier to deploy	Low graphics quality
		Easier to maintain	One-click functionality from a browser
		Adheres to Internet standards	
		Requires standard Web browser	
		Low bandwidth required	
Client side Internet GIS		Vector data can be used	Difficult to develop
		Better image quality	Requires additional software
		Enhanced GUI	Longer download times
			No adherence to standards
			Platform/browser incompatibility

When a user sends a request to the server, all the state variables and command(s) are extracted from the HTTP request. The command is then executed on the server, a map is generated reflecting the user's current state, the new map image is wrapped in HTML (including the current state variables), and the Web page is returned to the client. State maintenance on the server side is accomplished by maintaining map and database objects on the server for the life of a user session. The state maintenance variables are directly accessed as properties of the objects that are maintained on the server. The only state variables required to be maintained on the client are those that describe the interface (*i.e.* last command, active layer, etc.) A comparison of the advantages and disadvantages of server-side state maintenance is given in Table 12.4.

A typical Internet system may consist of three distinct components: a server application, a client interface, and a data repository (Fig. 12.7).

Table 12.4 Advantages and disadvantages of server-side state maintenance

Advantage	Disadvantage
Easier to develop	Not highly scalable
Less server processing required	Requires implementation of Session Management
Allows for complex applications	

Future Trends 263

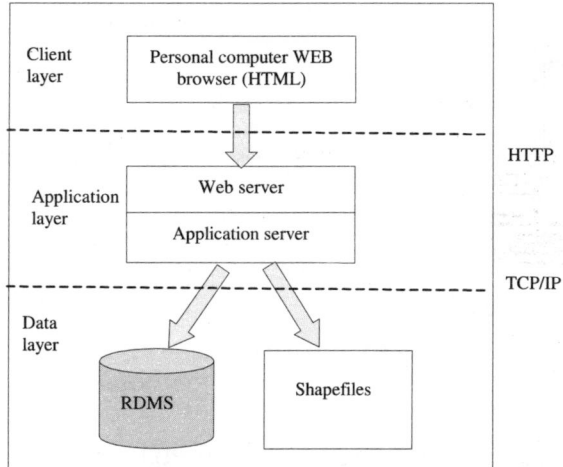

Fig. 12.7 System architecture

The client layer consists of a personal computer running a Web browser. This layer provides the user interface and operates by generating requests to the application server via HTTP and displays the resulting HTML file in a Web browser. The middle layer is itself a layered system consisting of a Web server layered on an application server. The Web server receives requests from the client which are processed by the application server's Web administration module and then passed to the application server. The application server makes requests to the data layer via TCP/IP and ODBC. The data layer is a data repository consisting of a relational SQL compliant database, and one or more directories of flat files in shapefile format. The data repository is built and maintained through an off-line data migration process that involves updating the data tables with new data, and geocoding new shapefiles. Although an off-line process, data migration is an integral part of the system and is included in the overall system design (Fig. 12.8).

In this design, data in the repository are updated via a migration process. The data repository is then accessed by the ESRI map application by means of an ODBC-TCP/IP connection. The map application processes data and generates HTML files which are in turn served to a client PC running a Web browser.

In order to support the above system, there are several hardware configurations. The configurations are: single computer configuration, two computer configuration, and multiple computer configurations. In the single computer configuration, the Web server, application server, and database server are installed on a single computer. In a two-computer configuration the Web server is installed on one machine, and the application server and database server are installed on a separate machine.

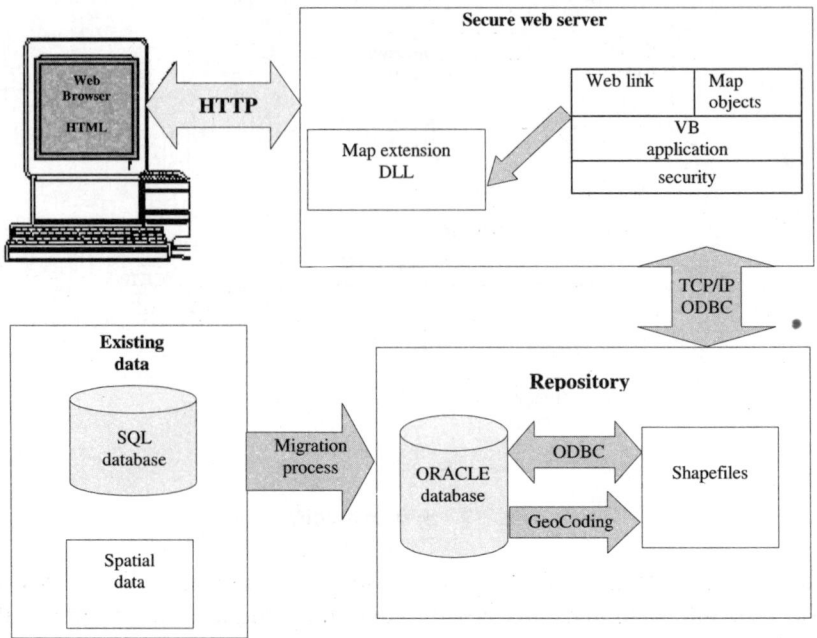

Fig. 12.8 System design diagram

In the multiple computer configurations, each component is installed on a separate computer. The ideal configuration for a particular deployment depends on the anticipated number of users visiting the site each day, and number of maps served (Table 12.5).

Table 12.5 Number of anticipated daily use

Configuration	Anticipated number of users
Single computer	100 - 1000
Two computer	1,000 – 1,500
Multiple computer	1,500 and more

Further, the server must have the following system specification.

  (*i*) Dual processor – 500 Mhz
  (*ii*) 512 Mb RAM
  (*iii*) Dual hard drive storage @ 13 GB each
  (*iv*) T1 (or better) Internet connection

Similarly, the system is required to have the following software for system development and deployment (Table 12.6).

A server-side Internet GIS application is composed of four distinct components: Web browser/client interface, Web application/server, GIS application/Map server, and a relational database management system. Although the components are integrated into a single system, each component is distinct and should be considered separately.

Table 12.6 Software requirement

Application	Software
Operating system	Window NT or higher
RDMS	Any SQL
Internet Browser	Internet Explorer or Netscape
Map/application	ESRI or any Internet map server
Geocoding engine	Arc view or any other compliant
Application development	Visual Basic, Map object
Client development	MS Visual Interdev

Client Interface

The client interface for an Internet GIS application is typically a Web browser implementing HTML form elements or implementing a Java applet. The client component can consist of a series of static and dynamic HTML pages that may or may not be implemented using HTML frames. By using HTML frames, the following advantages can be obtained.

(*i*) the entire interface need not be transmitted with every request
(*ii*) frames can be resized and scrolled individually
(*iii*) provides functionality similar to a stand alone application.

A Web application interface can be divided into number of functional area. Fig. 12.9 shows an example of such an interface.

Here Frame 1 is a non-resizeable, non-scrollable frame used to display a static HTML page consisting of the HTML form elements that compose the pulldown menu, the images that compose the image link button bar, and the JavaScript functions that process and submit users' actions. Frame 2 is a resizeable, scrollable frame used to display a dynamically generated HTML page consisting of the map's legend, image links that provide functionality to adjust the legend, and JavaScript functions used to process and submit user actions. Frame 3, also a resizeable, scrollable frame, is used to display a dynamically generated HTML page containing the map image, and JavaScript functions used to process and submit users' actions.

The GIS application/Map server may be designed and developed with appropriate software such as MS Visual Basic, ESRI Map Objects. The database design may be based on SQL or MS Access.

The success of Internet GIS application depends upon the requirement of the project and software. Developing GIS applications for the Internet is a situation where the best solution depends on the application requirements. By carefully analyzing requirements and planning an Internet GIS application, a software developer can greatly simplify the development process. The first step a developer must accomplish is to gain a thorough understanding of the application requirements. An understanding of the application requirements will allow the developer to make the right architectural choice for the application. Typically a server-side application is a good choice for developing an Internet GIS application because of advantages that include ease of development and standardization.

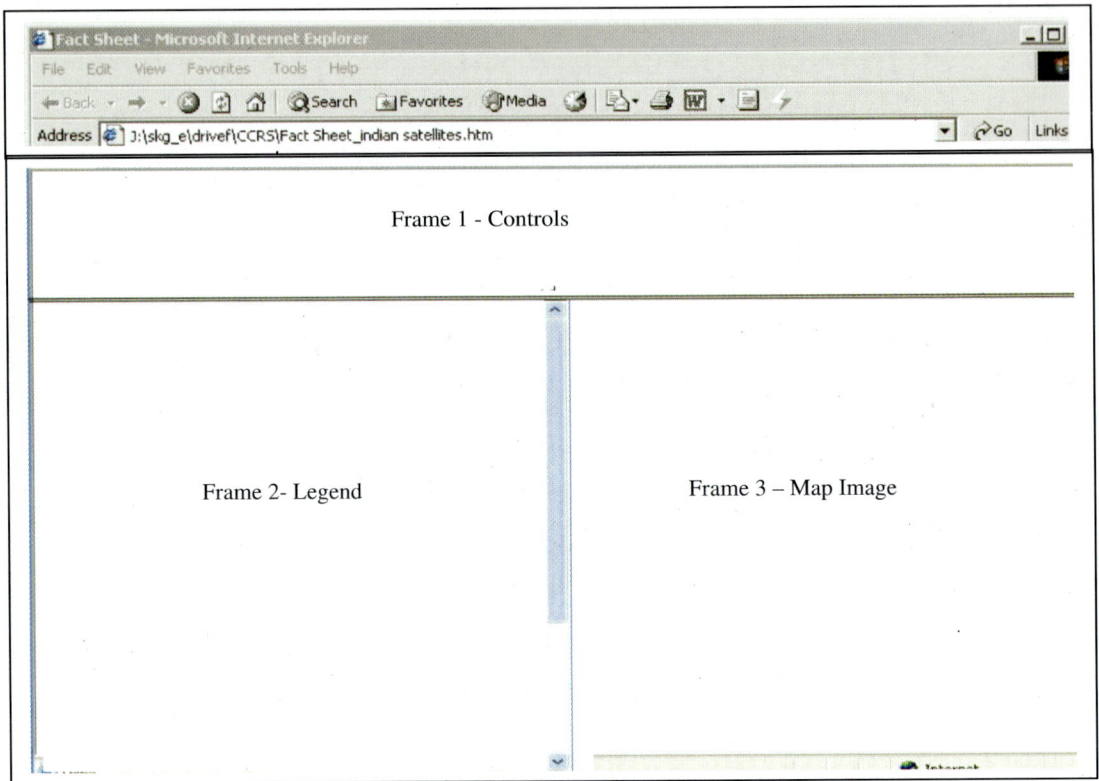

Fig. 12.9 Client interface layout

The structure of the spatial data, as well as the structure of a relational database is a critical factor that influences the performance of a GIS application. In an Internet environment, performance is usually the most important factor, thus a developer should make performance the number one priority when designing the database. A de-normalized database will provide the best performance for an Internet GIS application.

12.6 MOBILE GIS

The development of versatile computer hardware and software alongwith the successful implementation of Wireless Application Protocol (WAP) in communication network is a new concept of *work on the move*. It is easy to see that the integration of geospatial information and mobile internet is inevitable. The integrated system is designed to work on mobile intelligent terminals and brings new dimension and at any time and at any place to access geospatial and attribute information in GIS, which is known as *Mobile GIS*.

Traditionally, GIS mainly focused its attention on static spatial entity (also known as static GIS or SGIS). In SGIS, the analysis is correlated to position and its attribute, but it does not consider the moving nature of the world, *i.e.*, it cannot record the change of a piece of land in terms of its change in the boundary of the field or the ownership of a field. In 1988, Langram and Chrisman introduced the concept of Temporal GIS (TGIS) where the attention is focused on to a moving spatial object/entity. TGIS adds a new dimension in its analysis space *i.e.*, position, attribute, and time. Interestingly, both SGIS and TGIS consider a spatial entity that has a geographical aspect such as a road, mountain top, building, but does not consider a non-geographic entity such as a car, desk or a book.

Mobile GIS is a kind of GIS where it considers a non-geographic moving object in geographic space. It identifies a relationship between moving object and geographic entity, or a moving object and another moving object. For example, by integrating GIS, GPS, wireless internet to build a mobile GIS to monitor cars, the GIS analyst is interested in studying the moving car in a geographic entity space, and the moving car is a non-geographic entity.

Mobile GIS is not a modified conventional GIS to operate on a smaller computer, but a system built using a fundamentally new paradigm. It extends unlimited information on the Internet and powerful service functions of GIS to mobile devices to provide mobile users with geospatial information service. Mobile GIS has opened new avenues for business practice and thousands of potential application.

The general architecture of a mobile GIS is an integrated system of mobile client, a server, a wireless network, a mobile client position recording system, such as a GPS (Fig. 12.10). A mobile client can be a moving car equipped with GPS that can send information regarding the geographic position to the server by SMS. Alternately, a PDA equipped with GPS can show a digital map, and communicate with the server through a wireless network having GSM, CDMA, CDPP, or GPRS that can support digital data transmission. Mobile GIS can be simply divided into two categories, depending upon the manner by which access to Mobile Internet is done. The one is based on Short Message Service (SMS)/Multimedia Message Service (MMS), while the other is based on Wireless Application Protocol (WAP).

SMS/MMS-based Mobile GIS can only be suitable for mobile phones with simple system functions, unfriendly graphical user interface (GUI), poor information presentation and restricted application fields, because of the limitations of SMS/MMS such as restricted carrying information, time lag, unfriendly interactive mode, and so on.

On the contrary, WAP is a bear-independent international standard protocol that has been optimized for mobile devices with limited display and small keyboards of mobile handsets and low bandwidths of wireless networks, and permits applications and services to operate over all existing and foreseeable wireless networks such as GSM, CDMA, PHS, TDMA and WCDMA. The WAP specification encompasses a relatively simple and compact version of XML (eXtendable Markup Language) called WML (Wireless Markup Language), that makes it possible to request to a mobile service from a mobile terminal and return a map in the form of an embedded bitmap (*e.g.*, WBMP).

Hence, WAP-based Mobile GIS has richer information presentation, friendlier GUI, more system functions, and more application fields than the former. Moreover, it can work on a wide range of mobile devices with a WAP microbrowser only, from Personal Digital Assistants (PDAs), mobile phones, and in-car computers to other small mobile devices.

WAP-based Mobile GIS can be described as mobile users (with a WAP mobile terminal only). It can perform almost the same functionality as of Internet GIS but in a mobile environment at any time, any place and without the limitation of operating system and wired link. It is expected that because of the advantages of WAP, that WAP-based Mobile GIS will play a leading role in mobile information services markets.

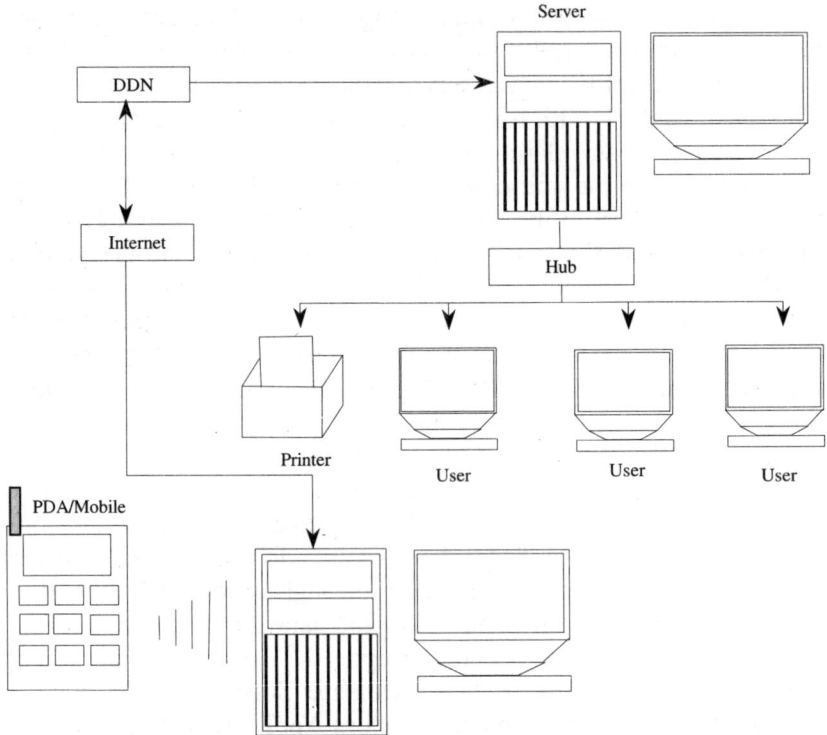

Fig. 12.10 Typical Architecture of Mobile GIS (*Source*: Luqun, et al, 2002)

WAP-based Mobile GIS is a "thin client" distributed system via mobile Internet. It is an open, extendable, stable and cross-platform distributed system, because of the increasing application demand and the diversity of mobile terminals.

Currently, popular system architecture solutions for distributed systems are all based on distributed object technologies. There are three mainstream industry standards: Microsoft's windows Distributed Network Architecture (Windows DNA)/.NET Architecture, Object Management Group's Common Object Request Broker Architecture (CORBA), Sun's Java 2 Enterprise Edition (J2EE) Architecture. A system based on Windows DNA/.NET Architecture solution can only use Microsoft's platforms from development, deployment to running, including developing platforms and operating system. It cannot cross operating system platform especially, which is the fatal weakness of Windows DNA/.NET Architecture. CORBA is too huge and complicated, and its technologies and standards are updated relatively slowly.

J2EE is a specification and standard created by Sun and her industry partners. J2EE provides support for the technologies such as Enterprise JavaBeans (EJB), Java Servlets API and Java Server Pagers (JSP), and so on. J2EE solution reduces the cost and complexity of developing a multi-tier distributed system which can be rapidly developed and deployed, and can enhance the portability, security, load balancing, and extensibility of a distributed system. The advantages of building distributed GIS based on J2EE are (*i*) cross-platform, (*ii*) multi-tier separation for performing complicated tasks (*iii*) component reusing, and (*iv*) module developing.

Fig. 12.11 illustrates a J2EE-based distributed architecture which is composed of four logic tiers from the client side to the server side: presentation tier, WAP service tier, application tier and data service tier.

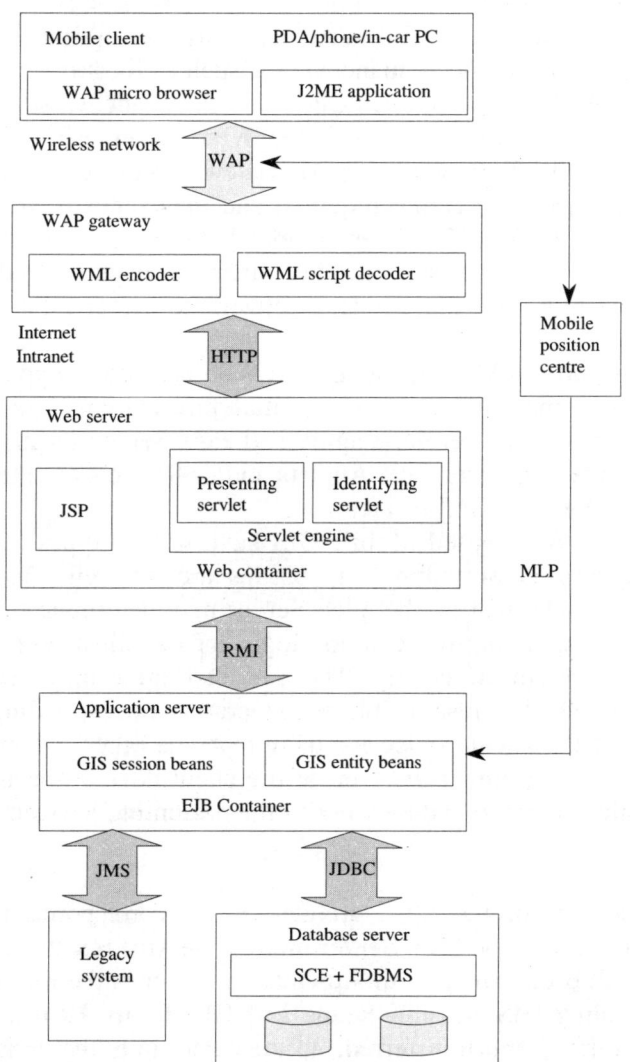

Fig. 12.11 A distributed architecture of WAP-based Mobile GIS (*Source*: Fangxiong and Zhiyong, 2003)

Presentation Tier

Presentation tier is a carrier of the client of WAP-based Mobile GIS, and is mainly responsible for implementing the presentation logic of GIS data. Generally, the client without local-storage data is a WAP microbrowser that controls the GUI and is analogous to a standard Web browser. The WAP microbrowser need not perform any GIS business logic nor it directly connects to a back-end database server or stores any state information, thus it is a really thin client. The client also may be a J2ME (Java 2 Micro Edition) application.

WAP Service Tier

WAP service tier includes a WAP Proxy (often referred to as a WAP Gateway) and a Web Server. The WAP Gateway is required to handle the protocol of interworking between the client and the Web Server. As depicted in Fig. 12.11, the WAP Gateway consists of WML encoders and WML-Script decoders. The WAP Gateway can optimize the communication process and may offer mobile service enhancements, such as location, privacy, and presence based services. The WAP Gateway communicates with the client (WAP microbrowser) using the WAP protocols and it communicates with the Web Server using the standard Internet protocols such as HTTP/HTTPS. Once the WAP Gateway receives WAP requests from the client, it translates the requests to HTTP requests, and then sends them to the Web Server. Once the WAP Gateway receives HTTP responses (Web contents) from the Web Server, it translates the Web contents to compact encoded binary formats for reducing the size and number of packets traveling over the wireless network to the client for displaying and/or processing.

The Web Server includes a Web container and Web protocols support, security support, and so on. Web caching mechanism for Geospatial Information, *The Web Container*, is responsible for managing a Java Servlet Engine and Java Server Pagers (JSP). An Internet GIS system based on Java often uses Java Applets and/or Servlets to extend the dynamically displaying functionality for the Web browser.

The client devices of WAP-based Mobile GIS have several types (*e.g.*, PDA and WAP phone) whose displaying and presenting capabilities are different. So there has to be a mechanism at the server side, that is, the Web Server may determine the type of the client device and generate corresponding presentation logic for the client. Fortunately, the Servlet Engine can solve the problem effectively. The Servlet Engine may provide two kinds of Servlets: the one Servlet which is responsible for generating corresponding presentation logic according to the type of the client device for the client; the other is identifying the Servlet which is responsible for determining the type of the client device (through ID of the client device accessing mobile networks) and then notify the presenting Servlet.

Application Tier

Application tier is the core of the J2EE architecture. It corresponds to GIS Application Servers that communicate with the Web Server in WAP service tier through Remote Method Invocation (RMI). An EJB container at an Application server is the runtime environment of EJB components including GIS Session Beans and GIS Entity Beans, and controls these components to be performed and transferred. At the same time, the container also provides these components with all required services for distributed computing environments. Thus, these EJB components could more efficiently execute in the Application Server. The EJB components can use JDBC (Java Database Connection) technology to access to database servers, and use JMS (Java Message Service) technology to connect to back-end legacy systems. The Application Server has a special Locating Entity Bean to communicate the Mobile Position Centre (MPC), a Server providing geographic location information, with Mobile Location Protocol (MLP). The mobile position technologies for real-time capturing the location information of mobile users generally are GPS, Cell Of Origin (COO), Time of Arrival (TOA), Angle Of Arrival (AOA), Enhanced Observed Time Difference (E-OTD), and so on.

Data Service Tier

Data service tier corresponds to Database Servers that are used to manage and store geospatial and attribute data of the whole system. Object oriented database management system (OODBMS) is the most desired database server for a GIS system, but OODBMS is immature and very costly at present, and hence it is not popular and commonly used (Gong, 2001). At present, the mainstream solution is that large object-relation database systems such as DB2, Oracle, Sybase, SQL Server, and so on are used to manage and store GIS data, at the same time, spatial data engine (SDE) also can be developed to build the communication between data service tier and application tier. SDE is an open standards-based middleware such as ArcSDE, Spatial Ware and Oracle Spatial.

At present, low bandwidths are still the main bottleneck of all mobile applications (Wei, *et al*, 2003). So researches on the organization of geospatial data at the fat server side and on the presentation of geospatial information at the thin client side should be done farther. Fortunately, the mobile network is towards the development of 3G. In 3G age, when the mobile terminal moves at the same speed as vehicle, the transmission speed is 144 kbps, when the mobile terminal moves at the walking speed or un-moves in outdoor, the transmission speed is 384 kbps, when the mobile terminal is in the room, the transmission speed is up to 2 Mbps. The 3G bandwidths will satisfy with the requirements of geospatial information wireless transmission. Therefore, it is not hard to understand that WAP-based Mobile GIS will have better development and application perspectives and considerable business value.

12.7 OPEN GIS CONSORTIUM (OGC)

Geographic data or geodata of the earth are being collected by satellites in digital form for the last 30 years. This data collection will increase rapidly with high resolution sensors and with growing number of people and organization. Geodata formats tend to be complex in comparison to other kinds of data as they are able to represent a range of information. The complexity is further enhanced by the use of particular software or the method of data acquisition.

Integrating geodata from various sources is increasingly becoming important as these are being used as input in GIS based solutions. However, one problem which all GIS analysts face is *interoperability of data i.e.*, easy to share data across system and software. To illustrate the problem of interoperability and data sharing, let us consider an example of three agencies, say soil survey agency (*A*), a research organization (*B*), and remote sensing application based company (*C*). Organization *A* collects soil samples and prepares maps for the whole country, while organization *B* provides and undertakes research studies related to soil maps using remote sensing and other data, and organization *C* undertakes analysis of remote sensing data. Now if the three organizations are asked to coordinate a project, wherein a GIS based soil database is to be prepared, the different levels of problems and coordination which may be required are

GIS Software: All the three organizations may be using different softwares. On conversion from one software to another, there is a likelihood that some data loss may occur as there may be no perfect union between the two system representational capabilities.
Operating System: Even if three organizations are using the same version of the GIS software, the hardware and operating system of the computers may not be same. This may lead to some variation in the storage and retrieval of data.
Definition of attributes: The data attributes may vary with the problem at hand and may not be consistent. Also data collection methods may be different, the sampling methods and parameters may vary. While geocoding the informational layers, the reference coordinate system definition and the acceptable error limits may vary.
Legal aspects: Since each organization is working independently, they may like to levy a price on data usuage, when working in an integrated program. Further information related to metadata (data about its data) has to be maintained alongwithresponsibility regarding its authenticity.
Security: The data available should not be tampered with. Illegal usuage or modification has to be accounted for.

In order to address to the above problems and issues, a non-profit trade association known as Open GIS Consortium Inc. (OGC) was founded in 1994. The mandate of this consortium is to promote new technical and commercial approaches to interoperable geoprocessing. The Open GIS frame work includes

  (*i*) a common means for digitally representing Earth and its phenomena, mathematically and conceptually.
 (*ii*) a common model for implementing services for access, management, manipulation, representation and sharing of geodata between information communities, and
(*iii*) a framework for using Open Geodata Model and Open GIS services Model to solve not only the technical but also the institutional non-interoperability problem.

Fig. 12.12 shows a schematic representation of Open GIS concept. OGC will enable software vendors to provide their software with *plug and play* interfaces of geoprocessing tools that GIS analyst can use to build the own customized geoprocessing functions into their information system.

OGC software specification is based on Open GIS Specification, which provides a *framework for software developers* to create software that enables their users to *access and process* geographic data *from a variety of sources across a generic computing interface within an open information technology foundation.* A critical analysis of the above statement reveals

(i) framework for software developers: It implies that Open GIS specification is a detailed software specification based on a comprehensive common plan for interoperable geoprocessing as formulated by industry and user community,

(ii) access and process: This will allow geodata users to query remote databases and control remote processing resources, and also take advantage of other distributed computing technologies,

(iii) from a variety of sources: This will allow the users to have access to data acquired in a variety of ways, and stored in a wide variety of relational and non-relational databases.

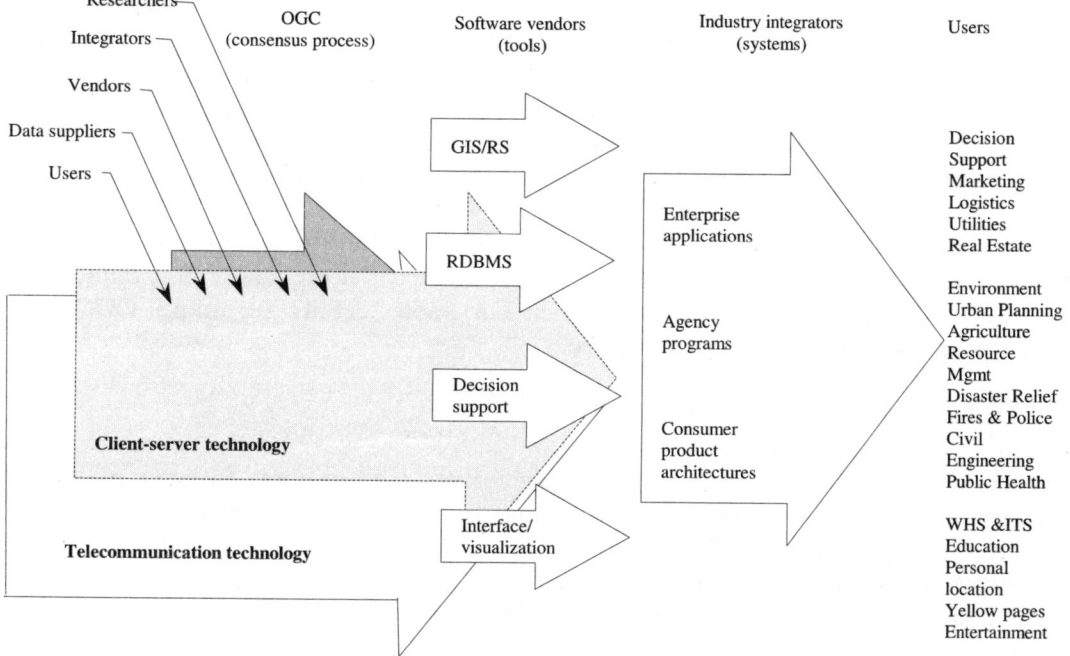

Fig. 12.12 Structure of Open GIS Specification

(iv) across a generic computing interface: It means that Open GIS interfaces will provide reliable communication between otherwise disparate software resources that are equipped to use such interfaces, and

(v) within an open information technology environment: It implies that Open GIS specification enables geoprocessing to take place outside the closed environment of a monolithic GIS, remote sensing that control and restrict database, user interface, network, and data manipulation functions.

The Open GIS specification has defined the manner in which the interoperability of data can be achieved. The basic components are

(i) Open Geodata Model which is a general and common set of basic geographic information type that canbe used to model the geodata needs of more specific application domains using object-based and/or conventional programming methods.

(ii) Open GIS Services are a set of services which will be required to an access and process the geographic types as defined in the Open GIS Model and provide capabilities to share geodata within different user communities having a common interest.

(iii) Information Community Model that employs the Open Geodata Model and Open GIS Service in a scheme that establishes a link for the community of geodata producers and users who share a common set of geographic feature definitions to efficiently and effectively maintain these definitions, and to catalog and share data sets conforming to these definitions. It will also provide an efficient and optimally accurate way for different communities of geodata users and producers to share geodata despite their dissimilar sets of geographic feature definitions. For example, well engineer, geologist and agronomist may share soil data despite the fact that each of them may characterize soil types differently according to their different professional objectives. The Information Communities Model defines a scheme for automated translation between different geographic feature lexicons.

To achieve the goals as described above, OGS has planned to bring about key companies and institutions to organize technical committees to arrive at consensus on technical issues. Fig. 12.13 shows the structure of OGC. For more details regarding OGC website www.opengeospatial.org may be refered to.

12.8 DECISION SUPPORT SYSTEM

For more than three decades, computers have been used as tools to support managerial decision making process. Some of the common computerized tools are Expert System (ES), Artificial Neural Networks (ANN), and Decision Support System (DSS), and out of these the Decision Support System has found its use in GIS.

Decision Support System is computer based software, which supports in taking decisions for an ill-structured problem. It can be defined as follows:

"A DSS is an interactive, flexible and adaptable computer based information system, specially developed for supporting the solution of a particular ill-structured problem for improved decision making. It utilizes data, provides easy user interface and it allows for the decision maker's own insights. Most sophisticated DSS also utilizes models (either standard and/or custom-made). It is build by an iterative process, supports the phases of the decision making, and includes a knowledge base".

The main objectives of Decision Support Systems are

(i) Effective generation of information on the decision problem from available data and ideas,
(ii) Effective generation of solutions (alternatives), and
(iii) To provide a good understanding of the structure and content of a decision problem.

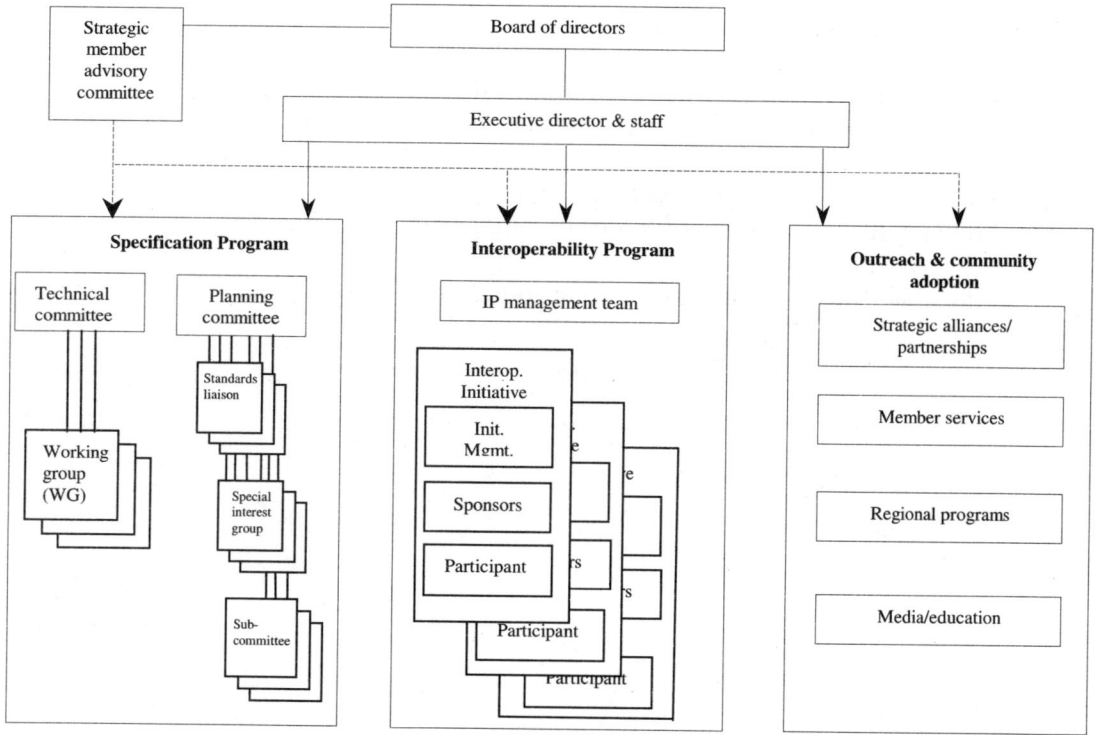

Fig. 12.13 Structure of Open GIS Consortium (*Source: www.opengeospatial.org*)

Characteristics of Decision Support System

In general, a decision support system requires an online access to data base through interactive programs and proper data communications. As the informational needs are not known, thus it requires an incremental design method. So, in decision support system, *decision* emphasizes its attention on decision making rather providing a simple information retrieval. *Support* highlights the role of computer as an aid and not as an replacement to the decision making, while *system* identifies the integrated nature of users, machine and decision making environment.

The main characteristics of DSS are

(*i*) to assist managers in their Decision Making Process for un-structured/semi-structured problems,
(*ii*) to support and enhance rather than replace managerial judgments,
(*iii*) to improve the use of models or analytical techniques with data access functions,
(*iv*) to emphasize flexibility and adaptability to respect changes in the context of decision making process, and

(v) to focus on features, which make them easy to be used interactively by non-experienced users.

Supplementary characteristics of the DSS are

(i) enabling an intuitive approach towards a solution,
(ii) helping in tentative procedures, as they could be supported by fast proto type environment,
(iii) including trial and error procedures, and
(iv) allowing the introduction of Subjective Judgments.

Fig. 12.14 illustrates the characteristics and capabilities of a DSS.

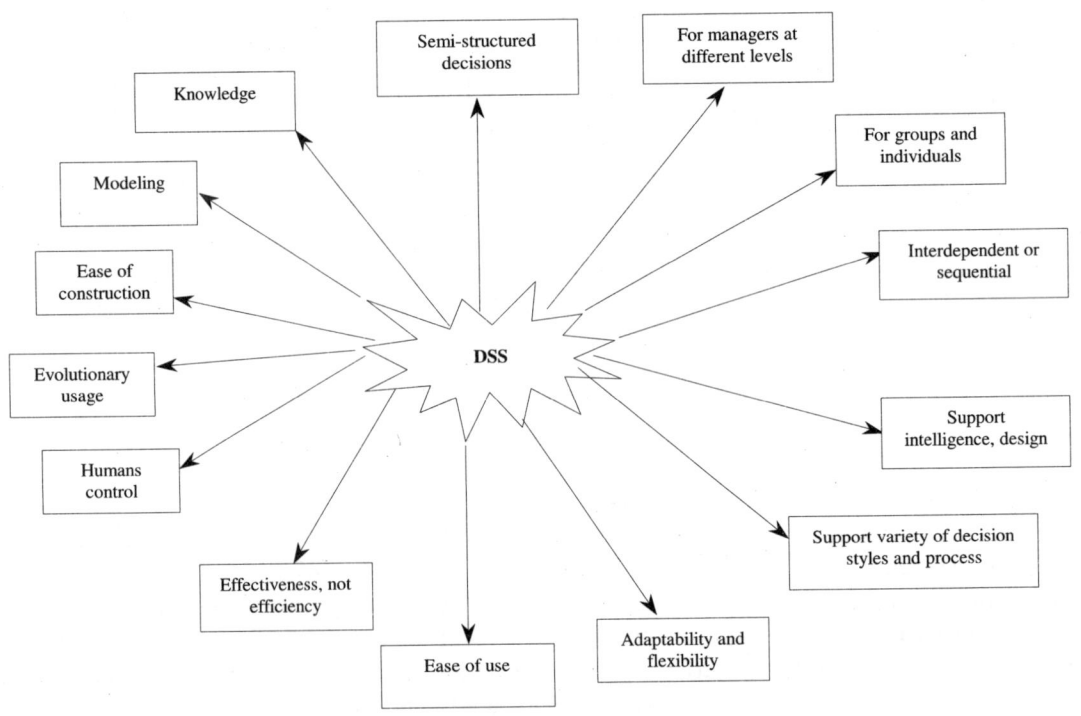

Fig. 12.14 Capabilities of a DSS.

Architecture of a DSS

The conceptual architecture of a DSS can be explained through the following three different approaches.

(i) A functional approach
(ii) A tool based approach
(iii) A combined approach

Functional Approach

In functional approach, conceptual architecture is explained as a combination of (*a*) Language system (*b*) Knowledge system and (*c*) Problem processing System.

Language System deals with the basic interaction of computers and human beings. It deals with menus, command languages, fill-in-forms and Natural Language Interface, while Knowledge System deals with the data inputting procedures and methods. The process of data input can be through databases, text, rules, sets, procedural models, frames, spreadsheets, and vocabularies. Problem Processing System deals with management of inputs and analytical methods, such as database management, text processing, model execution, interface engine and reasoning, spreadsheet analysis, and statistical analysis.

Tool Based Approach

In tools based approach the architecture of a typical DSS contains a Database, Model base and a Dialogue Module. Fig. 12.15 shows a typical DSS architecture on Tool based approach. Here the user communicates through a dialogue module, which then accesses the database and the model to provide a solution.

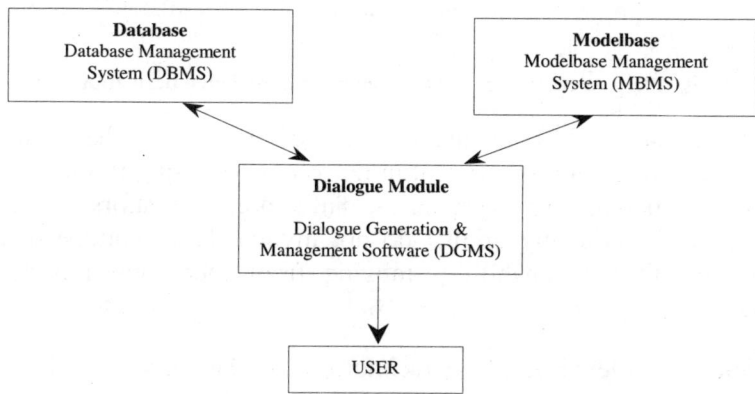

Fig. 12.15 DSS architecture on tool based approach

Knowledge based approach

In this approach, the user interacts through a well defined user interface which sends requests to system controller. Depending upon the type of query, access to information to either database, model or a knowledge base is made. Fig. 12.16 shows the typical architecture of a DSS using knowledge based approach.

Operational Structure of a DSS

The final product or finished DSS that accomplishes a task is called a Specific DSS (SDSS). It is used to support a specific application, for example, identifying wasteland in order to provide best possible routes within a given network of roads etc.

DSS generator: A generator is an integrated package of software that provides a set of capabilities to build a specific DSS quickly, inexpensively, and easily. One such popular PC based computer generator is Lotus 1-2-3. The term DSS generator emerges from the concept of application for program generator. Application generators are tools used by programmers and system analysts to expedite programming and systems development. For example, an application generator can be used to build an inventory control system.

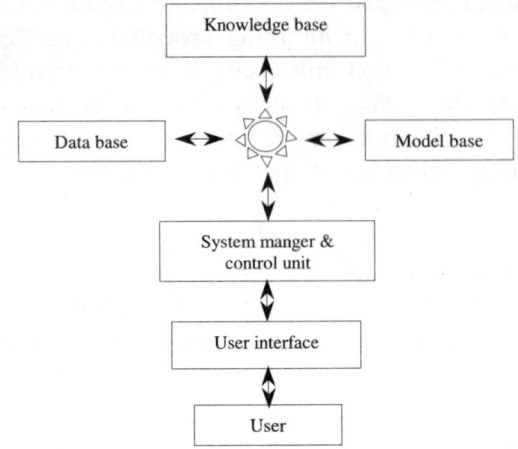

Fig. 12.16 Architecture of DSS based on knowledge base approach

Application generators add convenience and reduce costs for the creation of programs. The programs produced are not as efficient in terms of processing throughout as those coded from a scratch by experienced programmers. Therefore, generators are more suitable for applications that run infrequently or that do not involve large volume data processing. A generator possesses diverse capabilities ranging from modeling, report generation, and graphical display.

DSS tools: At the lowest level of a DSS technology are the software utilities or tools. These elements facilitate the development of either a DSS generator or a specific DSS. Examples of DSS tools are graphics (hardware and software), editors, query systems, random number generators, and spreadsheets.

The relationships among the three levels are presented in Fig. 12.17. The tools are used to construct generators, which in turn are used to construct specific DSS. However, tools can also be used directly to construct specific DSS. Sometimes, there may be simpler tools for constructing more complicated tools.

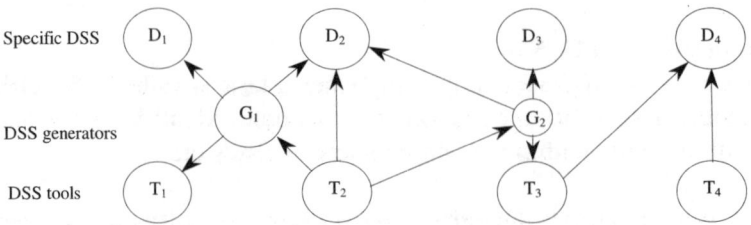

Fig. 12.17 Relationship between different operational levels of a DSS

DSS has a potential to be used along with GIS in order to make better understanding of the process involved in diverse natural resources applications. But it is till in an emerging stage for a day to day application oriented problems. DSS can be classified according to the type of support it provides, type of organizational utilization, degree of non-procedure, degree of dependency and type of support. Detailed study of different DSS will be of much use for research and developmental activities.

REFERENCES

Books and Journals

1. Adams, J.B., D.E.Sabol, V. Kapos, R.A. Filho, D.A.Roberts, M.O. Smith and A.R. Gillespie (1995): *Classifcation of Multispectral Images Based on Fractions of End Members: Applications to Land Cover in Brazilian Amazon*, Remote Sensing of Environment, Vol. 52, 137-145.
2. Arora, M. and S.K.Ghosh (1998): *Classification Accuracy Indices: Definitions, Comparisons and a Brief Review*, Asian Pacific Journal on Remote Sensing and GIS, Vol. 10, 1-9.
3. Baijal, R. (2002): *Development of a Knowledge Based System for Military Use*, M.Tech Thesis, Deptt of Civil Engg. IIT Roorkee, Roorkee.
4. Barrett, E.C. and L.F.Curtis, (1982): *Introduction to Environmental Remote Sensing*, Chapman and Hall, 2nd Edition.
5. Bezdek, J.C. (1981): *Pattern Recognition with Fuzzy Objective Function Algorithms*, Plenum, New York, USA.
6. Bezdek, J.C., R. Ehrlich and W. Full (1984): *FCM: The Fuzzy C-Means Clustering Algorithm*, Computers and Geosciences, Vol 1, 191-203.
7. Binaghi, E., R. Ehrlich and W. Full (1999): *A Fuzzy Set Based Accuracy Assessment of Soft Classification*, Pattern Recognition Letters, 20, 935-948.
8. Burrough, P.A. (1986): *Principles of Geographical Information System for Land Resources Assessment,* Claredon Press, Oxford.
9. Burrough, P.A. and R. A. McDonnell (2000): *Principles of Geographical Information System*, Oxford University Press, Great Clarendon Street, Oxford.
10. Campbell, J.B. (1978): *A Geographical Analysis of Image Interpretation Methods,* The Professional Geographer, Vol 30, pp 264-269.
11. Campbell, J.B. (1987),: *Introduction to Remote Sensing*, The Guilford Press, New York.
12. Chandra, A.M. (2002a): *Plane Surveying*, New Age International Pvt. Ltd., Publishers, New Delhi, India.
13. Chandra, A.M. (2002b): *Higher Surveying*, New Age International Pvt. Ltd., Publishers, New Delhi, India.
14. Checkland, P.B. (1981): *System Thinking – Systems Practice*, John Wiley, Chichester.
15. Clark, K. (2001): *Getting Started with Geographic Information Systems*, Prentice Hall, Upper Saddle River, NJ. Congalton, R.G., R.G.
16. Crist, E.P. and R.C. Cicone (1984): *Application of the Tasseled Cap Concept to Simulated Thematic Mapper Data*, Photogrammetric Engineering and Remote Sensing, Vol 50, pp 343-352.
17. Curran, P.J. (1986), *Principles of remote sensing,* Longman, London.
18. Date, C. J. (1986): *An Introduction to Database System*, 2nd edn., Addison-Wesley, Reading, Mass.
19. Deutsch M. and F. Ruggles (1974) : *Optical Data Processing and Project Application of Erts-1 Imagery Covering the 1973 Mississippi River Valley Flood*, Water Resources Bulletin, Vol. 10, No.5.

20. Dubey, R.P., N. Ajwani and R.R. Navalgund (1991): *Relation of Wheat Yield with Parameters Derived from a Spectral Growth Profile,* Journal of the Indian Society of Remote Sensing, Vol 1, No. 1.
21. ESRI (Earth System Research Incorporated, USA)
22. Eumetsat (2003): *Current Status of Insat And Kalpana-1 (Metsat) Geostationary Satellites for Meteorological Applications* (www.eumetsat.de/en/area2/cgms/cgms_xxxi/ CGMS_Working_Papers/ CGMS- XXXI_India_WPs/ CGMS_XXXI _IND _ WP_03.pdf.
23. Fangxiong, W. and J. Zhiyong (2004): *Research on Distributed Architecture on Mobile GIS based on WAP*, (URL: http://www.isprs.org/istanbul2004/comm2/papers/220.pdf)
24. Foody, G.M. (1992): *On the Compensation for the Chance Agreement in Image Classification Accuracy Assessment*, Photogrammetric Engineering and Remote Sensing, Vol. 58, 1459-1460.
25. Foody, G.M (1995): *Cross_entropy for the Evaluation of the Accuracy of a Fuzzy Land Cover Classification with Fuzzy Ground Data,* ISPRS Journal of Photogrammetry and Remote Sensing, 50, 2-12.
26. Foody, G.M. (1996) *Relating the Land Cover Composition of Mixed Pixels to Artificial Neural Network Classification Output*, Photogrammetric Engineering and Remote Sensing, Vol. 62, 491-499.
27. Foody, G.M and M.Arora (1996): *Incorporating Mixel Pixels in the Training, allocation and Testing stages of Supervised Classification*, Pattern Recognition Letters, 17, 1389-1398.
28. Friend, P.F. and R. Sinha (1993): *Braiding and Meandering Parameters*, from Best, J.L. and C. Bristows (eds.), Braided Rivers, Geological Society Spl. Publication No. 75.
29. Fung, T. and E. Le Drew (1988): *The Determination of Optimal Threshold Levels for Change Detection Using Various Indices*, Photogrammetric Engineering and Remote Sensing, Vol. 54, 1449-1454.
30. Ghosh, S.K. (1991): *River Basin Management using Remote Sensing Data Analysis – An Assessment of Data Input to Hydrological Model as Deriverd from Satellite Data*, Ph. D. Thesis, University of Strathclyde, Glasgow, U.K.
31. Gibson, P.J. (2000): *Introductory Remote Sensing-Principles and Concepts*, Routledge, 11 New Fetter Lane, London.
32. Gong Jianyi (2001): *Concept and Development Trends of Spatial Database Management System*, Science of surveying and Mapping, Vol 26(3), pp. 4-9.
33. Goodchild, M.F. (1991): *Geographical Information Systesm-Principles and Applications*, Longman London, Vol. 1.
34. Haralick, R.M. and K. Fu (1983): *Pattern Recognition and Classification*, Chapter 18. in the *Manual of Remote Sensing*, R.N. Colwell, Ed. Vol 1, pp 793-805.
35. Hutchinson, M.F. (1993): *Development of a Countrywide DEM with Applications to Terrain and Climate Analysis*, In Environmental Modeling with GIS by Goodchilds, M.F. Parks, B.O., and Stayert, L.T., eds. Pp. 392-399, New York: Oxford University Press.
36. Ibrahim, M.A (2004): *Evaluation of Soft Classifiers for Remote Sensing Data*, Ph.D Thesis, DEpartment of Civil Engineering, IIT Roorkee, Roorkee.
37. Jensen, J.R. (1986): *Introductory Digital Image Processing*, Prentice-Hall, New Jersey.

38. Jaganniwas (2004): *Development of Worldwide GIS Based Earthquake Information System*, M. Tech Thesis, Department of Civil Engineering, IIT Roorkee, Roorkee.
39. Kauth, R.J. and G.S Thomas (1976): *The Tasseled Cap – A Graphic Description of the Spectral-Temporal Development of Agricultural Crops as Seen by Landsat*, Proc. of Symp. On Machine processing of remotely sensed data, Purdue University, West Lafayette.
40. Kavzoglu, T. and P.M. Mather (2003): *The Use of Back Propagation Artificial Neural Network in Land Cover Classification*, International Journal of Remote Sensing, Vol. 24, 4907-4938.
41. Krishnapuram, R. and J.M. Keller (1993): *A Possibilistic Approach to Clustering*, IEEE Transaction of Fuzzy Systems Vol. 1, 98-108.
42. Krishnapuram, R. and J.M. Keller (1996): *The Possibilistic C-Means Algorithm: Insights and Recommendations*, IEEE Transaction of Fuzzy Systems Vol. 4, 98-108.
43. Krishnaswamy, K. and S. Kalyanaraman (2002): *Indian Remote Sensing Satellite Cartosat–1: Technical Features and Data Products* (URL: http://www.gisdevelopment.net/ technology /rs/ techrs023pf.htm)
44. Kumar, R. and L.F. Silva (1977): *Seperability of AgriculturalCover Types by Remote Sensing in the Visible and Infrared Wavelenght regions*, IEEE Transactions on Geoscience Electronics, Vol GE-15, pp42-49.
45. Luqun, L., L. Chenming and L. Zongjian (2002): *Investigation on the Concept of Model of Mobile GIS*, Symposium of Geospatial Theory, Processing and Applications, Ottawa, 2002.
46. Ma, Z. and R.L.Redmond (1995): *Tau Coefficient for Accuracy Assessment of Classification of Remote Sensing Data*, Photogrammetric Engineering and Remote Sensing, Vol. 61, 435-439.
47. Mather, P.M. (1999): *Computer Processing of Remotely-Sensed Images-An Introduction*, 2nd Ed., John Wiley & Sons, Chichester, U.K.
48. Lilliesand, T.M. and R.W.Kiefer (1999): Remote Sensing and Image Interpretation, 4th Edition, John Wiley and Sons, New York.
49. Lo, C.P. and A.K.W. Yeung (2002): *Concepts and Techniques of Graphic Information Systems*, Prentice Hall of India Pvt. Ltd., New Delhi.
50. Mannan, B. and A.K.Ray (2003): *Crisp and Fuzzy Competitive Learning Networks for Supervised Classification of Multispectral IRS Scenes*, International Journal of Remote Sensing, Vol. 24, 3491-3502.
51. Mark, D.M. (1984): *Automated Detection of Drainage Networks from Digital Terrain Models*, Cartographica, No. 21, pp. 168-178.
52. Maselli, F. A. Rodulf and C. Conese (1996): *Fuzzy classification of Spatially Degraded Thematic Mapper for the Estimation of Sub-pixel Components*, International Journal of Remote Sensing, 17, 537-551.
53. Massone, A.M., Masulli and A. Petrosino (2000): *Fuzzy Clustering Algorithms on Landsat Images for Detection of Waste Areas: A comparison in Advances in Fuzzy Systems and Intelligent Technologies*, (eds) F. Masulli, R. Parenti and G. Pasi, Shaker, Maastricht, NL. 165-175.

54. Mather, P.M. (1999): *Computer Processing of Remotely-Sensed Images- An Introduction*, John Wiley & Sons, New York, ed. II.
55. Mehrotra, K., C.K. Mohan and S. Ranka (1997): *Elements of Artificial Neural Networks*, Penram International Publishing, Mumbai, India.
56. Mishra, K.K. (1997): *Hydromorphological Study of River Kosi in a Selected Reach Using External Hypothesis and Satellite Data*, M.Tech Thesis, WRDTC, IIT Roorkee, Roorkee.
57. Moore, I.D., R.B. Grayson and A.R. Ladson (1991): *Digital Terrain Modelling: A Review of Hydrological, Geomorphologica, and Biological Applications*, Hydrological Process, Vol. 5, pp. 3-30.
58. Naesset, E. (1995): *Tests for Conditional Kappa and Marginal Homogeneity to Indicate Differences between User's and Producer's Accuracy*, International Journal of Remote Sensing, Vol. 16, 3147-3159.
59. Naidu, K.V.K (2000): *Case Study of Drainage Problem for a Plantation Area using GIS Techniques*, M.Tech Thesis, Department of Civil Engineering, IIT Roorkee, Roorkee.
60. Navalgund, R.R., J.S. Parihar, Ajai and P.P.Nageshwara Rao (1991): *Crop Inventory Using Remotely Sensed Data*, Current Science, Vol. 61, No. 3 & 4.
61. NWDM, (1987): *Wastelands definitions and Classification*, Technical Task Group Report, Nationational Development Board, Ministry of Forests and Environment, Govt. of India.
62. NRSA, 1997, *IRS-ID Handbook*, Report No. IRS-1D/NRSA/NDC/HB-12/97
63. Oderwald and R.A. Mead (1983): *Assessing Landsat Classification Accuracy Using Discrete Multivariate Analysis Statistical Techniques*, Photogrammetric Engineering and Remote Sensing, Vol. 49, pp 1671-1678.
64. Ozkan, C. and F.S. Erbek (2003): *The Comparison of Activation Functions for Multispectral Landsat TM Image Classification*, Photogrammetric Engineering and Remote Sensing, Vol. 69, 1225-1234.
65. Pearlman, J.S., P.S.Barry, C.C.Segal, J. Shepanski, D.Beiso and S.L.Carman (2003): *Hyperion, A Space-Based Imaging Spectrometer,* Transactions on Geosciences and Remote Sensing, Vol. 41, No.6, June 2003, 1160-1173.
66. Ramamoorthi, A.S., S. Thiruvengadachari and A.V. Kulkarni (1991): *IRS-1A Application to Hydrology and Water Resources*, Current Science, Vol. 61, No. 3&4.
67. Rao, D.R., P. Lalitha, P.P. Nageshwara Rao, A.K. Gupta and G. Vijayalakshmi (1983): *Disasters in India – Can Remote Sensing do something*, ISRO Technical report No. ISRO-EOS-TR-27-83, Indian Space Research Organization, Bangalore.
68. Rao, D.P., G. Behera, R.R. Navalgund, R.L.Karale and K.S Rao (1991), *IRS -1A Applications for District Level Planning*, Current Science, Vol. 61, No. 3&4.
69. Rao, D.P., N.C Gautam and B, Sahai (1991): *IRS-1A Application on Wasteland Mapping*, Current Science, Vol. 61, No. 3 &4.
70. Reddy, C.M. (2003): *Comparison of Classifiers in Various Image Processing Softwares*, M. Tech Thesis, Dept. of Civil Engineering., IIT Roorkee, Roorkee.
71. Richardson, A.J. and C.L. Wiegand (1977): *Distinguishing Vegetation from Soil Background Information*, Remote Sensing of Environment, Vol 8, pp 307-312.

72. Rosenfield, G.H. and K. Fitzpatrick-Lins (1986): *A Coefficient of Agreement as a Measure of Thematic Classification Accuracy*, Photogrammetric Engineering and Remote Sensing, Vol. 52, 397-399.
73. Sahai, B., A. Bhattacharya and V.S. Hedge (1991): *IRS-1A Application for Groundwater Targeting*. Current Science, Vol. 61, No. 3&4.
74. Schalkoff, R.J. (1997): *Artificial Neural Network*, McGraw-Hill, New York, U.S.A.
75. Settle, J.J. and N.A. Drake (1993): *Linear Mixing And Estimation of Ground Cover Proportions*, International Journal of Remote Sensing, Vol. 23, 1159-1177.
76. Skidmore, S. and B. Wroe (1988): *Introducing Systems Analysis,* NCC Publications, Manchester.
77. Soller, S.C. (1974): *The Use of Remote Sensing and Natural Indicators to Delineate Flood Plain – Preliminary Findings*, Proc. of the Ninth International Symposium on Remote Sensing of Environment, Vol. 1.
78. Sirisha, U. (2003): *Evaluation of Irrigation Performance Parameters Using Remote Sensing and GIS*. M. Tech Thesis, Department of Civil Engineering, IIT Roorkee, Roorkee.
79. Shalan, M.A., M. Arora and S.K. Ghosh (2003): *An Evaluation Of Fuzzy Classifications from IRS-1C LIS III Data*, International Journal of Remote Sensing, Vol. 23, 3179-3186.
80. Sharma, T. (1991), *Procedures for Wheat Yield Prediction Using Landsat and IRS-1A Data*, International Journal of Remote Sensing.
81. Smith, M.O, P.E. Johnson and J.B. Adams (1985): *Quantitative Determination of Mineral Types and Abundances from Reflected Spetra Using Principal Component Analysis*, Journal of Geophysical Research, Vol. 90, 797-804.
82. Soni, V.K. (2003): *GIS for Site Suitability in Urban Planning*, M. Tech Thesis, Department of Civil Engineering, IIT Roorkee, Roorkee.
83. Switzer, P., W.S. Kowalik and R.J.P. Lyon (1981): *Estimation of Atmospheric Path Radiance by the Covaraince Matrix Method*, Photogrammetric Engineering and Remote Sensing, Vol 47, 1469-1476.
84. Thomas, G., S.E. Hobbs and M. Dufour (1996): *Woodland Area Estimation By Spectral Mixing: Applying Goodness-of-fit Solution Method*. International Journal of Remote Sensing, Vol. 17, 291-301.
85. Tobler, W.R. (1987): *Measuring Spatial Resolution*, Proc. International Workshop on Geographic Information Systems, in Chen, Shupeng, *et al.*, Beijing.
86. Ungar, S.G., J.S. Pearlman, J.A. Mendanhall and D. Reuter (2003): *Review of the Earth Observing One (EO-1) Mission*, Transactions on Geosciences and Remote Sensing, Vol. 41, No.6, June 2003, 1149-1159.
87. Uraon, J.K. (2003): *Road Accident Analysis for Dehradun City within GIS Environment*, M.Tech Thesis, Department of Civil Engineering, IIT Roorkee, Roorkee.
88. Venkateswarlu, K. (2003): *Tourism GIS for Uttaranchal state*, M. Tech Thesis, Department of Civil Engineering, IIT Roorkee, Roorkee.
89. Wei Zhongya, Xu Suning, Wu Lun (2003): *A WAP – Based Geographic Information Mobile Services*, International conference on Communication Technology Proceedings, ICCT 2003, Vol. 2, 1455-1460.

90. Zhang, J. and G.M. Foody (1998): *A Fuzzy classification of sub-urban land cover from Remote Sensing Imagery*, International Journal of Remote Sensing, Vol. 19, 2721-2738.

Web Pages

1. www.telsat.belspo.be/beo/en/satellite/spot.htm - *Satellite And Sensors - SPOT (Satellite Pour l'Observation de la Terre)*
2. USGS, (2004): *Continuing The Legacy* (URL:http://ldcm.usgs.gov/legacy.php)
3. NRSA (a): *IRS-P3*. (URL: http://www.nrsa.gov.in/engnrsa/satellites/irsp3.html)
4. NRSA (b): *IRS-P3 MOS Sensor* (URL: http:// www.nrsa.gov.in/ engnrsa/ satellites /mosp3.html)
5. NRSA (c): IRS-P4 (URL: http:// www.nrsa.gov.in/ engnrsa/ satellites /irsp4.html)
6. NRSA (d): Resourcesat-1 (URL: http:// www.nrsa.gov.in/ engnrsa/ ebrochure /index.html)
7. NRSA (e): LISS - IV (URL: http:// www.nrsa.gov.in/ engnrsa/ ebrochure /liss4.html)
8. NRSA (f): LISS - IV (URL: http:// www.nrsa.gov.in/ engnrsa/ ebrochure /awifs.html)
9. www.infoterra-global.com/banddescrips.htm - *Band Descriptions - Satellite Band Designation And Principal Applications.*
10. SISEA (a): *IKONOS Satellite* (URL: http://www.sisea.com/kionos/overview.html)
11. SISEA(b), *IKONOS Characteristics* (URL: http://www.sisea.com /kionos/ characteristics.html)
12. www.tec.army.mil/tio/IKONOS.htm - *IKONOS.*
13. www.imagery-central.com/files/sensors.asp?sensornum=2–*Sensor Information – QuickBird – Digital Globe.*
14. www.nl.gim.be/p/316D77DDB5208B62C1256B6D005464F9 - *Quickbird Technical - Detailed Technical Product Information.*
15. www.spotimage.fr/html/_167_171_181_182_184_.php – *Quickbird.*
16. ORBIMAGE, 2004(a): *Orbview-1 (Mission Completed)* (URL: http://www. orbimage.com/ corp/ orbimage_system/ov1/index.html)
17. ORBIMAGE, 2004(b):*Orbview-2* (URL:http://www.orbimage.com/corp/ Orbimage_system/ ov2/ index.html)
18. ORBIMAGE,2004(c):*Orbview-3* (URL:http://www.orbimage.com/corp/Orbimage _system/ ov3)
19. www. spaceflightnow.com/delta/d282/001120eo1.html – *Earth Observing – 1.*
20. eo1.gsfc.nasa.gov (a), *Earth observing - 1 Baseline Mission Overview,* (URL: www.eo1. gsfc.nasa.gov/overview/eo1Overview.html).
21. eo1.gsfc.nasa.gov (b): *Advanced Land Imager* (URL:www.eo1.gsfc.nasa.gov/Technology/ ALIhome.html)
22. eo1.gsfc.nasa.gov (c): Atmospheric Corrector (URL: www.eo1.gsfc.nasa.gov/Technology/ AtmosCorr.htm)

INDEX

Absolute
    temperature, 10
    temporal resolution, 23
    zero, 10
Absorbed, 15
Absorbers solar radiation, 12
Absorptance, 10
Absorption, 12
Accuracy, 165
Acreage Estimation, 117
Activation levels, 251
Active sensors, 25
Advanced
    Land Imager (ALI), 48
    Very High Resolution Radiometer (AHVRR), 54
    Wide Field Sensor (AWIFS) sensor, 36
Aerial Photographs, 155
Aerosols, 14
Affine transformation, 87
Agriculture, 8
Agro-economic data, 156
Airborne, 25
Aircraft, 25
Algorithm, 187
Allocation, 207
Altitude, 86
AMEOBA, 109
Analog stereoplotter, 157
Analogue-to-digital, 23
Analysis of scheme, 214
Analysis, 146, 148
Analytical, 209
Anchor point, 200
Angle of Arrival (AOA), 271
Animated maps, 151
A-norms, 244
Anthocyanin, 17
Applicability, 165
Application tier, 269
ArcSDE, Spatial Ware 271
Area, 176, 187
Areal features, 204

Artificial Neural Networks (ANN), 242, 274
Aspatial query, 190
Aspect 204
Association, 74
Atmospheric, 12
    Corrector, 51
    effect, 84
    haze, 28
    moisture, 13
    Scattering, 14
    windows, 12
Attribute, 187
    Data, 153, 155
Automated shading method, 209
Average
    accuracy, 114
    divergence, 105
    Transformed Divergence, 116
    User's, 259
AWIFS, 34
AWIFS-A, 36
AWIFS-B, 36

Back Propagation Neural Network (BPNN), 248
Back-end GIS, 260
Badge of Tree, 94
Band Interleaved by
    Line (BIL), 62
    Pixel (BIP), 62
Band
    Sequential (BSQ), 62
    width, 23
    width 3G, 271
Bar charts, 151
Bare Soil, 19
Basic
    imagery, 44
    Stereo products, 67
Bhattacharyya distance, 105
Bias, 165
Bilinear interpolation, 88
Binomial distribution, 111

Black and White film, 23
Blackbody, 10
Block kriging, 203
Boolean 106
    Operators, 190
Brightness
    axis, 93
    values (BV), 61
Buffering function, 190
Carbon dioxide, 12
Carotheses, 17
Cartographic modeling, 214
Categorical data, 154
CCT, 59
CDMA, 267
CDPP, 267
CD-ROM Products, 71
Cell of Origin (COO), 271
Cell, 187
Central Processing Unit, 147
Centre National ï Études Spatiales (CNES), 26
Chains, 149
Change
    detection, 23, 92
    matrix, 124
Characteristics of Decision Support System, 275
Charged Coupled Device (CCD), 55
Chlorophyll absorptions band, 17
Choropleth maps, 150
Circular filter, 202
Classification 73
    Accuracy, 99, 109
    scheme, 99
Clay, 19
Client-side Internet GIS, 260
Cloud, 14
CLUSTER, 108
Clustering algorithms, 99
Coarse resolution, 22
Coastal Zone Colour Scanner (CZCS), 57
Co-kriging, 203
Colour, 15
    film, 23
Colour, 15
Combined
    accuracy, 114
    approach, 276
Commission, 111
Common Object Request Broker Architecture (CORBA), 268
Compatibility, 165

Competitive
    Layers 252
    Learning Neural Network, 252
Completeness, 165
Components, 96
Composite
    Land Development Unit' (CLDU), 135
    suitability map, 219
Computer
    Compatible Tapes (CCT'), 69
    hardware, 146
Conceptual
    error, 166
    model, 214
Conditional
    Kappa, 114
    Tau, 114
Confidence limit test, 111
Confusion matrix, 110
Coniferous trees, 16
Consistency, 165
Contingency table, 110
Continuous surface, 196
Contour maps, 151
Convolution, 196
Correlation
    coefficient (CC), 83, 255, 257
    matrixes, 81
Covariance matrix method (CMM), 85
Cowardin Wetland classification scheme, 100, 115
Crop
    condition assessment, 119
    identification, 115
    inventory, 116
Cross entropy, 255, 256
Cubic, 203
    convolution, 88

Dasymetic maps, 151
Data
    acquisition, 156
    analysis error, 166
    conversion error, 166
    conversion, 148
    editing error, 166
    editing, 159
    encoding error, 166
    filtering, 191
    layer, 187
    Management and Analysis Procedures, 148
    merging and GIS integration, 80

Index

model, 169, 214
    output errors, 166
    processing error, 166
    Relay Transponder (DRT), 55
    service tier, 269
    verification, 159
Database management system (DBMS), 148
DB2, 271
Deciduous, 16
    versus coniferous, 16
Decision Support System (DSS), 274
Decision, 275
Defense Meteorological Satellite Program (DMSP), 56
Definition of attributes, 272
Degree
    of fuzziness, 244
    of separability, 104
Delaunay triangulation, 200
Delineation, 73
DEM data products, 37
Demographic data, 156
Densitometer, 78
Densitometry, 73
Detection, 73
Deterministic
    Interpretation, 76
    or stochastic, 199
Diagonal, 244
Dichotomous, 154
Difference image, 92
Diffuse (or Lambertian), 15
Diffuse, 15
Digital
    counts, 61
    elevation model (DEM), 32, 157
    image classification, 97
    images, 23
    interprétation, 79
    numbers (DN), 61
    Orthophotographs, 155
    photogrammetric workstation, (DPWS), 158
    Terrain Data by GPS, 158
    Terrain Data from Internet/World Wide Web, 159
    terrain data, 156
    terrain visualization, 208
Digitization, 148
Digitizer, 146, 157
Direct files, 180
Directional, 92
Director Recognition, 75
Disaster management, 115, 135
Discretised surface, 196
Display of spatial data, 146
Distance matrix calculation, 207
Distributed Network Architecture, 268
District level planning, 115, 135
Divergence, 104
    Average, 105
    Average transformed, 116
Divide-and-conquer, 179
Domain, 149
Dot maps, 151
Drainage Problem in Tea Plantation Area, 225
DSS generator, 277
DSS tools, 277
DTM, 67
Dust, 14

Earth
    Observing-1(EO-1), 48
    Resources Technology Satellite (ERTS-1), 26
    rotation, 86
Edge, 200
    detection filters, 92, 192
Editing, 148
Eigen
    values, 97
    vectors, 97
EJB container, 270
Electromagnetic Spectrum, 10
Elevation 204
Embedded bitmap (e.g., WBMP), 267
Emissivity, 10
End member spectra, 243
Energy balance equation, 15
Energy interactions, 15
Enhanced
    Observed Time Difference (E-OTD), 271
    Thematic Mapper (ETM), 26
    Thematic Mapper Plus (ETM+), 26
Enterprise javabeans (EJB), 269
Entity, 187
Enumeration, 73
Environment, 8
Environmental
    impact assessment, 152
    phenomenon, 23
EOSAT, 59
Equipment for image interpretation, 77
ERDAS Imagine, 116

Error
    function E, 249
    Matrix, 110
    of commission, 103
    of omission, 103
ESRI Map Objects, 265
Euclidean, 244, 255
Exact
    objects, 171
    approximate, 199
Expected, 113
Expert System (ES), 274
Exponential, 91
Extendable Markup Language (XML), 267
Extrapolation, 196
Far infrared, 11
Fast format, 69
Feature selection, 103
Feature, 187
Field – based model, 170
Field Observation, 75
Field spectrometer, 17
File
    handling, 180
    management, 180
Filtering, 192
Flood
    inundation, 126
    plain mapping, 115, 125
Flow path length, 204
Fore-and-aft PAN cameras, 37
Forestry, 8
FORGY, 109
Format, 59
Frequency, 10
Function, or Operation, 187
Functional approach, 276, 277
Fuzzy, 242
    c-Means (FCM) 243
    entities, 171
    error matrix (FERM), 256, 257
    Set Based Method, 242, 243

Gain parameter, 249
Gamma rays, 10
Gaseous molecules, 14
Gaussian, 90
    histogram, 91
    stretch, 91
Generalization, 165
Geocoded products, 67

Geographic
    Coordinate System, 161
    Information System (GIS), 145
    referencing system, 148
    space, 151
Geographical data, 146
Geology, 8
Geometric corrections, 64
Geostationary Operational Environmental Satellite
    (GOES), 52
Geostationary satellites, 25
GIS
    Database, 179
    Entity Beans, 270
    Session Beans, 270
    Software, 272
Global Positioning System (GPS), 156
GMS satellites, 57
GPRS, 267
Gradient (Slope), 176
Gradual or abrupt, 199
Gram–Schmidt Sequential Orthogonalization
    technique, 93
Graphical user interface (GUI), 267
Gravity modeling, 207
Gray tones, 79
Greenness axis, 93
Greenwich meridian, 161
Ground
    Control Points (GCP), 86
    Receiving Station (GRS), 59
    swath, 30
    water mapping, 115, 121
    based platforms, 25
    sample distance (GSD), 42
    profile approach, 119
GSM, 267

Hard classifiers, 242
Hardware Components, 146
Heads-up digitizing, 157
Hidden layer unit, 249
Hierarchical database structure, 182
High
    boost filter, 91
    Resolution Geometric (HRG), 31
    Resolution Picture Transmission (HRPT), 47
    Resolution Stereoscopic (HRS), 32
    resolution, 22
    texture, 193
High-pass filters, 91, 192

Histogram, 81, 151
    Equalization Method, 90
    Minimum Method (HMM), 84
HTML, 261
HTTP, 261
Hydro morphological studies, 115, 128
Hydrothermal Mapping, 28
Hyperion, 48, 50
Hyperspectral Imaging Instrument (HIS), 50

Idealised sequence for
    selecting training data, 102
    digital image processing, 80
Identification, 73
IDENTITY, 195
IKONOS, 40
IKONOS-2, 42
Image, 60, 187
    addition, 92
    classification, 80, 81
    data file, 70
    division, 92
    enhancement, 80, 81, 88
    histogram, 89
    interpretation keys, 77
    interpretation strategies, 75
    Map Data Products (IMDP), 37
    Multiplication, 92
    processing, 79
    Rectification and Restoration, 81
    rectification, 80
    registration or image-to-image registration, 87
    subtraction, 92
    Support Data (ISD), 44
    Transformations, 81, 92
Imaging and sounding, 52
Impedance, 177, 178
Implementation problems, 216
Incident, 15
Index inverted file, 180
Indexed files, 179
Indistinguishable, 15
Indivisibility, 193
Inexact objects, 171
Information
    classes, 98
    Community Model, 274
    system, 145
Infrastructure data, 156
Initial Data Statistics, 81
Input layer, 249

INSAT Co-ordination Committee (ICC), 55
INSAT, 55
INSAT-1 series, 55
INSAT-1A, 55
INSAT-1B, 55
INSAT-1C, 55
INSAT-1D, 55
INSAT-2 series, 55
INSAT-2A, 55
INSAT-2B, 55
INSAT-2E, 55
INSAT-3 series, 55
INSAT-3A, 55
INSAT-3E, 55
Instantaneous Field of View (IFOV)
Intensity Interpolation, 88
INTERESECT, 195
Internal 154
INTERNET GIS, 260
Interoperability of data, 271
Interpolation, 196
Interpretation A, 127
Interpretation B, 127
Interpretation by Inference, 75
Interpretation C, 127
Interpretation of Flood Area, 127
Interval, 153
Inverted file, 180
Iron oxide, 19
Irradiance, 16
Irrigation Water Management, 221
IRS Satellite Systems, 32
IRS, 26
IRS-1A, 32
IRS-1B, 32
IRS-ID LISS-3, 34
IRS-P3 MOS-A/B/C, 34
IRS-P4 (OCEANSAT-2), 34
IRS-P5 (CARTOSAT-1), 34, 37
IRS-P6 (Resourcesat-1), 34, 36
IRS-P6 MX, 137
Iterative self Organizing Data Analysis Technique A (ISODATA), 109

Java 2 Enterprise Edition (J2EE), 268
Java
    applet, 265
    Database Connection (JDBC), 271
    Message Service (JMS), 271
    Server Pagers (JSP), 269
    Servlets API, 269

Javascript functions, 265

Kappa (k), 113
Kappa coefficient of agreement, 114
Karhunen-Loeve Transform, 96
Key, 183
Kinematic GPS surveying, 158
Knowledge based
    approach, 277
        Based System for Military Use, 225
Kriging, 203

L_1 distance, 255, 256
Land
    cover mapping, 115
    form maps, 151
    use mapping, 8
    use, 115
LANDSAT, 26
Land-use activity, 100
Large-scale map, 149
Latitude, 149
Leader file, 70
Leaf-area index (LAI), 116
Learning rate, 251
Least-square error objective function, 244
Legal aspects, 272
Length, 187
LGSOWG Superstructure format, 69
Light tables, 77
Light, 12
Lightning imaging instrument, 46
Line maps, 151
Line of sight analysis.207
Line, 149
Linear, 203, 249
    Contrast Enhancement, 89
    Etalon Imaging Spectral Array (LEISA)
        Atmospheric Corrector (LAC), 48
    features, 204
    Imaging Self Scanner (LISS-4) sensor, 36
    interpolation, 89
    methods 89
    Mixture Model (LMM), 243
Line-in-polygon, 194
LISS (Linear Imaging Self-scanning), 32
LISS IV AWIFS, 137
Local or global, 199
Location, 102
Location-allocation Modeling, 207, 208

Logarithmic models, 91
Logical volume file, 70
Longitude, 149
Look-Up-Table (LUT), 64
Low
    resolution, 22
    texture, 193
    variability, 193
    pass filter, 91, 192

Magnifiers, 78
Mahalonobis, 244
Man-made features, 152
Map
    Animated , 151
    Choropleth, 150
    Composite suitability, 219
    Contour, 151
    Dasymetic, 151
    Dot, 151
    ESRI Objects, 265
    Land form, 151
    Large-scale, 149
    Line, 151
    Overlay, 193
    Reference, 109
    Small-scale, 149
    Thematic, 150
    server, 264
Marine Observation Satellite (MOS), 57
Maximum
    Likelihood, 107, 116
    values, 82
Mean, 82
Membership function, 243
Mensuration, 73
Merging clusters, 108
Message, 185
Meteorological
    communication satellites, 25
    satellites, 52
Meteosat, 57
METSAT (Kalpana-1), 55
Michigan
    Classification System, 100, 119
    Land Use Classification (MLUC), 100
Microwave, 12
    Scanning Radiometer (MSR), 58
Middle Infra-Red (MIR), 35
Mie Scattering, 14

Minimum
    Distance to Means, 106, 116
    values, 82
Min-Max stretching, 89
Mirror scan velocity, 86
Mixed pixel, 23, 241
MOBILE
    GIS, 265
    Location Protocol (MLP), 271
    Position Centre (MPC), 271
Modeling
    surfaces, 176
    networks, 177
Moisture content, 19
Momentum factor, 251
MOS-1, 57
MOS-1b, 57
MS
    Access, 265
    Visual Basic, 265
Multi-
    band images, 4
    concept of, 4
    date images, 4
    dimensional Visualization of Terrain, 210
    disciplinary analysis, 4
    enhancement images, 4
    frequency Scanning Microwave Radiometer (MSMR), 35
    media Message Service (MMS), 267
    polarization images, 4
    spectral Electronic Self-Scanning Radiometer (MESSR), 57
    spectral mode (Mx), 36
    spectral mode (XS mode), 28
    Spectral Scanner (MSS), 26
    stage images, 4
    temporal, 23

National
    Drinking Water Mission (NDWM), 122
    Resources Management System (NNRMS), 32
    Wastelands Development Board, 130
Natural, 152
    resource data, 156
Near-diffuse, 15
Nearest neighbour, 88
Nearest-neighbourhood regions, 200
Near
    Infra Red (NIR), 92

polar orbits, 28
specular, 15
Neighbourhood functions, 190
Network, 177
    analysis, 207
    data structures, 182, 183
    features, 204
Networking, 148
Neural Network Based Methods, 248
New Millennium Program (NMP), 48
Nimbus-7, 57
Nitrogen (N_2), 14
NOAA AVHRR, 53
NOAA, 26
Nodes, 149
Nominal, 153
Non
    systematic, 85
    Coordinate System, 161
    homogeneous, 24
    linear methods, 89
    parametric classification, 106
    selective Scattering, 14
    spatial, 153
    such axis, 93
    systematic distortions, 86
Normalized Difference Vegetation Index (NDVI), 92
NRSA, 59
Null Volume file, 70
Number of training areas, 102

Object, 170
    class, 185
    oriented database management system (OODBMS), 271
    based model, 170
    oriented approach, 261
    oriented Database structure, 185
    Oriented Programming (OOP), 185
Observed, 113
    spectral response, 243
Ocean Colour Monitor (OCM), 35
Oceanography and marine, 8
Omission, 111
One-to many, 182
Open Geodata Model, 272, 274
OPEN GIS CONSORTIUM (OGC), 271, 272
Open GIS services, 274
    Model, 272
    Specification, 273

294 Index

Operating System, 146, 272
Operational
    Linescan System (OLS), 57
    Structure of a DSS, 277
Oracle Spatial, 271
Oracle, 271
Orbit, 25
Orbview, 40
Orbview-1 46
Orbview-2, 46
Orbview-3, 47
Ordered sequential files, 179
Ordinal, 153, 154
Ordinary kriging, 203
Organic matter, 19
Orientation (aspect), 176
Ortho Ready Standard imagery, 44
Orthoimage, 69
Ortho-rectified
    Imagery 44
    Imagery Products, 45
Other Sources, 155
Overall
    accuracy, 114
    classification, 110
Overlaying, 148
Oxygen (O_2), 14
Ozone, 12

PAN sensor, 32
Panchromatic mode (P mode), 28
Panoramic distortion, 86
Paper stretch, 160
Parallax bar, 78
Parametric classification, 106
Parent-child, 182
Passive sensors, 25
Path and Row, 65
Path/Row products, 67
Pathfinding, 207
Pattern, 74
People or activity, 100
Percentile stretching, 90
Perimeter, 187
Perpendicular Vegetation Index (PVI), 93
Personal
    Digital Assistants (PDAs), 267
    Operating GIS, 148
Perspective, 86
Phenomena, 170

Photograph, 60
Photographic/Image Interpretation, 73
Photomorphic Analysis, 77
Photons, 9
Physical model, 84, 214
Pie charts, 151
Piecewise enhancement, 90
PIN CODE, 163
Pipe-filter approach, 261
Pixel pixel, 23
Pixels, 156
Placement, 102
Plan curvature 204
Planck's constant, 10
Plane of
    Soil, 94
    Vegetation, 94
Planning Committee of NNRMS (PC-NNRMS), 32
Platform, 25
    velocity, 86
Point, 148
    directory, 174
Pointable optics, 28
Point-in-polygon, 194
Polar Satellite Launched Vehicle (PSLV), 35
Polygon, 149
Polygon-on-polygon, 194
Polymorphism, 186
Positional data, 151
Possibilistic c-Means (PCM), 243, 246
Postal address, 163
Preprocessing, 81
Presentation tier, 269
Primary
    attributes, 204
    colours, 12
Prime meridian, 161
Principal
    Component Analysis, 81, 96
    Conservation of energy, 15
Priori probabilities, 107
Probabilistic Interpretations, 76
Probability, 107
Producer's accuracy, 114, 259
Profiles curvature value, 204
Profiles, 193
Programme region analysis, 135
Project
    Evaluation, 216
    management, 215

Prototype approach, 215
Proximal, 151
    regions, 200
    calculation, 192
    function, 190
Proxy, 75
Pure, 24
    pixel, 24

Quadrant products, 67
Quadratic, 203
Quantitative analysis, 79
Quantization levels, 23
Quaternary Triangular Mesh, 161
Queries, 190
Quick Bird 1, 40, 42, 45

R, radius of the cluster, 108
Radiance, 16
Radiometric
    Corrections, 64, 84
    Resolution, 22, 23
Raster
    approach, 176
    data model, 171
    Data structures, 172
    Overlay, 196
Ratio, 153, 154
Rayleigh scattering, 14
Real material, 10
Real-time GPS surveying, 158
Reclassification, 190
Recognition, 73
Records, 155
Rectangular Coordinate System, 161
Recursive DEM, 205
Reference map, 109
Referencing scheme, 65
Reflectance, 15
    Characteristics, 16
Reflected, 15
Regression method, 84
Regularizing term, 246
Relational database structure, 182
Remote
    Method Invocation (RMI), 270
    Sensing System, 2
    sensing, 1
Resampling, 88
Resolution 22
    Absolute temporal, 23

    cell, 22
    Coarse, 22
    High, 22
    Low, 22
    Radiometric, 22, 23
    Spatial, 22
    Spectral, 22
    Temporal, 22, 23
Resource, 100
    region analysis, 134
Restoration, 80
Retrieval, 146
Return Beam Vidicon (RBV), 26
Revisit period, 26
Rich picture, 213
Ring pointer structures, 183
Road Accident Analysis, 220
Root Definition, 213
Root Mean Square Error (RMSE), 44, 87
Rough, 91
Route Tracing, 208
Rubber sheeting, 160
Rulers, 77

Salesman Travel Problem, 208
Sand, 19
Satellite Image Map Products
    2.5-D, 39
    2-D, 39
    3-D, 39
    Imagery, 154
Satellites, 25
    IKONOS, 40
    IKONOS-2, 42
    INSAT, 55
    INSAT-1 series, 55
    INSAT-1A, 55
    INSAT-1B, 55
    INSAT-1C, 55
    INSAT-1D, 55
    INSAT-2 series, 55
    INSAT-2A, 55
    INSAT-2B, 55
    INSAT-2E, 55
    INSAT-3 series, 55
    INSAT-3A, 55
    INSAT-3E, 55
    IRS, 26
    IRS-1A, 32
    IRS-1B, 32
    IRS-ID LISS-3, 34

Index

IRS-P3 MOS-A/B/C, 34
IRS-P4 (OCEANSAT-2), 34
IRS-P5 (CARTOSAT-1), 34, 37
IRS-P6 (Resourcesat-1), 34, 36
IRS-P6 MX, 137
Geostationary Operational Environmental Satellite (GOES), 52
GMS satellites, 57
LANDSAT, 26
Marine Observation Satellite (MOS), 57
Meteorological communication satellites, 25
Meteorological satellites, 52
Meteosat, 57
METSAT (Kalpana-1), 55
MOS-1, 57
MOS-1b, 57
Nimbus-7, 57
NOAA, 26
Orbview, 40
Orbview-1 46
Orbview-2, 46
Orbview-3, 47
Quick Bird 1, 40, 42, 45
Sun synchronouse, 25
System Pour ¼ Observation de la Terre (SPOT), 26
Television and Infrared Observation Satellite -1 (TIROS-1), 51
Tracking and Data Relay Satellite System (TDRSS), 59
Scale, 149
Scan skew, 86
Scatter Plots, 151
Scatter-grams, 81
Scattering, 12
Screen digitizing, 157
Seastar, 58
Sea-viewing Wide-Field-of View Sensor (seawifs), 58
Secondary attributes, 204
Security, 272
Sensors, 25
    Active, 25
    Advanced Land Imager (ALI), 48
    Advanced Very High Resolution Radiometer (AHVRR), 54
    Advanced Wide Field Sensor (AWIFS), 36
    AWIFS, 34
    AWIFS-A, 36
    AWIFS-B, 36
    Coastal Zone Colour Scanner (CZCS), 57
    Enhanced Thematic Mapper (ETM), 26
    Enhanced Thematic Mapper Plus (ETM+), 26
    Linear Etalon Imaging Spectral Array (LEISA) Atmospheric Corrector (LAC), 48
    Linear Imaging Self Scanner (LISS-4) sensor, 36
    LISS (Linear Imaging Self-scanning), 32
    LISS IV AWIFS, 137
    Ocean Colour Monitor (OCM), 35
    PAN, 32
    VEGETATION, 30, 31
    Very High Resolution Radiometer (VHRR), 55
    Visible and Infrared High-Resolution (HRVIR), 30
    Visible and Near Infra Red (VNIR), 36
    Visible Thermal Infrared Radiometer (VTIR), 58
    High Resolution Visible (HRV), 28
Separability measure, 104
Server-side Internet GIS, 261
Shaded contours method, 209
Shadow, 74
Shape, 74, 102
Shift Along Track products, 67
Short Message Service (SMS), 267
Shortest Path Problem, 207
Sigmoid, 249
Signal data spectral approach, 119
Silt, 19
Simple
    kriging, 203
    list, 179
Site Suitability in Urban Planning, 217
Site, 74
Site-specific accuracy, 110
Size, 74, 102
Slope 204
Slope transformation, 193
Small-scale map, 149
Smooth, 91
Socio-economic data, 156
Soft classification methods, 242
Soft Systems Approach, 214
Soft, 242
Software, 146, 148
Source data error, 166
Spaceborne, 25
Spatial
    Data engine (SDE), 271
    Data errors, 164
    Data Model, 169
    Data structures, 172
    Data, 148, 153
    Database, 145

Filtering, 91
Interaction, 207
Interpolation, 196
Moving Average, 202
Object, 148
Pattern recognition, 82
Query 190
Referencing system, 160
Referencing, 148
Resolution, 22
Specific catchment area, 204
Spectral classes, 98
  mixture analysis, 242
  pattern recognition, 82, 97
  ratioing, 92
  reflectance curve, 16
  Reflectance, 16
  resolution, 22
Specular, 15
  Reflection, 15
Speed of light c, 9
SQL Server, 271
Square filter, 202
Standard
  Deviation, 82
  Imagery Products, 44
  Imagery, 44
  Land Use Coding Manual, 100
  Products, 67
  Query Language (SQL), 185
Static GIS (SGIS), 267
Static GPS, 158
Statistical Information, 81
Stereoscopes, 78
Stereoscopic, 28
Storage, 146
Sub-pixel classification, 242
Sun synchronous satellite, 25
Sun, 12
Super mode panchromatic images, 32
Supervised, 99
Supply and demand, 177, 178
Support, 275
Surface, 171
  analysis, 202
  roughness, 15
  specific points, 204
Surrogate, 75
Survey Data, 155
SWIR band, 32
Sybase, 271

System Pour Observation de la Terre (SPOT), 26
System, 275
Systematic, 85
  distortions, 86
Systems life cycle, 215

Tangent hyperbolic, 249
Tape Formats, 62
Tassel Cap Transformation, 93
Tau coefficient, 114
Television and Infrared Observation Satellite -1
  (TIROS-1), 51
Temporal GIS (TGIS), 267
Temporal resolution, 22, 23
Terrain Data
  Collection by Photogrammetry, 157
  from Existing Maps, 157
  from Field Surveying Methods, 158
  from Satellite Remote Sensing, 156
Tessellation, 171
Texture, 74
  filters, 192
  transformation, 192
The Parallelepiped, 106
The Web Container, 270
Thematic
  layers, 148
  Mapper (TM), 26
  maps, 150
  infrared, 12
  Polygon Method, 199
  Polygon, 165, 200
Thin client, 268
Three-Dimensional Visualization of Terrain, 210
Time of Arrival (TOA), 271
Tone, 74
Tool based approach, 276, 277
Topological data structure, 174
Tourism Information System, 223
Tracing, 207
Tracking and Data Relay Satellite System (TDRSS), 59
Trailer file, 70
Training
  Area, 99
  Site Selection and Statistics Extraction, 100
Transformed divergence, 105
Transmission, 12
  Control Protocol/Internet Protocol (TCP/IP), 159
Transmitted, 15
Trend surface analysis, 203
Triangular facets, 200

Triangulated Irregular Network (TIN), 40, 157, 171, 200
Tsunami, 136
Tuples, 184
Two-and-a-Half-Dimensional Visualization of Terrain, 209
Two-dimensional Visualization of Terrain, 209

U.S. Geological Survey Land Use/Land Cover Classification System, 100, 119
Uniformity, 102
Unimodal, 90
UNION, 195
Universal kriging, 203
Universal Transverse Mercator Projection (UTM), 162
Unsupervised, 99
Unsystematic errors, 83
Urban growth studies, 115, 124
User's accuracy, 114, 259
USGS Land cover/Land Use classification, 115

Variance, 82
Variance-covariance 81
Variation, 165
Vector
    approach, 176
    data model, 171
    Data Structure, 173
    GIS, 206
    Overlay, 194
Vectorising, 148
Vegetation, 17
    Index (VI), 92, 119
    Index Number (VIN), 120
VEGETATION, 30, 31
Vertices, 200
Very High Resolution Radiometer (VHRR), 55
Viewpoints, 206
Viewshed analysis, 207
Violet portion, 11
Visibility
    analysis, 206
    region, 207

Visible and Infrared High-Resolution (HRVIR), 30
Visible and
    Near Infra Red (VNIR) sensor, 36
    Thermal Infrared Radiometer (VTIR), 58
    High Resolution (HRV), 28
    Red (VR), 92
Visual
    Display Unit, 147
    Interprétation, 79
Volume director file, 70
Voronoi polygon, 200

WAP service tier, 269
Wasteland mapping, 115, 130
Water, 21, 121
    resources, 8
    vapour, 12
Waterfall model, 215
Wave band, 23
Wavelength 1, 10
Web
    browser, 159, 262
    pages, 159
    server, 159
Weighted Kappa, 114
Weighting exponent, 244
Wetness, 94
Wide Field Sensor (Wifs), 32
Wireless Application Protocol (WAP), 266
Wireless Markup Language (WML), 267
World
    Wide Web (WWW), 159
    Earthquake Information System, 225
Wrapping, 160

Xanthophylls, 17
X-rays, 11

Yellowness axis, 93
Yield Forecasting, 117

Zoom transferscope, 78